The GABA Receptors, Third Edition

THE RECEPTORS

KIM A. NEVE, SERIES EDITOR

The GABA Receptors, Third Edition, EDITED BY S. J. Enna and Hanns Möhler, 2007

The Serotonin Receptors: From Molecular Pharmacology to Human Therapeutics, EDITED BY Bryan L. Roth, 2006

The Adrenergic Receptors: In the 21st Century, EDITED BY Dianne M. Perez, 2005

The Melanocortin Receptors, EDITED BY Roger D. Cone, 2000

The GABA Receptors, Second Edition, EDITED BY S. J. Enna and Norman G. Bowery, 1997

The Ionotropic Glutamate Receptors, EDITED BY Daniel T. Monaghan and Robert Wenthold, 1997

The Dopamine Receptors, EDITED BY Kim A. Neve and Rachael L. Neve, 1997

The Metabotropic Glutamate Receptors, EDITED BY P. Jeffrey Conn and Jitendra Patel, 1994

The Tachykinin Receptors, EDITED BY Stephen H. Buck, 1994

The Beta-Adrenergic Receptors, EDITED BY John P. Perkins, 1991

Adenosine and Adenosine Receptors, EDITED BY Michael Williams, 1990

The Muscarinic Receptors, EDITED BY Joan Heller Brown, 1989

The Serotonin Receptors, EDITED BY Elaine Sanders-Bush, 1988

The Alpha-2 Adrenergic Receptors, EDITED BY Lee Limbird, 1988

The Opiate Receptors, EDITED BY Gavril W. Pasternak, 1988

The GABA Receptors

THIRD EDITION

Edited by

S. J. Enna

Departments of Molecular and Integrative Physiology
and of Pharmcology, Toxicology, and Therapeutics
University of Kansas
Kansas City, KS

and

Hanns Möhler

Institute of Pharmacology
University of Zurich
Department of Chemistry and Applied Biosciecnes
Swiss Federal Institute of Technology (ETH)
and
Collegium Helveticum
Zurich, Switzerland

HUMANA PRESS
TOTOWA, NEW JERSEY

Softcover re-print of the Hardcover 3rd edition 2007

999 Riverview Drive, Suite 208
Totowa, New Jersey 07512
www.humanapress.com

Cover Illustration: Chapter 6, Fig. 16, "$GABA_A$-Receptor Mutations Associated With Idiopathic Generalized Epilepsies and Febrile Seizures," by Robert L. Macdonald, Jing-Qiong Kang, Martin J. Gallagher, and Hua-Jun Feng.

Cover design by Karen Schulz.

Production Editor: Amy Thau

This publication is printed on acid-free paper. ∞
ANSI Z39.48-1984 (American National Standards Institute) Permanence of Paper for Printed Library Materials
For additional copies, pricing for bulk purchases, and/or information about other Humana titles, contact Humana at the above address or at any of the following numbers: Tel.: 973-256-1699; Fax: 973-256-8341; E-mail: orders@humanapr.com; or visit our Website: www. humanapress.com.

ISBN 978-1-59745-465-0 (eBook)

Preface

This volume is the third edition of a monograph series that was first published in 1983. The demand for this work is a testament to the impact of studies on γ-aminobutyric acid (GABA) receptors on the basic understanding of synaptic transmission and on defining the clinical importance of the neurotransmitter system. Chronicled in *The GABA Receptors, Third Edition*, are the advances made in understanding the molecular and pharmacological properties of $GABA_A$ and $GABA_B$ receptors since the topic was last reviewed in 1996. Particular emphasis is placed on describing the assembly, structure, and function of $GABA_B$ sites, the first heterodimeric G protein-coupled receptors identified in vivo. In addition, there are reports dealing with the subunit composition, trafficking, and pharmacological selectivity of $GABA_A$ receptors. Aside from providing insights into the fundamental properties of ligand-gated ion channels and second messenger systems, the findings detailed in this work point the way for developing novel therapeutics capable of more selectively manipulating these transmitter sites. Chapters in this volume contain descriptions of new agents, including allosteric modulators, capable of activating or inhibiting GABA receptors. Descriptions are provided of potential clinical candidates for treating disorders as diverse as insomnia and cognitive impairments. The reports contained herein also detail new evidence directly linking $GABA_A$ and $GABA_B$ receptor dysfunctions to a host of neuropsychiatric conditions, including epilepsy, anxiety disorders, affective illness, and pain syndromes. These data provide a biological framework for understanding the clinical utility of GABAergic drugs as treatments for neurological and psychiatric disorders, and for their use as hypnotics and anesthetics.

Numbered among the contributors to *The GABA Receptors, Third Edition*, are many who have worked in this area for decades. All of the senior authors have been actively engaged in studying GABA receptor systems and are recognized for making seminal contributions to the field. In addition to highlighting advances over the past 10 years, the authors provide opinions on the implications of these findings and suggestions on fruitful avenues for future research. As was the case for the previous two editions, the aim of this volume is to not only serve as an information source, but as a stimulus for further advances in the field. This offering should be of particular value to basic and clinical neuroscientists in general, and neuropharmacologists, psychiatrists, and neurologists in particular.

S. J. Enna, PhD
Hanns Möhler, PhD

Contents

Contributors

BERNHARD BETTLER, PhD • *Pharmazentrum, Institute of Physiology, University of Basel, Basel, Switzerland*

NORMAN G. BOWERY, PhD • *Department of Pharmacology, Division of Neuroscience, The Medical School, University of Birmingham, Birmingham, UK*

NIGEL G. COOKE, PhD • *Novartis Pharma AG, Basel Switzerland*

ANDRÉS COUVE, PhD • *Program in Physiology and Biophysics, Centro de Neurociencias Integradas, Universidad de Chile, Santiago, Chile*

ANGELA N. DUKE, MA • *New England Primate Research Center, Harvard Medical School, Southborough, MA; and Neuroscience and Behavior Program, University of Massachusetts, Amherst, MA*

S. J. ENNA, PhD • *Department of Molecular and Integrative Physiology, Department of Pharmacology, Toxicology and Therapeutics, University of Kansas, Kansas City, KS*

DAVID H. FARB, PhD • *Department of Pharmacology and Experimental Therapeutics, Boston University School of Medicine, Boston, MA*

MARK FARRANT, PhD • *Department of Pharmacology, University College London, London, UK*

HUA-JUN FENG, PhD • *Department of Neurology, Vanderbilt University, Nashville, TN*

WOLFGANG FROESTL, PhD • *Department of Chemistry, AC Immune SA, Lausanne, Switzerland*

MARTIN J. GALLAGHER, MD, PhD • *Department of Neurology, Vanderbilt University, Nashville, TN*

TERRELL T. GIBBS, PhD • *Department of Pharmacology and Experimental Therapeutics, Boston University School of Medicine, Boston, MA*

MARIA C. GRAVIELLE, PhD • *Department of Pharmacology, Instituto de Investigacion Farmacologicas, Buenos Aires, Argentina*

JING-QIONG KANG, MD, PhD • *Department of Neurology, Vanderbilt University, Nashville, TN*

BERNHARD LÜSCHER, PhD • *Department of Biology, Department of Biochemistry and Molecular Biology, Pennsylvania State University, University Park, PA*

Robert L. Macdonald, MD, PhD • *Department of Neurology, Department of Molecular Physiology, and Department of Biophysics and Pharmacology, Vanderbilt University, Nashville, TN*

Stella C. Martin, PhD • *Department of Pharmacology and Experimental Therapeutics, Boston University School of Medicine, Boston, MA*

Stuart J. Mickel, PhD • *Novartis Pharma AG, Basel, Switzerland*

Hanns Möhler, PhD • *Institute of Pharmacology, University of Zurich; Department of Chemistry and Applied Biosciences, Swiss Federal Institute of Technology (ETH); and Collegium Helveticum, Zurich, Switzerland*

Stephen J. Moss, PhD • *Department of Neuroscience, University of Pennsylvania School of Medicine, Philadelphia, PA*

Menelas N. Pangalos, PhD • *Neuroscience Research, Wyeth Discovery Research, Princeton, NJ*

Donna M. Platt, PhD • *New England Primate Research Center, Harvard Medical School, Southborough, MA*

James K. Rowlett, PhD • *New England Primate Research Center, Harvard Medical School, Southborough, MA; and Neuroscience and Behavior Program, University of Massachusetts, Amherst, MA*

Shelley J. Russek, PhD • *Department of Pharmacology and Experimental Therapeutics, Boston University School of Medicine, Boston, MA*

Werner Sieghart, PhD • *Division of Biochemistry and Molecular Biology, Center for Brain Research of the Medical University Vienna, Vienna, Austria*

Janine L. Steiger, PhD • *CombinatoRx, Cambridge, MA*

Jim Yu-Hsiang Tiao, PhD • *Pharmazentrum, Institute of Physiology, University of Basel, Basel, Switzerland*

Xu Yuan, MS • *Department of Biology, Department of Biochemistry and Molecular Biology, Pennsylvania State University, University Park, PA*

1

The GABA Receptors

S. J. Enna

Summary

γ-Aminobutyric acid (GABA), an amino acid neurotransmitter, is widely distributed throughout the neuraxis. Two pharmacologically and molecularly distinct GABA receptors have been identified, $GABA_A$ and $GABA_B$. $GABA_A$ receptors are pentameric ligand-gated chloride-ion channels, whereas $GABA_B$ receptors are heterodimeric G protein-coupled sites. Although $GABA_A$ receptor subtypes can display pharmacological differences, the two molecularly distinct $GABA_B$ receptors have similar substrate specificities, limiting the ability to selectively manipulate this site. Gene deletion and point mutation studies have revealed the importance of GABA receptors in neural development and function, with subtle modifications in subunit amino acid composition having profound effects on behavioral phenotype and responses to drugs. The characterization of $GABA_A$ receptors has contributed substantially to the knowledge about allosteric regulation of ligand-gated ion channels. Such information is invaluable in defining precisely the mechanisms of action of numerous drugs, such as the benzodiazepines, and toxic agents. Research on $GABA_B$ receptors has proven the existence of dimeric metabotropic receptors and has provided the chemical tools necessary for defining such systems. The characterization of the pentameric $GABA_A$ and dimeric $GABA_B$ receptors has been crucial for understanding the neurobiological basis of some nervous system disorders. Given the importance of GABA in central nervous system function, further work on its receptors is likely to yield novel therapeutics for treating a host of neurological and psychiatric conditions.

Key Words: GABA; GABA pharmacology; $GABA_A$ receptors; $GABA_B$ receptors; GABA receptor subunits; GABA receptor function.

From: *The Receptors: The GABA Receptors, Third Edition*
Edited by: S. J. Enna and H. Möhler © Humana Press Inc., Totowa, NJ

1. Introduction

It has been nearly a quarter century since the publication of the first edition of *The GABA Receptors (1)*. That volume was devoted almost exclusively to reviewing studies on γ-aminobutyric acid-A ($GABA_A$) receptors, with only passing reference to the possibility of the existence of pharmacologically and molecularly distinct $GABA_B$ sites, evidence for which had only recently been reported *(2)*. The book consists of various chapters describing the biochemical and pharmacological properties of the $GABA_A$ receptor, with particular emphasis on the use of binding assays for characterizing this site and defining its relationship to the benzodiazepines.

Topics covered in the second edition, which appeared in 1996, reflected the pace of discoveries in the intervening 13 yr, with half of the chapters devoted to $GABA_A$ and half to $GABA_B$ *(3)*. By then the genes for $GABA_A$ receptor subunits were cloned, making it possible to more precisely determine the composition of the site. This in turn led to a better understanding of the way in which drugs interact with the $GABA_A$ system and pointed to the possibility of developing agents that affect only subsets of this receptor. As for the $GABA_B$ site, the second edition included a review of progress made in creating new receptor agonists and antagonists and of biochemical and electrophysiological studies showing that, unlike the ligand-gated $GABA_A$ receptor ion channel, the $GABA_B$ receptor is a G protein-coupled site. Although some reports suggested the existence of pharmacologically distinct $GABA_B$ subtypes, direct proof was lacking, as the gene for this receptor was proving difficult to clone. Predictions were made about potential advances in drug development once the $GABA_B$ receptor gene was isolated and the data on $GABA_A$ receptor subtypes fully exploited.

The offerings contained in this third edition review advances during the past decade. As with the last volume, approximately half the chapters deal with receptors for $GABA_A$ and half for $GABA_B$. Soon after publication of the second edition the genes associated with the $GABA_B$ receptor were cloned, revealing a heterodimeric G protein-coupled site *(4–6)*. As this was the first direct demonstration of dimeric seven transmembrane receptors, subsequent work was pioneering in describing how such sites are assembled and regulated. This discovery opened the way to identifying other G protein-coupled receptor dimers, indicating that the $GABA_B$ system is not unique in this regard *(7)*. There is also coverage in this latest edition of new $GABA_B$ receptor agonists and antagonists, including one that is undergoing a clinical trial (*see* Chapters 9 and 12).

As detailed in the present work, strides have also been made in defining the $GABA_A$ site. Examples include gene manipulation and deletion studies directly linking particular $GABA_A$ receptor subunits, indeed particular subunit amino acids, with the various pharmacological responses to drugs such as the benzodiazepines (*see* Chapter 2) *(8)*. Such findings make possible the design of more selective and less toxic hypnotics, anxiolytics, anticonvulsants, and muscle relaxants.

The aim of this chapter is to present an overview of GABA receptor systems. Though the subject is covered from an historical perspective to guide those new to the field, emphasis is placed on introducing and placing in context topics described in detail elsewhere in the text. Each chapter is authored by experts in the field, many of whom have labored for decades in this area. Those familiar with previous editions of this work will be impressed by the amount of progress made since the last volume, and by the impact GABA receptor research has made in defining chemical neurotransmission in general. As is made clear in this text, the ongoing efforts for characterizing the $GABA_A$ and $GABA_B$ receptors will undoubtedly continue to yield new insights into mechanisms that regulate synaptic transmission, the biological abnormalities associated with a host of neurological and psychiatric disorders, and the development of new drugs for treating these conditions.

2. General Overview

GABA, an amino acid neurotransmitter, is widely distributed throughout the neuraxis. While GABA is found in some peripheral tissues, and there is evidence it may regulate neuronal activity in the intestines, lungs, and bladder, its predominant effects are in the central nervous system. Because activation of neuronal GABA receptors generally results in hyperpolarization, this amino acid is considered an inhibitory neurotransmitter. Given the number of GABAergic neurons in the brain, and their widespread distribution, GABA appears to be the major inhibitory neurotransmitter in the central nervous system. It is this ubiquity that has hindered drug development because nonselective GABA receptor agonists and antagonists have generalized effects on central nervous system function.

Two pharmacologically and molecularly distinct GABA receptors have been identified, $GABA_A$ and $GABA_B$ *(3,4)*. $GABA_A$ receptors are ligand-gated chloride ion channels, whereas $GABA_B$ sites are heterodimers coupled to G proteins. Although many pharmacologically distinct $GABA_A$ receptors have been identified, the two molecularly distinct $GABA_B$ sites display similar pharmacological selectivity (*see* Chapter 11). However, it is possible that allosteric modulators might be able to distinguish between $GABA_B$ receptor subtypes (*see* Chapter 9) *(9)*.

3. $GABA_A$ Receptors

3.1. Molecular Pharmacology

As is the case with some other ligand-gated ion channels, the $GABA_A$ receptor is a pentameric structure made up of molecularly distinct subunits (Fig. 1)

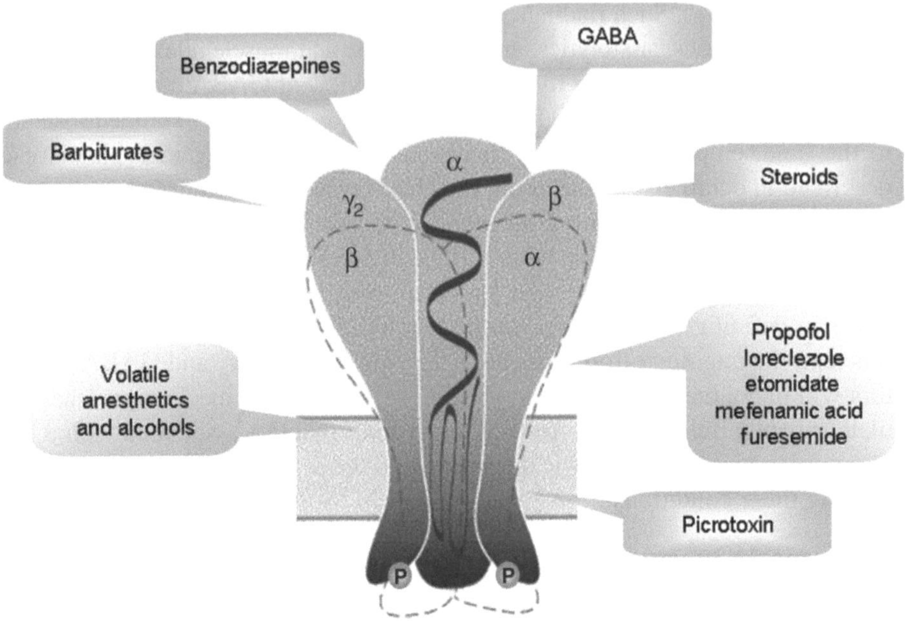

Fig. 1. Schematic representation of a $GABA_A$ receptor illustrating its pentameric structure and the different sites of action for drugs that interact with this site. The P designation represents a phosphorylation site. (Adapted from ref. *11*.)

(10,11). Generally, activation of a $GABA_A$ receptor increases the intraneuronal concentration of chloride ion, hyperpolarizing the cell (*see* Chapter 5) *(12,13)*. In select regions of the central nervous system, such as the hippocampus or dorsal root ganglia, or under certain circumstances, as during development, $GABA_A$ receptor activation causes neuronal depolarization *(12–15)*. In some circumstances this is because the intracellular concentration of chloride exceeds extracellular levels, with opening of the receptor-coupled channel resulting in a net efflux of this ion and neuronal excitation. In addition, whereas some $GABA_A$ receptors are phasically activated, others are tonically stimulated (*see* Chapters 5 and 8). The latter, such as those located in the cerebellar granule cells, can alter network excitability and can be selectively targeted by agents, such as the neurosteroids, that influence GABAergic transmission *(16–18)*.

As with other ligand-gated ion channels, the sensitivity and activity of $GABA_A$ receptors are modulated by drugs acting at distinct sites on individual subunits or subunit combinations (*see* Chapter 2) *(8,19)*. Whereas GABA itself activates the receptor by attaching to the recognition site, barbiturates, benzodiazepines, alcohol, neurosteroids, and fixed and general anesthetics facilitate

$GABA_A$ receptor transmission by acting on other components of the receptor complex (Figs. 1 and 2). The same is true for antagonists, with the convulsant bicuculline being a competitive antagonist at the $GABA_A$ receptor recognition site, whereas picrotoxin attaches elsewhere on the receptor complex (Figs. 1 and 2). The benzodiazepines, such as diazepam (Fig. 2), are the best characterized of the allosteric modulators of $GABA_A$ receptor function.

The subunits that assemble to form pentameric $GABA_A$ receptors are drawn from a pool of 19 distinct gene products (Table 1) (*see* Chapter 8). Given the variety of subunit proteins and their splice variants, the potential exists for an enormous number of molecularly distinct complexes (*see* Chapter 2) *(19,20)*. However, studies in various expression systems indicate that not all subunit combinations respond to GABA, with the estimated number of different $GABA_A$ receptors in the mammalian central nervous system believed to be less than 100, and probably as few as two dozen *(19,21,22)*.

The $GABA_A$ receptor subunits are widely and unevenly distributed throughout the brain and peripheral organs (*see* Chapter 4). Their function in peripheral tissues remains largely undefined, as GABAergic innervation is sparse outside the central nervous system. Whereas the precise stoichiometry of native $GABA_A$ receptors is unknown, subunit-labeling studies provide clues in this regard (Table 2). The largest single group of $GABA_A$ receptors appears to be made up of $\alpha_1\gamma_2$- and a β-subunit *(19)*. Indeed, it seems the majority of $GABA_A$ receptors possess either α_1- or α_2-subunit(s) in combination with a γ_2- and β-subunits(s). Subunit composition determines the biophysical and pharmacological properties of the site (*see* Chapters 2 and 5) (Table 2). For example, the α_1 or $\alpha_2\gamma_2\beta$ combination, designated as A1a- and A2a-GABA receptors, respectively, responds to benzodiazepine and nonbenzodiazepine anxiolytics and hypnotics. In contrast, receptors lacking γ-subunit, such as $\alpha_1\beta_x\delta$-, or $\alpha_1\beta_x\varepsilon$- (Table 2), or γ-subunits in combination with α_4 or α_6 are generally insensitive to the benzodiazepines and related drugs *(19)*. Trafficking and localization of $GABA_A$ receptors are determined in large measure by the subunit composition with, for example, γ_2 being important for routing receptors to synapses, whereas the δ-subunit is characteristic of $GABA_A$ receptors that accumulate at extrasynaptic sites (*see* Chapters 3–5 and 8) *(23,24)*.

Although the vast majority of $GABA_A$ receptors are heteromers, homomeric ρ_1-receptors have been identified, as have heterodimers containing only combinations of ρ_1-, ρ_2-, and ρ_3-subunits *(25,26)*. This ρ-containing family of $GABA_A$ receptors displays a unique pharmacological profile, being sensitive to the recognition site agonist *cis*-4-aminocrotonic acid (Fig. 2) but insensitive to bicuculline, benzodiazepines and, in some cases, picrotoxin (Table 2) *(26,27)*. This substrate selectivity led initially to their designation as $GABA_C$ sites, but subsequent

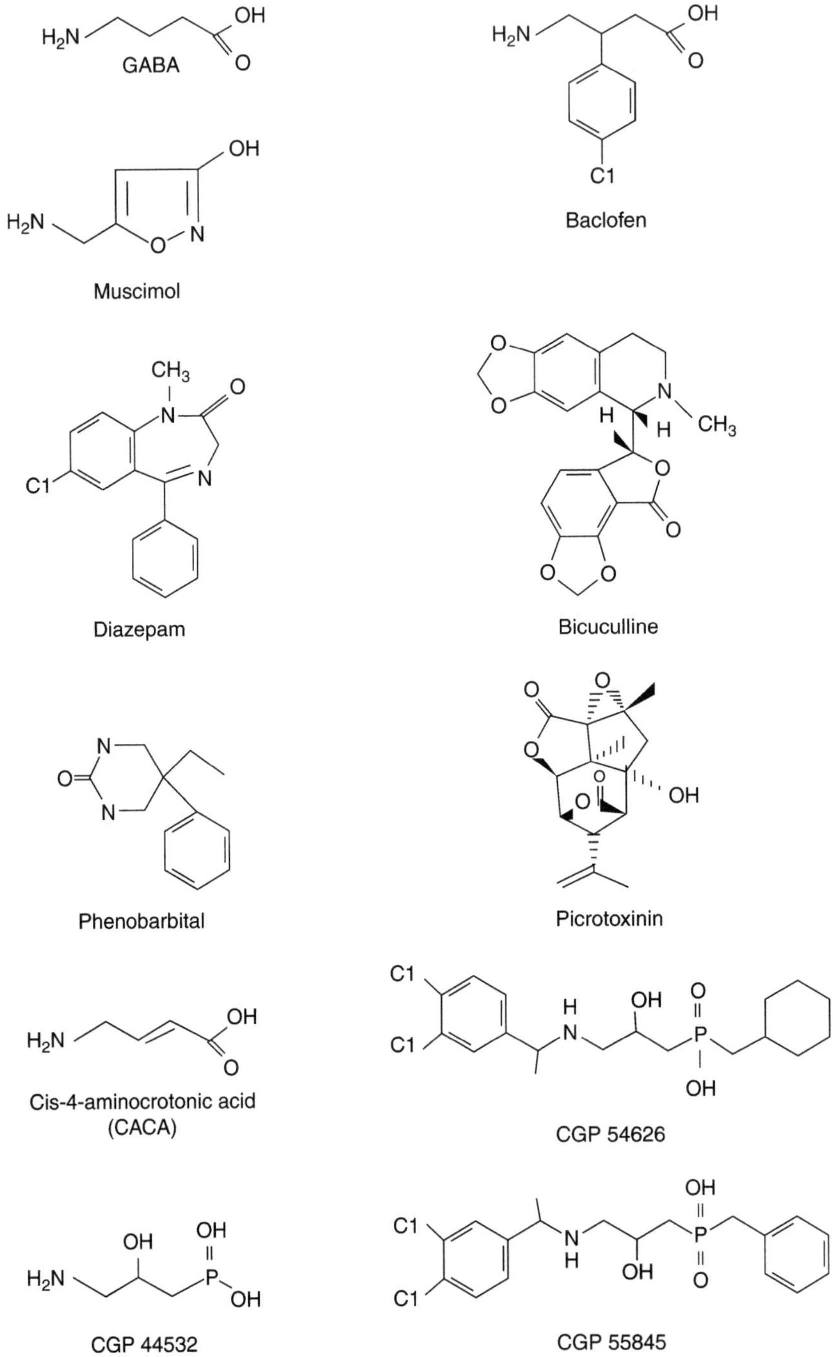

Fig. 2. Chemical structures of various $GABA_A$ and $GABA_B$ receptor agonists and antagonists.

Table 1
GABA Receptor Subunits

Receptor subtype	Subunits	GenBank accession no.	
		Human	Mouse
$GABA_A$	α_1	NM_000806	NM_010250
	α_2	BC022488	NM_008066
	α_3	BC028315	NM_008067
	α_4	NM_000809	BC094603
	α_5	BC111979	BC062112
	α_6	BC096241	NM_008068
	β_1	BC022449	NM_008069
	β_2	NM_000813[a]	NM_008070
	β_3	BC010641[b]	NM_008071[b]
	γ_1	NM_173536	NM_010252
	γ_2	BC069348[a]	NM_177408[a]
	γ_3	NM_033223	NM_008074
	ε	NM_004961[b]	NM_017369[b]
	θ	NM_018558	AK038859
	δ	BC033801	NM_008072
	π	NM_01421	BC023693
	ρ_1	NM_002042	NM_008075
	ρ_2	–	NM_008076
	ρ_3	XM_927388	–
$GABA_B$	$GABA_{B1}$	NM_021905[c]	NM_019439
	$GABA_{B1a}$	AF099148	AF114168
	$GABA_{B2}$	NM_005458	–

[a]Transcript variant 2.
[b]Transcript variant 1.
[c]Transcript variant 4.

characterization of their subunit composition and transduction mechanism revealed them to be a $GABA_A$ receptor subtype. Thus, the molecular characterization of the $GABA_A$ receptor has made possible a more precise definition of the mechanism of action of a number of drugs, and has provided new targets for designing novel therapeutics capable of activating or inhibiting select populations of $GABA_A$ sites.

3.2. Therapeutics

Most drugs that modify $GABA_A$ receptor function enhance the activity of this neurotransmitter system. Included among this group are the benzodiazepines and barbiturates (Fig. 2), zolpidem, and propofol. By attaching to the $GABA_A$ receptor, these agents either increase the frequency (benzodiazepines) or prolong

Table 2
Subunit Composition, Nomenclature, and Pharmacological Properties of Selected $GABA_A$ Receptor Subtypes[a]

		Pharmacology selectivity[b]				
Subunits	IUPHAR nomen-clature	Benzo-diazepine	Barbitu-rates	Bicuculline	Picro-toxinin	*cis*-4-amino-crotonic acid
$\alpha_1, \beta_x\gamma_2$	A1a	+	+	+	+	+
$\alpha_2, \beta_x\gamma_2$	A2a	+	+	+	?	?
ρ_1[c]	AOr1	–	–	–	+	+
$\rho_1\rho_2$[c]	AOr12	–	–	–	+	+
$\alpha_1, \beta_x\delta$	AO1	–	?	+	?	?
$\alpha_1, \beta_x\varepsilon$	AOle	–	?	+	?	?

[a]Adapted from ref. *25*.
[b](+), Responsive; (–), nonresponsive; (?), unknown.
[c]ρ-subunit homomers or heteromers were formerly classified as $GABA_c$ receptors.

the duration (barbiturates) of chloride channel opening in response to GABA. Although flumazenil attaches to the benzodiazepine-binding component of $GABA_A$ receptors, it differs from other benzodiazepines in being a competitive antagonist at this site. With no efficacy of its own, flumazenil is used clinically to reverse the effects of benzodiazepine agonists. Each of these drugs influences only those $GABA_A$ receptors with the requisite subunit composition, making them selective in this regard.

Some agents display differential effects at various $GABA_A$ receptor subtypes. An example is 4,5,6,7-tetrahydroisoxazolol[5,4-c]-pyridin-3-ol (THIP), a direct-acting drug. THIP is a partial agonist at the $\alpha_4\beta_3\gamma_2$-$GABA_A$ site, whereas it is a full agonist at the $\alpha_4\beta_3\delta$-receptor *(28–30)*. THIP is now undergoing clinical trials as a hypnotic agent *(29,30)*. Nonselective drugs that activate $GABA_A$ receptors do so indirectly by inhibiting the metabolism (vigabatrin) or reuptake (tiagabine) of this amino acid, thereby increasing its synaptic content and prolonging its action *(31)*. It has been suggested that the anticonvulsant GABApentin may, in some way, influence GABAergic transmission *(32,33)*, although the weight of evidence suggests its primary site of action is the neuronal calcium channel *(33,34)*. Diazepam and chlordiazepoxide are prescribed for the treatment of anxiety. As is the case for all central nervous system depressants, their long-term use can result in tolerance and physical dependence (*see* Chapter 7).

Other benzodiazepines, including flurazepam, alprazolam, and traizolam, as well as zolpidem, a nonbenzodiazepine that interacts with a subgroup of benzodiazepine receptors, are routinely used as hypnotics *(35)*. Although barbiturates are also available for this purpose, the benzodiazepines and zolpidem are preferred because of their greater margin of safety. Whereas the precise mechanisms of action of chloral hydrate and paraldehyde, two older hypnotics, are still unknown, there is evidence they influence GABAergic transmission, perhaps in a manner similar to ethanol (Fig. 1).

Both the barbiturates, in particular phenobarbital, and the benzodiazepines, such as chlorazepate and clonazepam, are used for the treatment of seizures *(31)*. Gene deletion and mutation studies indicate that alterations in $GABA_A$ subunits can dramatically influence seizure threshold, suggesting such changes may be responsible for some forms of epilepsy (*see* Chapter 6). However, the utility of the benzodiazepines is limited, because tolerance develops to their anticonvulsant effects. This is not the case for phenobarbital at doses used to control generalized tonic–clonic seizures, although tolerance develops to other effects of this agent. Clonazepam is used for the management of absence and myoclonic seizures, whereas chlorazepate is used as adjunctive therapy for complex partial seizures *(31)*. Intravenous diazepam or lorazepam are treatments of choice for terminating status epilepticus, an acute medical emergency. Tiagabine and vigabatrin are also used as antiepileptics *(31)*.

Some barbiturates, such as thiopental, are used as fixed or intravenous general anesthetics. Midazolam, a benzodiazepine, is also administered for this purpose, although benzodiazepines are less complete central nervous system depressants than barbiturates. Although propofol is neither a benzodiazepine nor a barbiturate, its mechanism of action is believed to be similar to the latter (Fig. 1) *(8)*. Benzodiazepines are also used as preanesthetic medications to speed the induction rate of inhalational agents. A number of GABAergic drugs are prescribed as muscle relaxants, and for treating neuropsychiatric conditions and certain types of pain. Whereas diazepam is used as a skeletal muscle relaxant, clonazepam is useful for treating bipolar disorder and for relieving dysesthetic and paroxysmal lancinating pain. Indeed, preclinical data suggest that activation of $GABA_A$ receptors, either directly with recognition site agonists or indirectly by inhibition of GABA uptake or metabolism, yields an antinociceptive response *(36–38)*.

The most common side-effects associated with drugs that enhance $GABA_A$ receptor activity are natural extensions of their pharmacological actions. These include sedation, ataxia, and motor incoordination, reflecting a diminution of central nervous system tone. Other adverse effects are anterograde amnesia and paradoxical excitement. While barbiturate overdose can be fatal owing to medullary depression and loss of respiratory drive, benzodiazepines are safer in this regard.

4. $GABA_B$ Receptors

4.1. *Molecular Pharmacology*

Although baclofen (Fig. 2), a muscle relaxant, was designed as a systemically active GABA receptor agonist, early studies revealed it does not stimulate bicuculline-sensitive sites, the contemporary criteria used for identifying such agents. Subsequently it was discovered that baclofen and GABA regulate the stimulated release of various neurotransmitters through activation of a receptor that is pharmacologically distinct from the bicuculline-sensitive sites *(39)*. These receptors, termed $GABA_B$, are located both pre- and postsynaptically and, unlike the $GABA_A$ site, are coupled to G proteins *(40,41)*. As the $GABA_B$ receptor is associated with G_i and G_o its stimulation reduces the activity of adenylyl cyclase or, through coincident signaling, enhances the production of cyclic AMP *(40,41)*. The predominant response to $GABA_B$ receptor activation is an increase in potassium conductance with a consequent hyperpolarization of the neuron *(42)*. Thus, as with the $GABA_A$ system, GABA serves as an inhibitory neurotransmitter when stimulating the $GABA_B$ site. Activation of $GABA_B$ receptors also reduces neuronal calcium conductance, an effect believed to be responsible for baclofen-induced inhibition of neurotransmitter release *(43)*. In all cases it appears the $GABA_B$ receptor-mediated changes in ion channel activities are owing to liberation of G protein subunits which, in turn, directly influence channel function and the generation of second messengers *(40,41)*. Persistent activation of $GABA_B$ receptors leads to desensitization through a GRK4-dependent process (*see* Chapter 10).

As with the ionotropic $GABA_A$ receptor, the metabotropic $GABA_B$ site is localized primarily to the central nervous system. Although there are data suggesting $GABA_B$ receptors regulate the release of acetylcholine in the enteric nervous system and lungs, and substance P in the pulmonary system *(44,45)*, the predominant responses to systemically administered $GABA_B$ receptor agonists and antagonists appear to be mediated by their effects in the central nervous system.

Biochemical and molecular cloning studies demonstrate the $GABA_B$ receptor is a class III metabotropic, G protein-coupled site (*see* Chapter 11) *(46)*. These experiments reveal this receptor functions as a heterodimer, being made up of $GABA_{B(1)}$ and $GABA_{B(2)}$ subunits, also referred to as $GABA_BR1$ and $GABA_BR2$, each of which is a seven transmembrane spanning protein (Table 1) (Fig. 3). Whereas the recognition site for GABA is located on the $GABA_{B(1)}$ component, the G protein-coupled effector system is selectively associated with the $GABA_{B(2)}$ subunit *(47,48)*. This explains why neither protein is capable of forming a fully functional receptor on its own, although there is some evidence that $GABA_{B(1)}$ homodimers may display some responsiveness to GABA *(49)*. Even though a number of $GABA_B$ receptor subunit isoforms have been

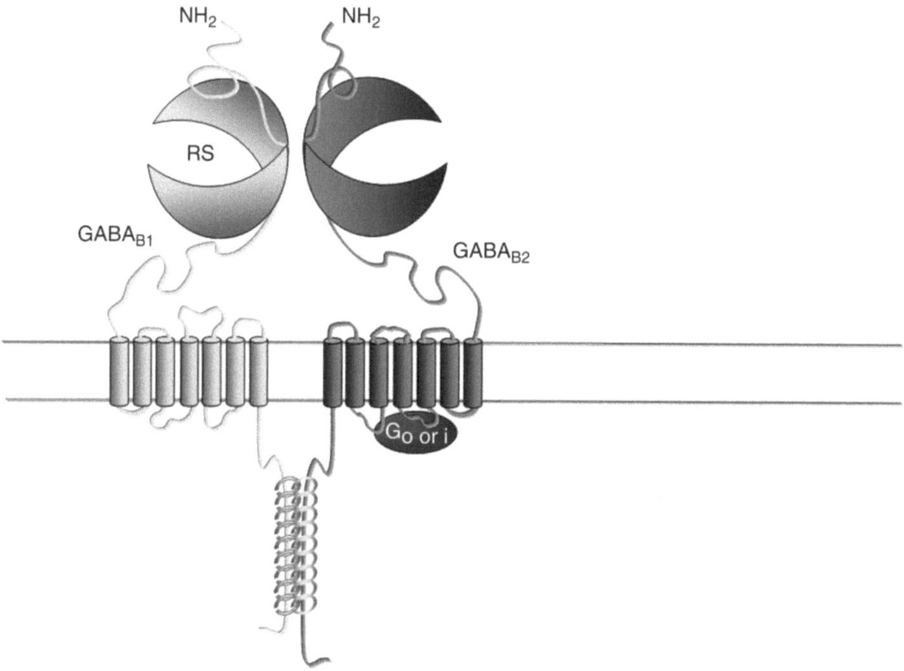

Fig. 3. Schematic representation of a $GABA_B$ receptor illustrating its dimeric structure. $GABA_{B1}$ and $GABA_{B2}$ designate the two subunits, RS the location of the recognition site on $GABA_{B1}$ and G_o or $_i$ the location of the G protein-binding site on $GABA_{B2}$. (Adapted from ref. *60*.)

identified, only $GABA_{B(1a)}$ and $GABA_{B(1b)}$ appear to be capable of forming an active site when either combines with $GABA_{B(2)}$ (*see* Chapter 8) *(5,6,41)*.

Although there is biochemical and electrophysiological evidence suggesting a multiplicity of pharmacologically distinct $GABA_B$ receptors *(50–54)*, it appears the amino acid sequence of the receptor recognition site is identical in all active forms of the $GABA_{B(1)}$ subunit *(41,55,56)*. In addition, gene deletion and mutation studies demonstrate that failure to express either $GABA_{B(1)}$ or $GABA_{B(2)}$ virtually eliminates $GABA_B$ receptor responses, supporting the conclusion that these subunits are uniquely capable of forming a fully functional receptor *(57,58)*. These findings suggest it may not be possible to synthesize $GABA_B$ receptor subtype-selective drugs, although there is evidence that allosteric agents may be capable of discriminating among different splice variants (*see* Chapters 9 and 11) *(9,59,60)*.

Numerous reports suggest a poor correlation between changes in $GABA_B$ subunit expression and receptor function *(61–63)*. For example, mild stress significantly increases $GABA_{B(1a)}$ gene expression in the spinal cord in the

Table 3
$GABA_{B1}$ Subunit Protein Partners

Partner	Function	References
CREB2/ATF4	Transcription factor	*68–70*
CHOP	Transcription factor	*71*
14-3-3	Scaffolding protein	*72*
Marlin-1	RNA-binding protein	*73*
msec7	Guanidine-nucleotide-exchange factor	*74*
Fibulin-2[a]	Extracellular matrix protein	*75*

[a]$GABA_{B1a}$ only.

CREB, cAMP response element binding; ATR4, activating transcription factor 4; CHOP, CCAT/ enhancer-binding protein (C/EBP) homologous protein.

absence of any apparent change in the responsiveness of the $GABA_B$ receptor system *(64)*. These findings indicate subunit proteins may serve functions in the cell, independent of their role as components of the $GABA_B$ receptor. Indeed, $GABA_{B(1)}$ is capable of partnering with a variety of cellular components, including transcription factors, scaffolding, RNA-binding, and extracellular matrix proteins (Table 3) (*see* Chapters 8 and 10) *(51,65–67)*. Some of these interactions are important for trafficking and anchoring the $GABA_B$ receptor heterodimer, or are involved in regulating the receptor-coupled effector system, whereas certain protein pairings could directly influence other cellular activities, such as gene transcription *(68–75)*.

A host of selective $GABA_B$ receptor agonists and antagonists have been developed (Fig. 2) (*see* Chapter 9) *(76,77)*. Included are phosphinic acid derivatives such as the agonist CGP 44532, and the antagonists CGP 54626 and CGP 55845 (Fig. 2). These and other agents were important for initially cloning the gene for the $GABA_{B(1)}$ subunit *(4)* and for characterizing the pharmacological properties of this site (*see* Chapter 12) *(78)*.

4.2. Therapeutics

Baclofen, a receptor recognition site agonist, is the only drug currently in use that directly influences $GABA_B$ receptor activity (*see* Chapter 12). Undoubtedly some of the responses to the nonselective GABAergic stimulants, such as tiagabine and vigabatrin, are the result of $GABA_B$ receptor stimulation, as these agents do not discriminate between GABA receptor subtypes.

Baclofen has for decades been used as a skeletal muscle relaxant. Although it is a primary treatment for spasticity, such as that associated with multiple sclerosis *(79)*, its effectiveness is enhanced by coadministration of diazepam or clonazepam, suggesting involvement of both $GABA_A$ and $GABA_B$ receptors in this condition. As continuous agonist administration desensitizes the $GABA_B$ receptor, tolerance develops to baclofen, limiting its clinical utility. Baclofen, usually in combination with certain anticonvulsants, is used for the treatment of neuropathic pain (*see* Chapter 12) *(38)*. As is the case with $GABA_A$, there are a significant amount of preclinical data indicating that baclofen and other $GABA_B$ receptor agonists display antinociceptive activity in a variety of animal models. It appears this response is a result, in part, of baclofen-induced inhibition of substance P and glutamate release in the spinal cord, interrupting the transmission of the pain impulse to higher centers *(38)*.

Laboratory animal studies suggest that $GABA_B$ receptor antagonists enhance learning and memory in both rodents and primates and that baclofen decreases cognition, although these effects might vary under different conditions (*see* Chapter 12) *(80,81)*. Based on these and other data, a clinical trial was initiated with SGS 742, formerly CGP 36742, a $GABA_B$ receptor antagonist, in patients with mild cognitive impairments to assess its effect on choice reaction time, visual information processing, and working memory *(78)*.

Because $GABA_B$ receptor activation contributes to the generation of abnormal synchronous discharges characteristic of absence epilepsy, it was speculated that $GABA_B$ receptor antagonists may represent a new approach for treating this condition. Studies in laboratory animal models of absence epilepsy indicate that $GABA_B$ receptor antagonists completely suppress these discharges and their behavioral manifestations *(82)*. In fact, numerous reports suggest that modifications in $GABA_B$ receptor expression or function may be responsible for certain forms of epilepsy, and that drugs acting at this receptor may be of benefit in treating these conditions (*see* Chapters 10 and 12).

Both clinical and preclinical data suggest that activation of the $GABA_B$ system reduces the reinforcing effects of addictive substances (*see* Chapter 12) *(83–85)*. Such findings suggest the possible use of baclofen, or other $GABA_B$ receptor agonists, for treating drug abuse. Gene deletion studies have provided insights into possible clinical uses for $GABA_B$ receptor agonists and antagonists (*see* Chapter 11) *(49,57,58,86,87)*. The results confirm that $GABA_{B(1)}$ is absolutely required for formation of a functional $GABA_B$ receptor because mice lacking this subunit gene are totally unresponsive to $GABA_B$ agonists. The phenotype displayed by these animals includes a reduced seizure threshold, retarded growth, hypothermia, hyperlocomotion, hyperalgesia, memory impairment, anxiety, and decreased immobility in the forced swim test (*see* Chapter 11). Some of these findings are perplexing given the results of pharmacological studies. For example, as noted earlier, whereas $GABA_B$ agonists cause absence seizures in laboratory animals, the gene deletion studies indicate that a reduction in $GABA_B$ tone leads to seizures as well, casting doubt on the clinical utility of $GABA_B$ receptor antagonists as anticonvulsants *(88–90)*. In fact, CGP

56999A, a potent $GABA_B$ receptor antagonist, induces seizures in mice *(91)*. It is also notable that, although $GABA_B$ receptor antagonists appear to enhance cognition in laboratory animals, elimination of $GABA_B$ receptor expression compromises memory. These apparently conflicting findings may be the consequence of developmental abnormalities resulting indirectly from the gene deletion rather than being directly associated with the loss of $GABA_B$ receptors.

However, a consistency between the gene deletion and pharmacological studies was found, with respect to the forced swim, analgesia, and locomotion results. Deletion of the $GABA_{B(1)}$ gene decreases immobility in the forced swim test, a result identical to that found with $GABA_B$ receptor antagonists *(92–95)*. This suggests that $GABA_B$ receptor antagonists may display antidepressant properties. Similarly, the hyperalgesia and hyperlocomotion noted in $GABA_{B(1)}$ null mice is consistent with reports that $GABA_B$ receptor agonists increase the pain threshold and reduce locomotor activity *(36,95–97)*.

Because $GABA_B$ receptor agonists influence neurotransmitter release from neurons in various types of smooth muscle, there has been interest in testing such agents as possible treatments for asthma and disorders of the gastrointestinal system and bladder. Although the results of preclinical in vivo and in vitro studies are promising, there are few clinical data to support this hypothesis. Indeed, questions remain about the relevance of $GABA_B$ receptors located on neurotransmitter terminals in peripheral tissues, as there is little, if any, GABA innervation to these sites. This suggests peripheral $GABA_B$ receptors may be of more pharmacological than physiological importance. Side-effects associated with the use of baclofen include sedation and confusion. Other problems encountered with this drug are constipation and urinary retention, both of which are probably secondary to baclofen-induced reductions in parasympathetic drive through inhibition of acetylcholine release. Preclinical and phase I clinical trials with the $GABA_B$ receptor antagonist SGS 742M, the fumarate salt of SGS 742, indicate it is relatively free of side effects and toxicities at the doses used in humans *(78)*.

5. Conclusions

It has been nearly 60 yr since Roberts and Frankel first identified GABA in the mammalian central nervous system *(98)*. The first 15 yr were needed to prove that GABA is a neurotransmitter substance, with studies during the following decade focused on defining its distribution, synthesis, storage, release, and electrophysiological properties. In the mid-1970s attention shifted to GABA receptor sites, which remains the focus to this day. As reviewed in the three editions of this volume, research on GABA receptors has yielded a wealth of information about neurotransmission in general, and GABAergic systems in particular. Characterization of $GABA_A$ receptors contributed substantially the knowledge about allosteric regulation of ligand-gated ion channels. The resulting information has been invaluable in defining precisely the mechanisms of action of a host of drugs that interact with such sites. Among these are the benzodiazepines and other sedative, hypnotic, and anesthetic agents.

Similarly, $GABA_B$ receptor studies have had a broad impact in the neurosciences. Although the existence of G protein-coupled receptor dimers was suspected for some time, they were difficult to demonstrate until the heterodimeric $GABA_B$ site was identified. This finding led subsequently to the discovery of a host of dimerized metabotropic sites, with significant implications regarding the design and development of drugs that interact with these complexes. The characterization of the pentameric $GABA_A$ and the dimeric $GABA_B$ receptors has also been crucial for understanding the neurobiological basis of some central nervous system disorders. Thus, gene deletion and point mutation studies have revealed the importance of GABA receptors in neural development and function. These experiments have also revealed that even subtle modifications in the amino acid composition of receptor subunits can have profound effects on the behavioral phenotype and the response to drugs. Such information has had a significant impact on the direction of research into the causes and treatment of neurological and psychiatric disorders. Given past experience it is likely that further research on GABA receptors will continue to yield information of value to neuroscientists in general, neuropharmacologists, neurologists, and psychiatrists in particular. As outlined in this chapter, and detailed in this volume, the recent studies on GABA receptors continue a tradition established during the past half a century in providing new perspectives on the functions of neurotransmitters and their receptor sites. It is hoped the information contained herein will not only broaden the perspective of those working in this field, but also provide new ideas and insights for scientists in related areas.

Acknowledgments

A special thanks to Ms. Lynn LeCount for her editorial assistance.

References

1. The GABA Receptors (1983) (Enna, S. J., ed.), Humana Press, Clifton, New Jersey, pp. 341.
2. Bowery, N. G., Doble, A., Hill, D. R., et al. (1981) Bicuculline-insensitive GABA receptors on peripheral autonomic nerve terminals. *Eur. J. Pharmacol.* **71,** 53–70.
3. The GABA Receptors (1996) (Bowery, N. G. and Enna, S. J., eds.), 2nd Ed., Humana Press, Clifton, New Jersey, pp. 332.
4. Kaupmann, K., Huggel, K., Heid, J., et al. (1997) Expression cloning of GABA(B) receptors uncovers similarity to metabotropic glutamate receptors. *Nature* **386,** 239–246.

5. Jones, K. A., Borowsky, B., Tamm, J. A., et al. (1998) $GABA_B$ receptor function as a heteromeric assembly of the subunits of $GABA_{BR1}$ and $GABA_{BR2}$. *Nature* **396,** 674–679.
6. Kaupmann, K., Malitschek, B., Schuler, V., et al. (1998) GABA(B)-receptor subtypes assemble into functional heteromeric complexes. *Nature* **396,** 683–687.
7. Angers, S., Salapour, A., and Bouvier, M. (2002) Dimerization: an emerging concept for G protein-coupled receptor ontogeny and function. *Annu. Rev. Pharmacol. Toxicol.* **42,** 409–435.
8. Möhler, H., Fritschy, J. M., Crestani, F., Hensch, T., and Rudolph, U. (2004) Specific $GABA_A$ circuits in brain development and therapy. *Biochem. Pharmacol.* **68,** 1685–1690.
9. Urwyler, S., Mosbacher, J., Lingenhoehl, K., et al. (2001) Positive allosteric modulation of native and recombinant gamma-aminobutyric acid(B) receptors by 2,6-Di-tert-butyl-4-(3-hydroxy-2,2-dimethyl-propyl)- phenol (CGP7930) and its aldehyde analog CGP13501. *Mol. Pharmacol.* **60,** 881–884.
10. Barnard, E. A. (1996) The transmitter-gated channels: a range of receptor types and structures. *Trends Pharmacol. Sci.* **17,** 305–308.
11. Benke, D. and Möhler, H. (2006) GABA-A receptors, in *xPharm* (Enna, S. J., Bylund, D. B., eds.), Elsevier, New York, http://www.xpharm.com/citation?article_ID=532. Last accessed on Jan. 3, 2007.
12. Payne, J. A., Rivera, C., Voipio, J., and Kaila, K. (2003) Catioin-chloride co-transporters in neuronal communication, development and trauma. *Trends Neurosci.* **26,** 199–206.
13. Rivera, C., Voipio, J., and Kaila, K. (2005) Two developmental switches in GABAergic signaling: the K^+-Cl^- cotransporter KCC2 and carbonic anhydrase CAVII. *J. Physiol.* **562,** 27–36.
14. Krnjevic, K. (2004) How does a little acronym become a big transmitter? *Biochem. Pharmacol.* **68,** 1549–1555.
15. Cherubini, E., Giairsa, J. L., and Ben-Ari, Y. (1991) GABA, an excitatory transmitter in early postnatal life. *Trends Neurosci.* **14,** 515–519.
16. Semyanov, A., Walker, M. C., Kullmann, D. M., and Silver, R. A. (2004) Tonically active $GABA_A$ receptors: modulating gain and maintaining the tone. *Trends Neurosci.* **27,** 262–269.
17. Scimemi, A., Semyanov, A., Sperk, G., Kullmann, D. M., and Walker, M. C. (2005) Multiple and plastic receptors mediate tonic $GABA_A$ receptor currents in the hippocampus. *J. Neurosci.* **25,** 10,016–10,024.
18. Adkins, C. E., Pillai, G. V., Kerby, J., et al. (2001) Alpha4beta3delta GABA(A) receptors characterized by fluorescence resonance energy transfer-derived measurements of membrane potential. *J. Biol. Chem.* **276,** 38,934–38,939.
19. Möhler, H., Benke, J., Benson, B., Lüscher, B., Rudolph, U., Fritschy, J. M. (1997) Diversity in structure, pharmacology, and regulation of $GABA_A$ receptors in *The GABA Receptors*, (Enna, S. J. and Bowery, N. G., eds.), 2nd Ed., Humana Press, Totowa, New Jersey, pp. 11–36.
20. Simon, J., Wakimoto, H., Fujita, N., Lalande, M., and Barnard, E. A. (2004) Analysis of the set of $GABA_A$ receptor genes in the human genome. *J. Biol. Chem.* **279,** 41,422–41,435.

21. Kittler, J. T., McAinsh, K., and Moss, S. J. (2002) Mechanisms of $GABA_A$ receptor assembly and trafficking-Implications for the modulation of inhibitory neurotransmission. *Mol. Neurobiol.* **26,** 251–268.
22. Lüscher, B. and Keller, C. A. (2004) Regulation of $GABA_A$ receptor trafficking, channel activity, and functional plasticity at inhibitory synapses. *Pharmacol. Therap.* **102,** 195–221.
23. Nusser, Z., Sieghart, W., and Somogyi, P. (1998) Segregation of different $GABA_A$ receptors to synaptic and extrasynaptic membranes of cerebellar granule cells. *J. Neurosci.* **18,** 1693–1703.
24. Wei, W., Zhang, N., Peng, Z., Houser, C. R., and Mody, I. (2003) Perisynaptic localization of d subunit-containing $GABA_A$ receptors and their activation by GABA spillover in the mouse dentate gyrus. *J. Neurosci.* **23,** 10,650–10,661.
25. Barnard, E. A., Skolnick, P., Olsen, R. W., et al. (1998) International Union of Pharmacology. XV. Subtypes of gamma-aminobutyric acid A receptors: classification on the basis of subunit structure and receptor function. *Pharmacol. Rev.* **50,** 291–313.
26. Johnston, G. A. (2002) Medicinal chemistry and molecular pharmacology of $GABA_c$ receptors. *Curr. Top. Med. Chem.* **2,** 903–913.
27. Bormann, J. (2000) The 'ABC' of GABA receptors. *Trends Pharmacol. Sci.* **21,** 16–19.
28. Wafford, K. A. and Ebert, B. (2006) Gaboxadol—a new awakening in sleep. *Curr. Opin. Pharmacol.* **6,** 30–36.
29. Krogsgaard-Larsen, P., Frolund, B., Liljefors, T., and Ebert, B. (2004) $GABA_A$ agonists and partial agonists: THIP (Gaboxadol) as a non-opioid analgesic and a novel type of hypnotic. *Biochem. Pharmacol.* **68,** 1573–1580.
30. Ebert, B., Wafford, K. A., and Deacon, S. Treating insomnia: current and investigational pharmacological approaches. *Pharmacol. Therap.* **12,** 612–629.
31. Fisher, R. S. (1998) Epilepsy, in *Pharmacological Management of Neurological and Psychiatric Disorders*, (Enna, S. J. and Coyle, J. T., eds.), McGraw-Hill, New York, pp. 459–503.
32. Honmou, O., Kocsis, J. D., and Richardson, G. B. (1995) Gabapentin potentiates the conductance increase induced by nipecotic acid in CA1 pyramidal neurons in vitro. *Epilepsy Res.* **20,** 193–202.
33. Bertrand, S., Ng, G. Y., Purisai, M. G., et al. (2001) The anticonvulsant, antihyperalgesic agent gabapentin is an agonist at brain gamma-aminobutyric acid type B receptors negatively coupled to voltage-dependent calcium channels. *J. Pharmacol. Exp. Ther.* **298,** 15–24.
34. Mortell, K. H., Anderson, D. J., Lynch, J. J. 3rd, et al. (2006) Structure-activity relationships of alpha-amino acid ligands for the alpha2delta subunit of voltage-gated calcium channels. *Bioorg. Med. Chem. Lett.* **16,** 1138–1141.
35. Winokur, A. (1998) Sleep disorders, in *Pharmacological Management of Neurological and Psychiatric Disorders*, (Enna, S. J., and Coyle, J. T., eds.), McGraw-Hill, New York, pp. 213–235.
36. Zorn, S. H. and Enna, S. J. (1985) The effect of mouse spinal cord transaction on the antinociceptive response to the GABA agonists THIP and baclofen. *Brain Res.* **338,** 380–383.

37. Zorn, S. H. and Enna, S. J. (1985) Uptake inhibitors produce a greater antinociceptive response in the mouse tail immersion assay than other types of GABAergic drugs. *Life Sci.* **37,** 1901–1912.
38. Enna, S. J. and McCarson, K. M. (2006) The role of GABA in the mediation and perception of pain. in *Advances in Pharmacology*, (Enna, S. J., ed.), Academic Press New York, Vol. 54, pp. 1–28.
39. Bowery, N. G. (1993) $GABA_B$ receptor pharmacology. *Annu. Rev. Pharmacol. Toxicol.* **33,** 109–147.
40. Enna, S. J. (2001) GABA-B receptor signaling pathways, in *Pharmacology of GABA and Glycine Transmission*, (Möhler, H., ed.) Springer, Heidelberg, pp. 329–342.
41. Enna, S. J. (2001) GABA-B mystery. The search for pharmacologically distinct $GABA_B$ receptors. *Mol. Interven.* **1,** 208–218.
42. Crunelli, V. and Leresche, N. (1991) A role for $GABA_B$ receptors in excitation and inhibition of thalamocortical cells. *Trends Neurosci.* **14,** 16–21.
43. Wall, M. J. and Dale, N. (1994) $GABA_B$ receptors modulate an omega-conotoxin-sensitive calcium current that is required for synaptic transmission in the *Xenopus* embryo spinal cord. *J. Neurosci.* **14,** 6248–6255.
44. Ong, J. and Kerr, D. I. (1990) GABA receptors in peripheral tissues. *Life Sci.* **46,** 1489–1501.
45. Chapman, R. W., Hey, J. A., Rizzo, C. A., and Bolser, D. C. (1993) $GABA_B$ receptors in the lung. *Trends Pharmacol. Sci.* **14,** 26–29.
46. Kubo, Y. and Tateyama, M. (2005) Towards a view of functioning dimeric metabotropic receptors. *Curr. Opin. Neurobiol.* **15,** 289–295.
47. Margeta-Mitrovic, M., Jan, Y. N., and Jan, L. Y. (2000) A trafficking checkpoint controls $GABA_B$ receptor dimerization. *Neuron* **27,** 97–106.
48. Robbins, M. J., Calver, A. R., Fillipov, A. K., Couve, A., Moss, S. J., and Pangalos, M. N. (2001) The $GABA_{B2}$ subunit is essential for G protein coupling of the $GABA_B$ receptor heterodimer. *J. Neurosci.* **21,** 8043–8052.
49. Gassmann, M., Shaban, H., Vigot, R., et al. (2004) Redistribution of GABAB(1) protein and atypical GABAB responses in GABAB(2)-deficient mice. *J. Neurosci.* **24,** 6086–6097.
50. Bowery, N. G., Bettler, B., Froestl, W., et al. (2002) International Union of Pharmacology. XXXIII. Mammalian gamma-aminobutyric acid (B) receptors: structure and function. *Pharmacol. Res.* **54,** 247–264.
51. Möhler, H. and Fritschy, J. M. (1999) GABAB receptors make it to the top as dimmers. *Trends Pharmacol. Sci.* **20,** 87–89.
52. Bonanno, G. and Raiteri, M. (1993) Multiple GABAB receptors. *Trends Pharmacol. Sci.* **14,** 259–261.
53. Cunningham, M. D. and Enna, S. J. (1996) Evidence for pharmacologically distinct GABAB receptors associated with cAMP production in rat brain. *Brain Res.* **720,** 220–224.
54. Yamada, K., Yu, B., and Gallagher, J. P. (1999) Different subtypes of GABAB receptors are present at pre- and postsynaptic sites within the rat dorsolateral septal nucleus. *J. Neurophysiol.* **81,** 2875–2883.

55. Brauner-Osborne, H. and Krogsgaard-Larsen, P. (2000) Sequence and expression pattern of a novel orphan G-protein-coupled receptor, GPRC5B, a family C receptor with a short amino-terminal domain. *Genomics* **65,** 121–128.
56. Robbins, M. J., Charles, K. J., Harrison, D. C., and Pangalos, M. N. (2002) Localization of the GPRC5B receptor in the rat brain and spinal cord. Brain Res. *Mol. Brain Res.* **106,** 136–144.
57. Schuler, V., Lüscher, C., Blanchet, C., et al. (2001) Epilepsy, hyperalgesia, impaired memory, and loss of pre- and postsynaptic GABA(B) responses in mice lacking GABA(B(1)). *Neuron.* **31,** 47–58.
58. Prosser, H. M., Gill, C. H., Hirst, W. D., et al. (2001) Epileptogenesis and enhance prepulse inhibition in GABA (B1)-deficient mice. *Mol. Cell. Neurosci.* **17,** 1059–1070.
59. Urwyler, S., Pozza, M. F., Lingenhoehl, K., et al. (2003) N,N'-Dicyclopentyl-2-methylsulfanyl-5-nitro-pyrimidine-4,6-diamine (GS39783) and structurally related compounds: novel allosteric enhancers of gamma-aminobutyric acidB receptor function. *J. Pharmacol. Exp. Ther.* **307,** 322–330.
60. Couve, A., Calver, A. R., Fairfax, B., Moss, S. J., and Pangalos, M. N. (2004) Unraveling the unusual signaling properties of the $GABA_B$ receptor. *Biochem. Pharmacol.* **68,** 1527–1536.
61. Sands, S. A., McCarson, K. E., and Enna, S. J. (2004) Relationship between the antinociceptive response to desipramine and changes in $GABA_B$ receptor function and subunit expression in the dorsal horn of the rat spinal cord. *Biochem. Pharmacol.* **67,** 743–749.
62. Sands, S. A., McCarson, K. E., and Enna, S. J. (2003) Differential regulation of $GABA_B$ receptor subunit expression and function. *J. Pharmacol. Exp. Ther.* **305,** 191–196.
63. Sands, S. A., Reisman, S. A., and Enna, S. J. (2004) Effect of antidepressants on $GABA_B$ receptor function and subunit expression in the rat hippocampus. *Biochem. Pharmacol.* **68,** 1489–1495.
64. McCarson, K. E., Duric, V., Reisman, S. A., Winter, M., and Enna, S. J. (2006) $GABA_B$ receptor function and subunit expression in the rat spinal cord as indicators of stress and the antinociceptive response to antidepressants. *Brain Res.* **1068,** 109–117.
65. Marshall, F. H., Jones, K. A., Kaupmann, K., and Bettler, B. (1999) GABAB receptors – the first 7TM heterodimers. *Trends Pharmacol. Sci.* **20,** 396–399.
66. Lehtinen, M. J., Meri, S., and Jokiranta, T. S. (2004) Interdomain contact regions and angles between adjacent short consensus repeat domains. *J. Mol. Biol.* **344,** 1385–1396.
67. Calver, A. R., Davies, C. H., and Pangalos, M. (2002) GABA(B) receptors: from monogamy to promiscuity. *Neurosignals* **11,** 299–314.
68. Nehring, R. B., Horikawa, H. P., El Far, O., et al. (2000) The metabotropic $GABA_B$ receptor directly interacts with the activating transcription factor 4. *J. Biol. Chem.* **275,** 35,185–35,191.
69. Vernon, E., Meyer, G., Pickard, L., et al. (2001) GABA(B) receptors couple directly to the transcription factor ATF4. *Mol. Cell. Neurosci.* **17,** 637–645.

70. White, J. H., McIlhinney, R. A., Wise, A., et al. (2000) The $GABA_B$ receptor interacts directly with the related transcription factors CREB2 and ATFx. *Proc. Natl. Acad. Sci. USA* **97,** 13,967–13,972.
71. Sauter, K., Grampp, T., Fritschy, J. M., ct al. (2005) Subtype-selective interaction with the transcription factor CCAAT/enhancer-binding protein (C/EBP) homologous protein (CHOP) regulates cell surface expression of GABA(B) receptors. *J. Biol. Chem.* **280,** 33,566–33,572.
72. Couve, A., Kittler, J. T., Uren, J. M., et al. (2001) Association of GABA(B) receptors and members of the 14-3-3 family of signaling proteins. *Mol. Cell. Neurosci.* **17,** 317–328.
73. Couve, A., Restituito, S., Brandon, J. M., et al. (2004) Marlin-1, a novel RNA-binding protein associates with GABA receptors. *J. Biol. Chem.* **279,** 13,934–13,943.
74. Restituito, S., Couve, A., Bawagan, H., et al. (2005) Multiple motifs regulate the trafficking of $GABA_B$ receptors at distinct checkpoints within the secretory pathway. *Mol. Cell. Neurosci.* **28,** 747–756.
75. Blein, S., Ginham, R., Uhrin, D., et al. (2004) Structural analysis of the complement control protein (CCP) modules of $GABA_B$ receptor 1a: only one of the two CCP modules is compactly folded. *J. Biol. Chem.* **279,** 48,292–48,306.
76. Froestl, W., Mickel, S. J., Von Sprecher, G., et al. (1995) Phosphinic acid analogues of GABA. 2. Selective, orally active $GABA_B$ antagonists. *J. Med. Chem.* **38,** 3313–3331.
77. Enna, S. J. (1997) γ –Aminobutyric acid-B ($GABA_B$) receptor agonists and antagonists: pharmacological properties and therapeutic possibilities. *Exp. Opin. Invest. Drugs* **6,** 1319–1325.
78. Froestl, W., Gallagher, M., Jenkins, H., et al. (2004) SGS742: the first $GABA_B$ receptor antagonist in clinical trials. *Biochem. Pharmacol.* **68,** 1479–1487.
79. Guarnaccia, J. B., Bollmer, T. L., and Waxman, S. G. (1998) Multiple sclerosis, in *Pharmacological Management of Neurological and Psychiatric Disorders*, (Enna, S. J. and Coyle, J. T., eds.), McGraw-Hill, New York, pp. 377–428.
80. Mondadori, C., Preiswerk, G., and Jaekel, J. (1992) Treatment with a $GABA_B$ receptor blocker improves the cognitive performance of mice, rats and rhesus monkeys. *Pharmacol. Comm.* **2,** 93–97.
81. DeSousa, N. J., Benninger, R., Jhamandas, K., and Boegman, R. J. (1994) Stimulation of $GABA_B$ receptors in the basal forebrain selectively impairs working memory of rats in the double Y-maze. *Brain Res.* **641,** 29–38.
82. Snead, O. C. (1995) Basic mechanisms of generalized absence epilepsy. *Ann. Neurol.* **37,** 146–157.
83. Corrigall, W. A., Coen, K. M., Adamson, K. L., Chow, B. L., and Zhang, J. (2000) Response of nicotine self-administration in the rat to manipulations of mu-opioid and gamma-aminobutyric acid receptors in the ventral tegmental area. *Psychopharmacology* **149,** 107–114.
84. Haney, M., Hart, C. L., and Foltin, R. W. (2006) Effects of baclofen on cocaine self-administration: opioid- and nonopioid-dependent volunteers. *Neuropsychopharmacology* **31,** 1814–1821.

85. Flannery, B. A., Garbutt, J. C., Cody, M. W., et al. (2004) Baclofen for alcohol dependence: a preliminary open-label study. *Alcohol Clin. Exp. Res.* **28,** 1517–1523.
86. Queva, C., Bremner-Danielsen, M., Edlund, A., et al. (2003) Effects of GABA agonists on body temperature regulation in GABA(B(1))-/- mice. *Br. J. Pharmacol.* **140,** 315–322.
87. Thuault, S. J., Brown, J. T., Sheardown, S. A., et al. (2004) The $GABA_{B2}$ subunit is critical for the trafficking and function of native $GABA_B$ receptors. *Biochem. Pharmacol.* **68,** 1655–1666.
88. Getova, D. P., Bowery, N. G., and Spassov, V. (1997) Effects of GABAB receptor antagonists on learning and memory retention in a rat model of absence epilepsy. *Eur. J. Pharmacol.* **320,** 9–13.
89. Hosford, D. A., Wang, Y., Liu, C. C., and Snead, O. C. (1995) Characterization of the antiabsence effects of SCH 50911, a GABA-B receptor antagonist, in the lethargic mouse, gamma-hydroxybutyrate, and pentylenetetrazole models. *J. Pharmacol. Exp. Ther.* **274,** 1399–1403.
90. Vergnes, M., Boehrer, A., Simler, S., Bernasconi, R., and Marescaux, C. (1997) Opposite effects of GABAB receptor antagonists on absences and convulsive seizures. *Eur. J. Pharmacol.* **332,** 245–255.
91. Badran, S., Schmutz, M., and Olpe, H. R. (1997) Comparative in vivo and in vitro studies with the potent GABAB receptor antagonist, CGP 56999A. *Eur. J. Pharmacol.* **333,** 135–142.
92. Nakagawa, Y., Sasaki, A., and Takashima, T. (1999) The GABA(B) receptor antagonist CGP36742 improves learned helplessness in rats. *Eur. J. Pharmacol.* **381,** 1–7.
93. Heese, K., Otten, U., Mathivet, P., Raiteri, M., Marescaux, C., and Bernasconi, R. (2003) GABA(B) receptor antagonists elevate both mRNA and protein levels of the neurotrophins nerve growth factor (NGF) and brain-derived neurotrophic factor (BDNF) but not neurotrophin-3 (NT-3) in brain and spinal cord of rats. *Neuropharmacology* **39,** 449–462.
94. Mombereau, C., Kaupmann, K., Froestl, W., Sansig, G., van der Putten, H., and Cryan, J. F. (2004) Genetic and pharmacological evidence of a role for GABA(B) receptors in the modulation of anxiety- and antidepressant-like behavior. *Neuropsychopharmacology* **29,** 1050–1062.
95. Vaught, J. L., Pelly, K., Costa, L. G., Settler, P., and Enna, S. J. (1985) A comparison of the antinociceptive responses to the GABA receptor agonists THIP and baclofen. *Neuropharmacology* **24,** 211–216.
96. Smith, G. D., Harrison, S. M., Birch, P. J., Elliott, P. J., Malcangio, M., and Bowery, N. G. (1994) Increased sensitivity to the antinociceptive activity of (±)-baclofen in an animal model of chronic neuropathic, but not chronic inflammatory hyperalgesia. *Neuropharmacology* **33,** 1103–1108.
97. Fromm, G. (1994) Baclofen as an adjuvant analgesic. *J. Pain Symptom Manage.* **9,** 500–509.
98. Roberts, E. and Frankel, S. (1950) Gamma-aminobutyric acid in brain: its formation from glutamic acid. *J. Biol. Chem.* **187,** 55–63.

2

Functional Relevance of $GABA_A$-Receptor Subtypes

Hanns Möhler

Summary

Oscillatory activities are hallmarks of neuronal network functions. They contribute to information processing in a multitude of brain activities including cognitive functions. A key requirement for the generation of network oscillations is regular and synchronized neuronal activity. The temporal structure and spatial coherence of network oscillations is largely controlled by GABAergic interneurons. To provide the appropriate response characteristics the synaptic specialization of these interneurons includes an extensive diversity of $GABA_A$ receptors. The cell- and domain-specific location of these receptor subtypes offers the possibility to gain functional insights into the role of defined neuronal circuits. Pharmacologically, neurons operating with α_1GABA_A receptors mediate sedation while the small population of α_2GABA_A receptors largely mediates anxiolysis. Schizophrenia-related phenotypes can be ameliorated through α_3GABA_A receptors. Memory and learning are enhanced by diminishing α_5GABA_A receptor function. Thus, by targeting selective neuronal circuits $GABA_A$ receptor subtypes provide a new circuit-specific GABA pharmacology for a multitude of CNS dieases.

Key Words: Benzodiazepines; anxiolytics; memory enhancers; schizophrenia; sedation; anaesthesia.

1. Introduction

The dynamics of neural networks are largely shaped by the activity pattern of interneurons, most of which are γ-aminobutyric acid (GABA)ergic *(1–6)*. The

From: *The Receptors: The GABA Receptors, Third Edition*
Edited by: S. J. Enna and H. Möhler

activity of these interneurons is believed to set the spatio-temporal conditions required for different patterns of network oscillations that may be critical for information processing *(2,6–14)*. The response properties of interneuron signaling are shaped by the type of $GABA_A$ receptor expressed synaptically or extrasynaptically. For instance, two types of basket cells innervate the soma of hippocampal pyramidal cells. The fast spiking parvalbumin-containing basket cells form synapses containing $\alpha_1 GABA_A$ receptors, which display fast kinetics of deactivation *(3,5,15,16)*. In contrast, the synapses of the regular spiking cholecystokinin-positive basket cells contain α_2-$GABA_A$ receptors, which display slower kinetics than α_1-receptors *(15,17,18–20)*. Axon initial segments of principal cells also contain α_2-receptors, which appear to be kinetically sufficient for simple on/off signaling. Furthermore, distinct $GABA_A$ receptors are segregated to synaptic and extrasynaptic membranes *(21,22)*. Thus, functionally specialized interneurons operate with the kinetically appropriate $GABA_A$-receptor subtypes to regulate network behavior (Figs. 1 and 2). As GABAergic interneurons are operative throughout the brain, a highly diverse repertoire of $GABA_A$ receptors is required.

2. Diversity of $GABA_A$ Receptors

The physiological significance of the structural diversity of $GABA_A$ receptors lies in the provision of receptors, which differ in their channel kinetics, affinity for GABA, rate of desensitization, and ability for transient chemical modification such as phosphorylation. In addition, $GABA_A$-receptor subtypes can show a cell-type specific expression and—in case of multiple receptor subtypes present in a neuron—a domain-specific location. Based on the presence of seven subunit families including at least 18 subunits in the central nervous system (α_{1-6}, β_{1-3}, γ_{1-3}, δ, ε, θ, ρ_{1-3}) the $GABA_A$-receptors display an extraordinary structural heterogeneity. Most $GABA_A$-receptors subtypes in vivo are considered to be heteropentamers made up of isoforms of α-, β-, and γ-subunits (Fig. 1). In certain extrasynaptic receptors the γ-subunit is replaced by the δ-subunit (for review *see* refs. *23–32*).

2.1. *Diazepam-Sensitive $GABA_A$ Receptors*

Receptors containing the α_1-, α_2-, α_3-, or α_5-subunits in combination with any of the β-subunits and the γ_2-subunit are most prevalent in the brain. These receptors are sensitive to benzodiazepine modulation. The major receptor subtype is assembled from the subunits $\alpha_1\beta_2\gamma_2$, with only a few brain regions lacking this receptor (e.g., granule cell layer of the olfactory bulb, reticular nucleus of the thalamus, spinal cord motoneurons) *(22,33,34)*.

Receptors containing the α_2- or α_3-subunit are considerably less abundant and are highly expressed in brain areas where the α_1-subunit is absent or present at low levels. The α_2- and α_3-subunits are frequently coexpressed with the

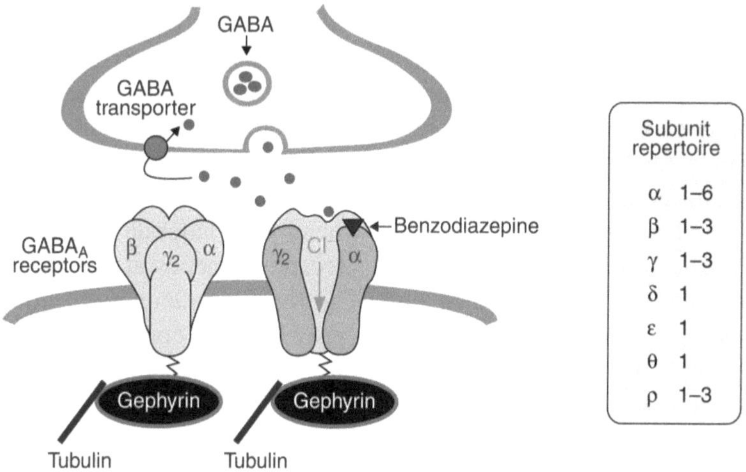

Fig. 1. Scheme of GABAergic synapse depicting major elements of signal transduction. The GABA$_A$ receptors are heteropentameric membrane proteins, which are linked by a yet unknown linker protein to the synaptic anchoring protein gephyrin and the cytoskeleton *(28)*.

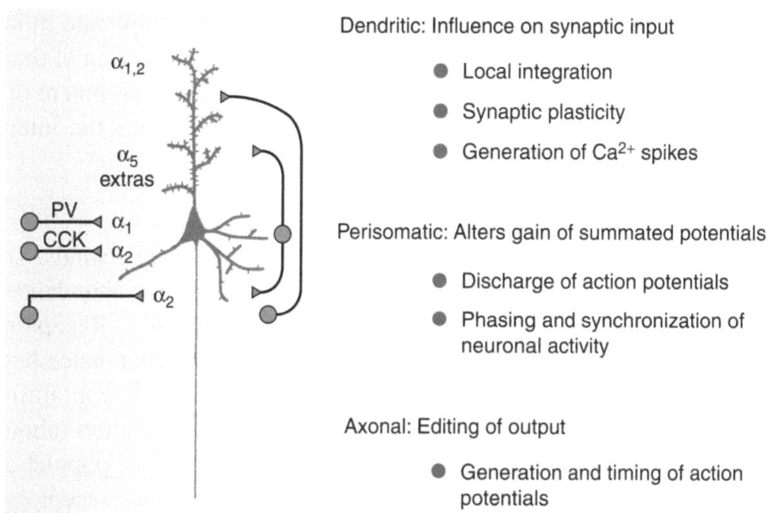

Fig. 2. Domain-specific inhibitory innervation of hippocampal pyramidal cell (center) as shown schematically for five types of GABAergic interneurons. The domain-specific postsynaptic GABA$_A$-receptor subtypes are indicated by the preponderant type of α-subunit. PV and CCK indicate parvalbumin and cholecystokinin containing GABAergic interneurons, respectively. extras, extrasynaptic.

β_3- and γ_2-subunits, which is particularly evident in hippocampal pyramidal neurons ($\alpha_2\beta_3\gamma_2$) and in cholinergic neurons of the basal forebrain ($\alpha_3\beta_3\gamma_2$). The α_3GABA$_A$ receptors are the main subtypes expressed in monoaminergic and basal forebrain cholinergic cells *(35),* and in addition, are strategically located in the thalamic reticular nucleus for modulating the thalamo-cortical circuit *(36)*. Marked differences in desensitization kinetics have been reported between synaptic α_2- and extrasynaptic α_3-receptors whereby the latter desensitize very slowly *(37)*. The factors regulating GABA$_A$ receptor kinetics at synaptic and extrasynaptic sites are yet unknown *(38)*. The ligand-binding profile of the α_2- and α_3-receptors differs from that of $\alpha_1\beta_2\gamma_2$ by having a considerably lower displacing potency for ligands such as βCCM, CL 218,872, and zolpidem.

Receptors containing the α_5-subunit are of minor abundance in the brain but are expressed to a significant extent in the hippocampus, where they include 15–20% of the diazepam-sensitive GABA$_A$ receptor population, predominately coassembled with the β_3- and γ_2-subunits. Pharmacologically, the α_5-receptors are differentiated from $\alpha_1\beta_2\gamma_2$-, $\alpha_2\beta_3\gamma_2$-, and $\alpha_3\beta_3\gamma_2$-receptors by a lower affinity to CL 218, 872 and near-insensitivity to zolpidem.

The subunits-γ_1 and -γ_3 characterize a small population of receptors that contain various types of α- and β-subunits. Owing to their reduced affinity for the classical benzodiazepines they do not appear to contribute to any great extent to their pharmacology in vivo. It should be kept in mind that complex benzodiazepine actions such as the development of tolerance can implicate more than a single receptor subtype. For instance, whereas the sedative action of diazepam is mediated by α_1GABA$_A$ receptors (*see* Heading 3.), the development of tolerance to this action under chronic diazepam treatment requires the interaction with both α_1GABA$_A$- and α_5GABA$_A$ receptors *(39)*.

2.2. *Diazepam-Insensitive GABAA Receptors*

GABA$_A$ receptors that do not respond to clinically used ligands, such as diazepam, flunitrazepam, clonazepam, and zolpidem are of low abundance in the brain and are largely characterized by the α_4- and α_6-subunits. Receptors containing the α_4-subunit are generally expressed at very low abundance but more prominently in thalamus and dentate gyrus *(34)*; those containing the α_6-subunit are restricted to the granule cell layer of the cerebellum (about 30% of all GABA$_A$ receptors in the cerebellum *[40]*). Both receptor populations are structurally heterogeneous, and the majority of the α_6-containing receptors are of the $\alpha_6\beta_2\gamma_2$ combination. Apart form the lack of affinity of classical benzodiazepines, the benzodiazepine-site profile of α_4- and α_6-receptors is characterized by a low affinity for flumazenil and bretazenil and an agonistic efficacy of Ro 15-4513 and bretazenil *(41)*. The δ-subunit is frequently coassembled with the α_4- or the α_6-subunit in benzodiazepine insensitive receptors *(24,26,29)*. Receptors containing the δ-subunit are located exclusively at extrasynaptic sites as shown in dentate gyrus and cerebellum. They are tailor made for tonic inhibition, owing to their high affinity for GABA and slow desensitization kinetics *(42,43)*.

In the retina homomeric receptors consisting of the ρ-subunit represent a particular class of GABA-gated chloride channels. Their GABA site is insensitive to bicuculline and baclofen and they are not modulated by barbiturates or benzodiazepines. Owing to these distinctive features the receptors are frequently termed GABA$_c$-receptors *(44)*, although they are a homomeric class of GABA$_A$-receptors *(23)*.

3. A New GABA$_A$-Receptor Subtype Pharmacology

Classical benzodiazepines do not differentiate among GABA$_A$-receptor subtypes and act indiscriminately at all subtypes. However, the selective pattern of expression of GABA$_A$-receptor subtypes opened the possibility to modulate distinct neuronal circuits, provided novel ligands were found, which displayed a differential interaction with GABA$_A$-receptor subtypes based on either selective affinity or selective efficacy (Table 1). Such agents would be expected to share with the classical benzodiazepines the excellent overall tolerability, but display therapeutic indications which are more selective than those of the classical benzodiazepines and go beyond their spectrum of activity.

As a prerequisite, it had to be determined which pharmacological effects were mediated by distinct GABA$_A$-receptor subtypes. The dissection of the receptor pharmacology was achieved experimentally by generating four lines of point-mutated mice in which the receptors containing the α_1-, α_2-, α_3-, or α_5-subunits, respectively, had been rendered diazepam-insensitive by replacing a conserved histidine residue (H) in the drug-binding domain by an arginine (R) *(45–47)*. In the respective point mutated mice the pharmacological action linked to the point-mutated receptor would be missing and thereby reveal the pharmacological relevance of the respective receptor in wild-type mice. As the subunit composition and distribution of GABA$_A$-receptor subtypes is largely conserved between rodents and nonhuman primates the results were expected to be relevant for the human condition *(48)*.

4. Modulation of Sleep

Frequently, sedation is taken as a surrogate marker for hypnotic activity. The sedative component of benzodiazepines, measured by the reduction of locomotor activity, has been attributed to neuronal circuits expressing α_1GABA$_A$ receptors, the most prevalent receptor subtype in the brain. Mice in which the α_1GABA$_A$ receptor had been rendered diazepam-insensitive by a point mutation (α_1[H101R]) failed to be sedated by diazepam *(45,49)*. Ligands with preferential affinity for α_1-receptors such as zolpidem or zaleplon are used as hypnotics. Similarly, the changes in the electroencephalogram (EEG) pattern

Table 1
$GABA_A$-Receptor Subtype Ligands

Drug	Main activity	Interaction with recombinant $GABA_A$ receptors[a,b]	References
Benzodiazepine site ligands			
Zolpidem	Hypnotic	Preferential affinity for α_1	*82*
Zaleplone	Hypnotic	Preferential affinity for α_1	*82*
Indiplon	Hypnotic	Preferential affinity for α_1	*83*
L-838 417	Anxiolytic	Comparable affinity at α_1-, α_2-, α_3-, α_5-subtype. Partial agonist at α_2-, α_3-, α_5-(not α_1-) subtype	*49*
Ocinaplon	Anxiolytic	Comparable affinity at α_1-, α_2-, α_3-, α_5-subtype. Partial agonist at α_2-, α_3-, α_5-subtype nearly full agonist at α_1	*84*
SL 651 498	Anxiolytic	Agonist at α_2, α_3, partial agonist at α_1- and α_5-subtype	*85*
TPA 023	Anxiolytic	Partial agonist at α_2-, α_3- subtypes, antagonist at α_1-, α_5-subtypes	*86*
TPA 003	Anxiolytic	Partial agonist at α_3-subtype	*59*
ELB 139	Anxiolytic	Selective receptor profile uncertain	*87*
L-655 708	Memory enhancer, anxiogenic	Partial inverse agonist with preference for α_5-subtype	*88–91*
α_3 IA	Anxiogenic	Weak inverse agonist at α_3	*92*
Modulatory site other than benzodiazepine site			
Ethanol	Anxiolytic sedative	High sensitivity (≥3 m*M*) at $\alpha_4(\alpha_6)\ \beta_3\delta^c$, medium sensitivity (≥30 m*M*) at $\alpha_4(\alpha_6)\ \beta_2\delta^c$, low sensitivity (≥100 m*M*) at $\alpha_4(\alpha_6)\ \beta_3\gamma_2$	*93*
Neurosteroids (e.g., 3α-, 5α-THDOC)	Anxiolytic, sedative, anesthetic	High sensitivity at δ-containing subtypes[c] and at α_1-, α_3-receptors in combination with β_1	*94*

Intravenous anesthetics (Etomidate and Propofol)	Sedative, anesthetic	Act on receptor subtypes containing β_3, i.e., mainly α_2- and α_3-subtypes	*70*
GABA site			
Gaboxadol	Hypnotic	Partial agonist at α_1-, α_3-subtypes, full agonist at α_5, and superagonist at $\alpha_4\beta_3\delta$-receptors[c]	*54*

This table is a modified version from ref. *32*.

Abbreviation: THDOC, 5α-pregnane3α,21-diol-20-one.

[a]Classical partial agonists which do not differentiate between $GABA_A$-receptor subtypes such as Bretazenil *(95)* or Pagoclone *(96)* are not considered in this review.

[b]Data should be treated with caution as properties of recombinant receptors that are expressed in foreign host cells might not give an accurate reflection of their neuronal counterparts.

[c]GABA is a weak partial agonist on δ-containing receptors, which largely explains the strong modulatory response of ligands acting on δ-containing receptors *(97)*.

induced by zolpidem in wild-type mice were almost exclusively mediated through α_1GABA$_A$ receptors *(50)*. However, the changes in sleep architecture (suppression of rapid eye movement [REM] sleep) and EEG-frequency profiles (reduction of slow-wave sleep, increase in fast β-frequencies) induced by classical benzodiazepines are largely owing to effects mediated by receptors others than α_1 *(51)*. The enhancement of α_2GABA$_A$ receptors by diazepam appears to have the most pronounced effect on the sleep EEG in wild-type mice. When the α_2GABA$_A$ receptor was rendered diazepam-insensitive by a point mutation [α_2(H101R)], the diazepam-induced suppression of δ-waves, the increase in fast β-waves in non-REM sleep (>16 Hz), and the diazepam-induced increase of θ-waves in REM sleep were strongly attenuated *(52)*. Thus, the hypnotic EEG fingerprint of diazepam can be dissociated from its sedative action. Future hypnotics might target changes in the EEG pattern, which are characteristic of physiological sleep and thereby aim at improving sleep quality. For instance, the GABA-mimetic gaboxadol (4,5,6,7-tetrahydroisoxazolo[5,4-c]pyridine-3-01 hydrochloride), which interacts preferentially with $\alpha_4\beta_3\delta$-GABA$_A$ receptors in vitro *(53,54)* was found to enhance slow-wave sleep in vivo *(55,56)*.

5. Anxiolytics Devoid of Sedation

As α_1GABA$_A$ receptors were found to mediate sedation but not anxiolysis *(45,49)*, the anxiolytic activity of benzodiazepines was expected to reside in one or several of the remaining benzodiazepine-sensitive GABA$_A$ receptors (α_2, α_3, α_5). The differentiation of GABA$_A$ receptors by knockin point mutations showed that it was the α_2-, but not the α_3- or α_5GABA$_A$ receptor, which mediated the anxiolytic activity of diazepam *(46,47)*. In α_2(H101R) mice, but not α_3(H102R) or α_5(H105R) mice, diazepam failed to induce anxiolytic activity (light-dark paradigm, elevated plus maze). With the α_2GABA$_A$ receptor a highly selective target for the anxiolytic activity of benzodiazepine tranquillizers had been identified. In keeping with this notion, the benzodiazepine site ligand L-838417, which showed efficacy at α_2, α_3, and α_5 but not α_1GABA$_A$ receptors, proved to be anxiolytic in wild-type rats (Table 1) *(49)*. Similarly, partial agonists of 3-heteroaryl-2-pyridones acting at the benzodiazepine site with efficacy at α_2-, α_3-, and α_5-receptors, but not at α_1 receptors, were found to show anxiolytic activity in rodents (Table 1) *(57)*. It remained to be clarified to what extent the α_3GABA$_A$ receptor component contributed to the anxiolytic activity of these ligands. In mice that lacked α_3GABA$_A$ receptors, the anxiolytic activity of diazepam was undiminished *(58)*. However, an α_3-selective inverse agonist was anxiogenic and proconvulsant in rodents (Table 1) *(57)*. In addition, TP003 with selective efficacy at α_3GABA$_A$ receptor was anxiolytic, although only at high receptor occupancy *(59)*. Classical benzodiazepines exert anxiolysis at low receptor occupancy, suggesting that the α_2GABA$_A$ receptors rather than the α_3GABA$_A$

receptors are the major mediators of this activity. The contribution of α_3GABA$_A$ receptors is unlikely to be of major relevance. Thus, the strategy to develop novel daytime anxiolytics, which are free of sedation, is clear *(27,28,60)*.

α_2GABA$_A$ receptors by their preponderant localization on the axon-initial segment of principal cells in cerebral cortex and hippocampus can control the output of these cells. In addition, α_2-receptors are the only GABA$_A$ receptors found in the central nucleus of the amygdala, a key area for the control of emotions *(33)*. Thus, by their strategic distribution in brain areas involved in anxiety responses, α_2GABA$_A$ receptors are key substrates for anxiolytic drug action.

6. Enhancement of Learning and Memory

Hippocampal pyramidal cells express various structurally diverse GABA$_A$ receptors in a domain-specific manner. Whereas α_1- and α_2GABA$_A$ receptors are largely synaptic, α_5GABA$_A$ receptors are located extrasynaptically at the base of the spines and on the adjacent shaft of the pyramidal cell dendrite. The α_5GABA$_A$ receptors are therefore in a privileged position to modulate the excitatory input arising at the spines through NMDA (*N*-methyl-D-aspartate) receptors. The introduction of a point mutation (H105R) in the α_5-subunit is associated with a specific reduction of the hippocampal α_5-subunit containing GABA$_A$ receptors, whereas the pattern of distribution is undisturbed *(47)*. Mice with a partial deficit of α_5GABA$_A$ receptors in hippocampus, showed an improved performance in trace fear conditioning, a hippocampus dependent memory task *(47)*. In addition, these mutants displayed a resistance to extinction of conditional fear for several days *(61)*. Similarly, in a mouse line in which α_5GABA$_A$ receptors were deleted in the entire brain *(60,62)* an improved performance in the water maze model of spatial learning was observed. Furthermore, a partial inverse agonist acting at α_5GABA$_A$ receptors enhanced the performance of wild-type rats in the water maze test *(63)* (Table 1). Thus, neuronal inhibition in the hippocampus mediated through α_5GABA$_A$ receptors is a critical element in the regulation of the acquisition and expression of associative memory.

It is striking that the behavioral consequences of an impairment of α_5GABA$_A$ receptors are opposite to those of an NMDA receptor deficit, as shown in spatial and temporal associative memory tasks. Whereas mice with a deficit in hippocampal NMDA receptors show a deficit in the formation of spatial and temporal memory *(64,65)*, the mice with a partial deficit of α_5GABA$_A$ receptors in hippocampus display an improvement in spatial and temporal memory performance. Thus, it appears that these two receptor systems play a complementary role in controlling signal transduction at the hippocampal principal cells *(31)*. Although the initial results with α_5-selective partial inverse agonist, described above, support a role in memory function, it has to be verified that such ligands do not interfere with other hippocampal functions such as sensorimotor gating.

7. Improving Sensorimotor Processing in Schizophrenia

A deficit in GABAergic inhibitory control is one of the major hypothesis underlying the symptomatology of schizophrenia *(66)*. A potential contribution of $GABA_A$-receptor subtypes was therefore investigated regarding the overactivity of the dopaminergic system, considered to be a major factor in schizophrenia. The dopaminergic system is under GABAergic inhibitory control mainly through α_3-containing $GABA_A$ receptors *(33,34)*. Their functional role was explored in mice lacking the α_3-subunit gene. α_3-Knockout mice displayed no adaptive changes in the expression of α_1-, α_2-, and α_5-subunits and anxiety-related behavior was normal. However, the mice displayed a marked deficit in prepulse inhibition of the acoustic startle reflex, pointing to a deficit in sensorimotor information processing *(58)*. This deficit in prepulse inhibition was normalized by administration of the antipsychotic D2 receptor antagonist haloperidol, suggesting that the phenotype is caused by hyperdopaminergia *(58)*. Attenuation of prepulse inhibition is a frequent phenotype of psychiatric conditions including schizophrenia. These results suggest that α_3-selective agonists may constitute an effective treatment for sensorimotor gating deficits in various psychiatric conditions. This view is supported by the observation that the partial benzodiazepine site agonist bretazenil in earlier open clinical trials displayed an antipsychotic activity similar to neuroleptic drugs *(67)*. It is conceivable the α_3-selective agonists would lack the sedative or extrapyramidal side-effects of classical neuroleptics and would thus be valuable agents.

Among various brain structures, the hippocampus is believed to play an important role in the modulation of prepulse inhibition. In α_5(H105R) point-mutated mice, the expression of the α_5-subunit containing $GABA_A$ receptors in the hippocampus is reduced (*see* above Heading 6 and ref. *47*). In these animals, prepulse inhibition was attenuated concomitant with an increase in spontaneous locomotor activity *(68)*. Thus, the α_5-subunit containing $GABA_A$ receptors, which are located extrasynaptically and are believed to mediate tonic inhibition, are important regulators of the expression of prepulse inhibition and locomotor exploration. Postmortem analysis of schizophrenia brains have consistently revealed structural abnormalities of developmental origin in the hippocampus *(66)*. Such abnormalities may include disturbances of $\alpha_5 GABA_A$ receptor function given that schizophrenia patients are known to exhibit a deficit in prepulse inhibition. Thus, agonists acting on both α_3- and $\alpha_5 GABA_A$ receptors might therefore be beneficial in overcoming this endophenotypic disease manifestation.

8. Anesthetic Actions

It is largely unknown which molecular targets mediate the clinical effects of general anesthetics *(69,70)*. Recent work focused on the role of $GABA_A$ receptors. The studies were based on the analysis of point-mutated knockin mice carrying point mutations in the β_3- and β_2-subunits of the $GABA_A$ receptor. These mutations rendered the GABA$_A$ receptors containing the respective subunits insensitive to modulation by etomidate, propofol, and certain volatile anesthetics, for example, enflurane. It was found that β_3-containing GABA$_A$ receptors mediate the immobilizing action of etomidate and propofol apparently in full (Table 1) *(71)* and of enflurane, isoflurane, and halothane in part *(71–73)*. In addition, they also mediate part of the hypnotic action of etomidate and propofol *(71)*, but apparently not of the volatile anesthetics *(71,72)*. In contrast, the hypnotic action of etomidate was found to be mediated by β_2-containing GABA$_A$ receptors *(74)*. Further studies revealed that the respiratory depressant action of etomidate and propofol is also mediated by β_3-containing GABA$_A$ receptors, whereas the heart rate depressant action and to a large part the hypothermic action of etomidate and propofol are mediated by other targets *(75,76)*. Thus, a β_3-selective agent would be predicted to be immobilizing and respiratory depressant, but largely lack the heart rate depressant and hypothermic actions of etomidate and propofol. The analysis of α-subunits involved in mediating the actions of general anesthetics is expected to result in further insights into the contribution of GABA$_A$ receptors to anaesthesia. Mutations in α-subunits have been identified in recombinant receptor studies that render $\alpha x\beta x\gamma_2$-GABA$_A$ receptors insensitive to specific volatile anesthetics but not to etomidate or propofol *(77,78)*. It is expected that studies using knockin mice carrying these mutations will yield information as to the contribution of individual GABA$_A$-receptors subtypes and to the GABA$_A$ receptor family as a whole to the action of volatile general anesthetics.

The role of GABA$_A$ receptors containing the ε-subunit remain somewhat controversial. Recombinant GABA$_A$ receptors containing the ε-subunit instead of the γ_2-subunit were initially considered to be insensitive to general anesthetics *(79)*. However, when the ε-subunit was expressed at low levels, ε-subunit-containing receptors were sensitive to general anesthetics *(80)*. In slices from rat-brainstem, adenoviral expression of the ε-subunit in cardiac parasympathetic preganglionic neurons rendered GABA$_A$ receptors insensitive to pentobarbital *(81)*. The functional role of the ε-subunit requires further clarification.

Acknowledgment

I would like to express my gratitude to my colleagues for their excellent contributions to the investigation of the GABA$_A$-receptor system over the years, in particular to Dietmar Benke, Florence Crestani, Jean-Marc Fritschy, Bernhard Lüscher, and Uwe Rudolph.

References

1. Buzsaki, G. and Chrobak, J. J. (1995) Temporal structure in spatially organized neuronal ensembles: a role for interneuronal networks. *Curr. Opin. Neurobio.* **5,** 504–510.

2. Paulsen, O. and Moser, E. I. (1998) A model of hippocampal memory encoding and retrieval: GABAergic control of synaptic plasticity. *Trends Neurosc.* **21,** 273–278.
3. Freund, T. F. and Buzsaki, G. (1996) Interneurons of the hippocampus. *Hippocampus* **6,** 345–470.
4. Miles, R. (2000) Diversity in inhibition. *Science* **287,** 244–146.
5. Klausberger, T., Roberts, J. D., and Somogyi, P. (2002) Cell type- and input-specific differences in the number and subtypes of synaptic $GABA_A$ receptors in the hippocampus. *J. Neurosci.* **22,** 2513–2521.
6. Klausberger, T., Magill, P. J., Marton, L. F., et al. (2003) Brain state-and cell type-specific firing of hippocampal interneurons in vivo. *Nature* **421,** 844–848.
7. O'Keefe, J. and Nadel, L. (1978) The hippocampus as a cognitive map. Clarendon Press, Oxford, pp. 477–543.
8. O'Keefe, J. and Recce, M. L. (1993) Phase relationship between hippocampal place units and the EEG theta rhythm. *Hippocampus* **3,** 317–330.
9. Skaggs, W. E., McNaughton, B. L., Wilson, M. A., and Barnes, C. A. (1996) Theta phase precession in hippocampal neuronal populations and the compression of temporal sequences. *Hippocampus* **6,** 149–172.
10. Engel, A. K., Fries, P., and Singer, W. (2001) Dynamic predictions: oscillations and synchrony in top-down processing. *Nat. Rev. Neurosci.* **2,** 704–716.
11. Harris, K. D., Henze, D. A., Hirase, H., et al. (2002) Spike train dynamics predicts theta–related phase precession in hippocampal pyramidal cells. *Nature* **417,** 738–741.
12. Metha, M. R., Lee, A. K., and Wilson, M. A. (2002) Role of experience and oscillations in transforming a rate code into a temporal code. *Nature* **417,** 741–746.
13. Bortos, M., Vida, I., and Jonas, P. (2007) Synaptic mechanisms of synchronited gamma oscillations in inhibitory networks. *Nature Rev. Neurosci.* **8,** 45–56.
14. Spencer, K. M., Nestor, P. G., Perlmutter, R., et al. (2004) Neural synchrony indexes disordered perception and cognition in schizophrenia. *Proc. Natl. Acad. Sci. USA* **101,** 17,288–17,293.
15. Nyíri, G., Freund, T. F., and Somogyi, P. (2001) Input-dependent synaptic targeting of α_2 subunit containing $GABA_A$ receptors in hippocampal pyramidal cells of the rat. *Eur. J. Neurosci.* **13,** 428–442.
16. Pawelzik, H., Hughes, D. I., and Thomson, A. M. (2002) Physiological and morphological diversity of immunocytochemically defined parvalbumin- and cholecystokinin-positive interneurons in CA1 of the adult rat hippocampus. *J. Comp. Neurol.* **443,** 346–367.
17. Brussaard, A. B. and Herbison, A. E. (2000) Long-term plasticity of postsynaptic $GABA_A$-receptor function in the adult brain: insights from the oxytocin neurone. *Trends Neurosci.* **23,** 190–195.
18. Hutcheon, B., Morley, P., and Poulter, M. O. (2000) Developmental change in $GABA_A$ receptor desensitization kinetics and its role in synapse function in rat cortical neurons. *J. Physiol.* **522,** 3–17.
19. Jüttner, R., Meier, J., and Grantyn, R. (2001) Slow IPSC kinetics, low levels of α_1 subunit expression and paired-pulse depression are distinct properties of neonatal

inhibitory GABAergic synaptic connections in the mouse superior colliculus. *Eur. J. Neurosci.* **13,** 2088–2098.
20. Vicini, S., Ferguson, C., Prybylowski, K., et al. (2001) GABA$_A$ receptor α_1 subunit deletion prevents developmental changes of inhibitory synaptic currents in cerebellar neurons. *J. Neurosci.* **21,** 3009–3016.
21. Nusser, Z., Sieghart, W., and Somogyi, P. (1998) Segregation of different GABA$_A$ receptors to synaptic and extrasynaptic membranes of cerebellar granule cells. *J. Neurosci.* **18,** 1693–1703.
22. Fritschy, J. M. and Brünig, I. (2003) Formation and plasticity of GABAergic synapses: physiological mechanisms and pathophysiological implications. *Pharmacol. Therapeutics* **98,** 299–323.
23. Barnard, E. A., Skolnick, P., Olsen, R. W., et al. (1998) Subtypes of γ-aminobutyric acid$_A$ receptors: classification on the bases of subunit structure and receptor function *Pharmacol. Rev.* **50,** 291–313.
24. Whiting, P., Wafford, K. A., and McKernan, R. M. (2000) Pharmacologic subtypes of GABA$_A$ receptors based on subunit composition, in *GABA in the Nervous System: the view at fifty years* (Martin, D. L., and Olsen, R. W., eds.), Lippincott, Philadelphia, pp. 113–126.
25. Sieghart, W. and Sperk, G. (2002) Subunit composition, distribution and function of GABA$_A$ receptor subtypes. *Curr. Top. Med. Chem.* **2,** 795–816.
26. Möhler, H., Benke, D., Fritschy, J. M., and Benson, J. (2000) The benzodiazepine site of GABA$_A$ receptors, in *GABA in the Nervous System: the view at fifty years* (Martin, D. L., and Olsen, R. W., eds.), Lippincott, Philadelphia, PA, pp. 97–112.
27. Möhler, H., Fritschy, J. M., and Rudolph, U. (2002) A new benzodiazepine pharmacology. *J. Pharm. Exptl. Ther.* **300,** 2–8.
28. Möhler, H., Fritschy, J. M., Vogt, K., et al. (2005) Pathophysiology and Pharmacology of GABA$_A$ Receptors, in *Anxiety and Anxiolytic Drugs, Handbook of Experimental Pharmacology* (Holsboer, F., and Ströhle, A., eds.), Springer Publishers, Vol. **169,** pp. 225–247.
29. Möhler, H. (2001) Functions of GABA receptors: Pharmacology and Pathophysiology, in *Pharmacology of GABA and Glycine Neurotransmission* (Möhler, H., ed.), Springer Publishers, New York, pp. 101–116.
30. Möhler, H. (2002) Pathophysiological aspects of diversity in neuronal inhibition: a new benzodiazepine pharmacology, in *Dialogues Clin. Neurosci.* **4,** 261–269.
31. Fritschy, J. M., Crestani, F., Rudolph, U., and Möhler, H. (2004) GABA$_A$ receptor subtypes with special reference to memory function and neurological disorders, in *Excitatory inhibitory balance: synapses, circuits and systems plasticity* (Hensch, T. K., and Fagiolini, M., eds.), Kluwer Academic / Plenum Publishers, New York, pp. 215–228.
32. Rudolph, U. and Möhler, H. (2006) GABA-based therapeutic approaches: GABA$_A$ receptor subtype functions. *Curr. Opin. Pharmacol.* **6,** 18–23.
33. Fritschy, J. M. and Möhler, H. (1995) GABA$_A$ receptor heterogeneity in the adult rat brain: differential regional and cellular distribution of seven major subunits. *J. Comp. Neurol.* **359,** 154–194.

34. Pirker, S., Schwarzer, C., Wieselthaler, A., et al. (2000) $GABA_A$ receptors: immunocytochemical distribution of 13 subunits in the adult rat brain. *Neuroscience* **101,** 815–850.
35. Gao, B., Fritschy, J. M., Benke, D., and Möhler, H. (1993) Neuron-specific expression of $GABA_A$ receptor subtypes: differential associations of the α_1- and α_3-subunits with serotonergic and GABAergic neurons. *Neuroscience* **54,** 881–892.
36. Huntsmann, M. M., Porcello, D. M., and Homanics, G. E., et al. (1999) Reciprocal inhibitory connections and network synchrony in the mammalian thalamus. *Science* **283,** 541–543.
37. Devor, A., Fritschy, J. M., and Yarom, Y. (2001) Spatial distribution and subunit composition of $GABA_A$ receptors in the inferior olivary nucleus. *J. Neurophysiol* **85,** 1686–1696.
38. Moss, S. J. and Smart, T. G. (2001) Constructing inhibitory synapses. *Nature. Rev. Neurosci.* **2,** 240–250.
39. van Rijnsoever, C., Tauber, M., Choulli, M. K., et al. (2004) Requirement of α_5 $GABA_A$ receptors for the development of tolerance to the sedative action of diazepam in mice. *J. Neurosci.* **24,** 6785–6790.
40. Nusser, Z., Sieghart, W., Stephenson, F. A., and Somogyi, P. (1996) The α_6 subunit of the $GABA_A$ receptor is concentrated in both inhibitory and excitatory synapses on cerebellar granule cells. *J. Neurosci.* **16,** 103–114.
41. Benson, J., Löw, K., Keist, R., et al. (1998) Pharmacology of recombinant $GABA_A$ receptors rendered diazepam-insensitive by point-mutated α-subunits. *FEBS. Lett.* **431,** 400–404.
42. Brickley, S. G., Cull-Candy, S. G., and Farrant, M. (1996) Development of a tonic form of synaptic inhibition in rat cerebellar granule cells resulting from persistent activation of $GABA_A$ receptors. *J. Physiol.* **497,** 753–759.
43. Brickley, S. G., Revilla, V., Cull-Candy, S. G., et al. (2001) Adaptive regulation of neuronal excitability by a voltage independent potassium conductance. *Nature* **409,** 88–92.
44. Bormann, J. (2000) The "ABC" of GABA receptors. *Trends Pharmacol. Sci.* **21,** 16–19.
45. Rudolph, U., Crestani, F., Benke, D., et al. (1999) Benzodiazepine actions mediated by specific γ-aminobutyric $acid_A$ receptor subtypes. *Nature* **401,** 796–800.
46. Löw, K., Crestani, F., Keist, R., et al. (2000) Molecular and neuronal substrate for the selective attenuation of anxiety. *Science* **290,** 131–134.
47. Crestani, F., Keist, R., Fritschy, J. M., et al. (2002) Trace fear conditioning involves hippocampal α_5 $GABA_A$ receptors. *PNAS.* **99,** 8980–8985.
48. Rudolph, U. and Möhler, H. (2004) Analysis of $GABA_A$ receptor function and dissection of pharmacology of benzodiazepines and general anaesthetics by mouse genetics. *Ann. Rev. Pharmacol. Toxicol.* 475–498.
49. McKernan, R. M., Rosahl, T. W., Reynolds, D. S., et al. (2000) Sedative but not anxiolytic properties of benzodiazepines are mediated by the GABAA receptor $\alpha 1$ subtype. *Nature Neurosci.* **3,** 587–592.
50. Kopp, C., Rudolph, U., and Tobler, I. (2004a) Sleep EEG changes after zolpidem in mice. *Neuroreport* **15,** 2299–2302.

51. Tobler, I., Kopp, C., Deboer, T., and Rudolph, U. (2001) Diazepam-induced changes in sleep: role of the α_1GABA$_A$ receptor subtype. *Proc. Natl. Acad. Sci. USA* **98,** 6464–6469.
52. Kopp, C., Rudolph, U., Löw, K., and Tobler, I. (2004b) Modulation of rhythmic brain activity by diazepam: GABA(A) receptor subtype and state specificity. *Proc. Natl. Acad. Sci. USA* **101,** 3674–3679.
53. Brown, N., Kerby, J., Bonnert, T. P., et al. (2002) Pharmacological characterization of a novel cell line expressing human $\alpha4\beta3\delta$ GABA$_A$ receptors. *Br. J. Pharmacol.* **136,** 965–974.
54. Storustovu, S. and Ebert, B. (2003) Gaboxadol: in vitro interaction studies with benzodiazepines and ethanol suggest functional selectivity. *Eur. J. Pharmacol.* **467,** 49–56.
55. Lancel, M. and Steiger, A. (1999) Sleep and its modulation by drugs that affect GABA$_A$ receptor function. *Angew. Chem. Int. Ed.* **111,** 2852–2864.
56. Huckle, R. (2004) Gaboxadol. Lundbeck/Merck. *Curr. Opin. Invest. Drugs.* **5,** 766–773.
57. Collins, I., Moyes, C., Davey, W. B., et al. 3-Heteroaryl-2-pyridones: benzodiazepine site ligands with functional delectivity for alpha 2/alpha 3-subtypes of human GABA(A) receptor-ion channels. *J. Med. Chem.* **45,** 1887–1900.
58. Yee, B. K., Keist, R., von Boehmer, L., et al. (2005) A schizophrenia-related sensorimotor deficit links α_3-containing GABA$_A$ receptors to a dopamine hyperfunction. *Proc. Natl. Acad. Sci. USA* **102,** 17,154–17,159.
59. Dias, R., Sheppard, W. F., Fradley, R. L., et al. (2005) Evidence for a significant role of alpha3-containing GABAA receptors in mediating the anxiolytic effects of benzodiazepines. *J. Neurosci.* **25,** 10,682–10,688.
60. Whiting, P. J. (2003) The GABAA receptor gene family: new opportunities for drug development. *Curr. Opin. Drug. Discov. Devel.* **6,** 648–655.
61. Yee, B. K., Hauser, J., Dolgov, V. V., et al. (2004) GABA receptors containing the α_5 subunit mediate the trance effect in aversive and appetitive conditioning and extinction of conditioned fear. *Eur. J. Neurosci.* **20,** 1928–1936.
62. Collinson, N., Kuenzi, F. M., Jarolimek, W., et al. (2002) Enhanced learning and memory and altered GABAergic synaptic transmission in mice lacking the α_5 subunit of the GABA$_A$ receptor. *J. Neurosci.* **22,** 5572–5580.
63. Chambers, M. S., Atack, J. R., Broughton, H. B., et al. (2003) Identification of a novel, selective GABA$_A$ α_5 receptor inverse agonist which enhances cognition. *J. Med. Chem.* **46,** 2227–2240.
64. McHugh, T. J., Blum, K. I., Tsien, J. Z., et al. (1996) Impaired hippocampal representation of space in CA1-specific NMDAR1 knockout mice. *Cell* **87,** 1339–1349.
65. Tsien, J. Z., Huerta, P. T., and Tonegawa, S. (1996) The essential role of hippocampal CA1 NMDA receptor-dependent synaptic plasticity in spatial memory. *Cell* **87,** 1327–1338.
66. Lewis, D. A., Hashimoto, T., and Volk, D. W. (2005) Cortical inhibitory neurons and schizophrenia. *Nature Rev. Neurosci.* **6,** 312–324.
67. Dellini-Stula, A. and Berdah-Tordjman, D. (1996) Antipsychotic effects of bretazenil, a partial benzodiazepine agonist in acute schizophrenia—a study group report. *J. Psychiatr. Res.* **30,** 239–250.

68. Hauser, J., Rudolph, U., Keist, R., et al. (2005) Hippocampal α_5 subunit containing $GABA_A$ receptors modulate expression of prepulse inhibition. *Mol. Psychiatry* **10,** 201–207.
69. Campagna, J. A., Miller, K. W., and Forman, S. A. (2003) Mechanisms of actions of inhaled anesthetics. *N. Engl. J. Med.* **348,** 2110–2124.
70. Rudolph, U. and Antkowiak, B. (2004) Molecular and neuronal substrates for general anaesthetics. *Nat. Rev. Neurosci.* **5,** 709–720.
71. Jurd, R., Arras, M., Lambert, S., et al. (2003) General anesthetic actions in vivo strongly attenuated by a point mutation in the GABA(A) receptor beta3 subunit. *Faseb. J.* **17,** 250–252.
72. Lambert, S., Arras, M., Vogt, K. E., and Rudolph, U. (2005) Isoflurane-induced surgical tolerance mediated only in part by beta3-containing GABA(A) receptors. *Eur. J. Pharmacol.* **516,** 23–27.
73. Liao, M., Sonner, J. M., Jurd, R., et al. (2005) Beta3-containing gamma-aminobutyric acidA receptors are not major targets for the amnesic and immobilizing actions of isoflurane. *Anesth. Analg.* **101,** 412–418.
74. Reynolds, D. S., Rosahl, T. W., Cirone, J., et al. (2003) Sedation and anesthesia mediated by distinct GABA(A) receptor isoforms. *J. Neurosci.* **23,** 8608–8617.
75. Zeller, A., Arras, M., Lazaris, A., et al. (2005) Distinct molecular targets for the central respiratory and cardiac actions of the general anesthetics etomidate and propofol. *Faseb. J.* **12,** 1677–1679.
76. Cirone, J., Rosahl, T. W., Reynolds, D. S., et al. (2004) Gamma-aminobutyric acid type A receptor beta 2 subunit mediates the hypothermic effect of etomidate in mice. *Anesthesiology.* **100,** 1438–1445.
77. Mihic, S. J., Ye, Q., Wick, M. J., et al. (1997) Sites of alcohol and volatile anaesthetic action on GABA(A) and glycine receptors. *Nature* **389,** 385–389.
78. Krasowski, M. D., Koltchine, V. V., Rick, C. E., et al. (1998) Propofol and other intravenous anesthetics have sites of action on the gamma-aminobutyric acid type A receptor distinct from that for isoflurane. *Mol. Pharmacol.* **53,** 530–538.
79. Davies, P. A., Hanna, M. C., Hales, T. G., and Kirkness, E. F. (1997) Insensitivity to anaesthetic agents conferred by a class of GABA(A) receptor subunit. *Nature* **385,** 820–823.
80. Thompson, S. A., Bonnert, T. P., Cagetti, E., et al. (2002) Overexpression of the GABA(A) receptor epsilon subunit results in insensitivity to anaesthetics. *Neuropharmacology* **43,** 662–668.
81. Irnaten, M., Walwyn, W. N., Wang, J., et al. (2002) Pentobarbital enhances GABAergic neurotransmission to cardiac parasympathetic neurons, which is prevented by expression of GABA(A) epsilon subunit. *Anaestesiology* **97,** 717–724.
82. Dämgen, K. and Lüddens, H. (1999) Zaleplon diaplays a selecitvity to recombinant $GABA_A$ receptors different from zolpidem, zopiclone and benzodiazepines. *Neurosci. Res. Comm.* **25,** 139–148.
83. Foster, A. C., Pelleymounter, M. A., Cullen, M. J., et al. (2004) In vivo pharmacological characterization of indiplon, a novel pyrazolopyrimidine sedative-hypnotic. *J. Pharmacol. Exp. Ther.* **311,** 547–559.

84. Lippa, A., Czobor, P., Stark, J., et al. (2005) Selective anxiolysis produced by ocinaplon, a GABA(A) receptor modulator. *Proc. Natl. Acad. Sci. USA* **102,** 7380–7385.
85. Griebel, G., Perrault, G., Simiand, J., et al. (2003) SL651498, a $GABA_A$ receptor agonist with subtype-selective efficacy, as a potential treatment for generalized anxiety disorder and muscle spasms. *CNS. Drug. Rev.* **9,** 3–20.
86. Atack, J. R., Wafford, K., Tye, S. J., et al. (2006a) TPA023, an agonist selective for α_2- and α_3-containing $GABA_A$ receptors, is a non-sedating anxiolytic in rodents and primates. *J. Pharm. Exp. Ther.* **316,** 410–422.
87. Langen, B., Egerland, U., Bernoster, K., et al. (2005) Characterization in rats of the anxiolytic potential of ELB139 [1-(4-chlorophenyl)-4-piperidin-1-yl-1,5-dihydro-imidazol-2-on], a new agonist at the benzodiazepine binding site of the $GABA_A$ receptor. *J. Pharmacol. Exp. Ther.* **314,** 717–724.
88. Sternfeld, F., Carling, R. W., Jelley, R. A., et al. (2004) Selective, orally active gamma-amonobutyric acidA alpha5 receptor inverse agonists as cognition enhancers. *J. Med. Chem.* **47,** 2176–2179.
89. Chambers, M. S., Atack, J. R., Carling, R. W., et al. (2004) An orally bioavailable, functionally selective inverse agonist at the benzodiazepine site of $GABA_A$ alpha5 receptors with cognition enhancing properties. *J. Med. Chem.* **47,** 5829–5832.
90. Navarro, J. F., Buron, E., and Martin-Lopez, M. (2002) Anxiogenic-like activity of L-655,708, a selective ligand for the benzodiazepine site of GABA(A) receptors which contain the alpha-5 subunit, in the elevated plus-maze test. *Prog. Neuropsychopharmacol. Biol. Psychiatry* **26,** 1389–1392.
91. Navarro, J. F., Buron, E., and Martin-Lopez, M. (2004) Behavioral profile of L-655 708, a selective ligand for the benzodiazepine site of $GABA_A$ receptors which contain the α_5 subunit in social encounters between male mice. *Aggress. Behav.* **30,** 319–325.
92. Atack, J. R., Hutson, P. H., Collinson, N., et al. (2005) Anxiogenic properties of an inverse agonist selective for α_3 subunit-containing $GABA_A$ receptors. *Br. J. Pharmacol.* **144,** 357–366.
93. Wallner, M., Hanchar, H. J., and Olsen, R. W. (2003) Ethanol enhances alpha 4 beta3 delta and alpha 6 beta 3 delta gamma-aminobutyric acid type A receptors at low concentration known to affect humans. *Proc. Natl. Acad. Sci. USA* **100,** 15,218–15,223.
94. Belelli, D. and Lambert, J. J. (2005) Neurosteroids: endogenous regulators of the GABA(A) receptor. *Nat. Rev. Neurosci.* **6,** 565–575.
95. Haefeli, W., Martin, J. R., and Schoch, P. (1990) Novel anxiolytics that act as partial agonists at benzodiazepine receptors. *TIPS.* **11,** 452–456.
96. Atack, J. R., et al. (2006b) The in vivo properties of pagoclone in rat are most likely mediated by 5′-hydroxy pagoclone. *Neuropharmacol.* **50,** 677–689.
97. Bianchi, M. T. and McDonald, R. L. (2003) Neurosteroids shift partial agonist activation of GABA(A) receptor channels from low- to high-efficacy gating patterns. *J. Neurosci.* **23,** 10,934–10,943.

3

Trafficking of Postsynaptic $GABA_A$ Receptors by Receptor-Associated Proteins

Xu Yuan and Bernhard Lüscher

Summary

γ-Aminobutyric acid receptors ($GABA_A$ Rs) are the principal receptors that mediate neural inhibition in the brain. Changes in the function of GABAergic transmission are implicated in activity-dependent adaptation of neural excitability. Of particular interest are mechanisms that control the size of the postsynaptic $GABA_A$-receptor pool, a major determinant of synaptic strength. Mechanisms of trafficking of postsynaptic $GABA_A$ Rs contribute to regulation of inhibitory synaptic transmission in response to changes in neural activity and extracellular stimuli. This review summarizes current information available on the receptor structures relevant for trafficking of $GABA_A$ Rs, the molecular composition of the submembrane cytoskeleton of inhibitory synapses, and the receptor interacting proteins that regulate the localization and trafficking of postsynaptic $GABA_A$ Rs during exocytosis, lateral diffusion and endocytic recycling, and degradative pathways.

Key Words: Trafficking; synaptogenesis; synaptic plasticity; inhibitory synapses; palmitoylation.

1. Introduction

Functional adaptation of γ-aminobutyric acid (GABA)ergic synapses can generally be achieved by changes in either the neurotransmitter release properties of GABAergic neurons or changes in gene expression, cellular distribution, and function of postsynaptic $GABA_A$ Rs. However, experimental evidence suggests

From: *The Receptors: The GABA Receptors, Third Edition*
Edited by: S. J. Enna and H. Möhler © Humana Press Inc., Totowa, NJ

that the synaptic efficacy of GABAergic synapses is tightly correlated with the number of postsynaptic $GABA_A$ Rs *(1,2)*, indicating that activity-dependent changes in trafficking of $GABA_A$ Rs to synapses represent one of the most important mechanisms underlying functional plasticity of GABAergic synapses. Moreover, changes in trafficking of $GABA_A$ Rs might contribute to the etiology and/or manifestation of a wide range of neurological and psychiatric disorders including epilepsy *(3–6)*, mood disorders such as anxiety and depression *(7–13)*, and alcoholism *(14,15)*.

Different subtypes of $GABA_A$ Rs can be distinguished based on their pharmacological profile, their subcellular localization, or simply their subunit composition (*see* review by W. Sieghart in Chapter 4). For the purpose of this review, the focus will be specifically on the trafficking mechanisms relevant for the major subtypes of $GABA_A$ Rs that are concentrated at postsynaptic membrane sites, and on the proteins that directly or indirectly interact with $GABA_A$ Rs to regulate their trafficking and accumulation at inhibitory synapses.

The surface expression and stability of $GABA_A$ Rs is ultimately regulated by neural activity *(5,16–20)*, growth factors, such as brain-derived neurotrophic factor *(21,22)*, insulin *(23)*, or tumor necrosis factor-α *(24)*, as well as $GABA_A$ R-modulating drugs such as the benzodiazepines *(25,26)* and alcohol *(15)*. However, the specific pathways through which these cues tie into the known protein–protein interactions at GABAergic synapses remain ill-defined and hence beyond the scope of this review.

2. The Structure of Postsynaptic and Extrasynaptic $GABA_A$ R Subtypes

$GABA_A$ Rs belong to the superfamily of heteropentameric ion channels encompassing nicotinic acetylcholine receptors, glycine receptors, and 5HT3 receptors, the subunits of which share a common structure including a large extracellular N-terminal domain followed by four transmembrane domains (TM1–4) and an extended cytoplasmic-loop domain between TM3 and -4 *(27)*. $GABA_A$ Rs are assembled from α_{1-6}-, β_{1-3}-, γ_{1-3}-, δ-, ε-, ρ-, and θ-subunits and the large majority of native receptors represent assemblies of two α-, two β-, and the γ_2-subunit *(28)*. The $\gamma_{1/3}$-, δ-, and ε-subunits can substitute for the γ_2-subunit in comparatively minor, and spatially and developmentally restricted receptor populations (Fig. 1).

The many different receptor subtypes formed in this way can be broadly divided into postsynaptic and extrasynaptic $GABA_A$ Rs (for review *see* refs. *29,30*). Postsynaptic receptors are distinguished from extrasynaptic and perisynaptic $GABA_A$ Rs, which mediate tonic inhibition of neurons. Structurally, these nonsynaptic $GABA_A$ Rs represent assemblies of α_4- or α_6-subunits

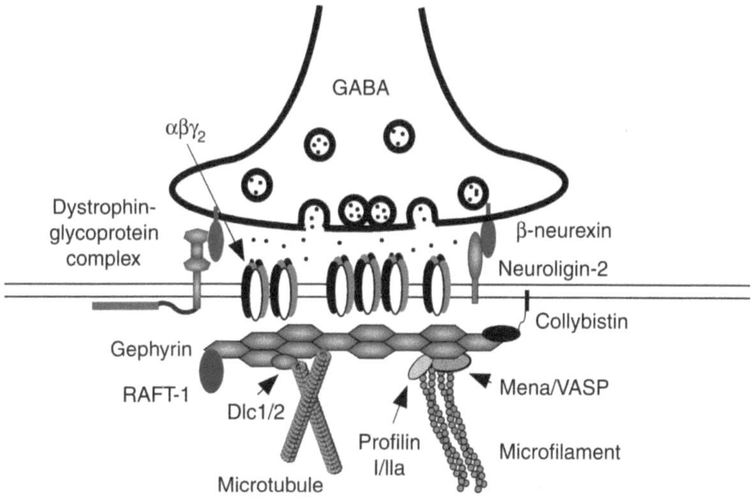

Fig. 1. Schematic representation of the protein composition of the postsynaptic apparatus of GABAergic synapses. Shown are a GABAergic terminal apposed to a postsynaptic membrane with clusters of γ_2-subunit containing $GABA_A$ Rs. Integral membrane proteins concentrated in the postsynaptic membrane include NL2 and β-dystroglycan, which together with the peripheral membrane protein α-dystroglycan and subsynaptic dystrophin forms the dystrophin-associated glycoprotein complex that also interacts with presynaptic β-neurexin. NL2 is believed to be part of a *trans*-synaptic complex with presynaptic β-neurexin(s). The postsynaptic density contains gephyrin and its associated proteins including the GEF collybistin, the rapamycin and FKBP12 target RAFT1, Dlc1/2 (dynein LC1 and -2), microtubules, Mena/VASP, and profilin I and IIa (for references *see* text).

together with any type of β-subunit and the δ-subunit and appear to be tailored to function in low-ambient concentrations of GABA in the submicromolar range *(31–33)*. Accordingly, they exhibit high-agonist affinity and significantly reduced desensitization, allowing them to function in the continuous presence of agonist *(33–36)*.

Postsynaptic $GABA_A$ R subtypes are concentrated in the plasma membrane apposed to presynaptic GABAergic terminals, where they mediate inhibitory postsynaptic currents (IPSCs) underlying the phasic form of neural inhibition. Consistent with comparatively high concentrations of GABA in the synaptic cleft estimated at 300 m*M* *(37)* these postsynaptic receptor subtypes generally exhibit low affinity for GABA *(35)* and rapid desensitization *(34,36)*. Structurally, the major postsynaptic $GABA_A$ R subtypes represent heteropentamers of α_1-, α_2-, or α_3-subunits together with β-subunits and the γ_2-subunit. In vivo, they are invariably colocalized with immunoreactive clusters of the putative subsynaptic scaffold protein gephyrin, a postsynaptic marker for

GABAergic synapses *(38–43)*. Consistent with the γ_2-subunit being part of all postsynaptic $GABA_A$ R subtypes, this subunit is essential for GABAergic miniature inhibitory synaptic currents and for clustering and targeting to the postsynaptic plasma membrane of all postsynaptic $GABA_A$ R subtypes identified to date, as well as gephyrin *(38,44,45)*. Importantly, whereas $\alpha_{1-3}\beta\gamma_2$-receptors are generally concentrated at synapses they are also abundant in the extrasynaptic membrane *(38,41,46,47)*, consistent with recent evidence for dynamic mobility and rapid exchange of γ_2-subunit-containing receptors between postsynaptic and extrasynaptic receptor pools *(48,49)*.

Receptors made up of $\alpha_5\beta\gamma_2$-subunits are atypical in that they contain a γ_2-subunit, yet are largely excluded from synapses, as seen by immunohistochemical and electrophysiological analysis of hippocampal brain slices *(41,50–52)*. However, there are conflicting reports with respect to the localization of these receptors in cultured hippocampal neurons *(41,53)*, indicating that α_5-subunit-containing receptors may under certain conditions also be synaptic. Moreover, $\alpha_3\beta\gamma_2$-receptors are typically concentrated at synapses of hippocampal pyramidal cells, but are specifically excluded from the postsynaptic membrane of neurons in the inferior olivary nucleus *(54,55)*. Thus, the α_5- and α_3-subunits and/or their associated proteins appear in some cases to interfere with postsynaptic localization of $GABA_A$ Rs. Alternatively, some γ_2-subunit-dependent trafficking mechanisms might be neuronal cell type specific.

3. Subunit Domains Implicated in Clustering and Postsynaptic Targeting of $GABA_A$ Rs

A direct approach to address the mechanism by which $GABA_A$ Rs are concentrated at synapses involves mapping of the subunits and subunit domains involved. Given the evidence mentioned earlier for a critically important role of the γ_2-subunit, Alldred et al. *(56)* used transfection of cultured γ_2-subunit-deficient neurons with chimeric α_2/γ_2-subunit constructs, to map the γ_2-subunit domains that could rescue postsynaptic accumulation of $GABA_A$ Rs and gephyrin in these mutant neurons. Surprisingly, these experiments revealed that trafficking of $GABA_A$ Rs to postsynaptic dendritic domains is critically dependent on the TM4 of the γ_2-subunit. Furthermore, the γ_2-subunit major intracellular loop domain was neither sufficient nor required for postsynaptic localization of subunit assemblies containing α_2/γ_2-chimeric constructs. Unlike the postsynaptic accumulation of $GABA_A$ Rs, the recruitment of gephyrin to synapses and rescue of GABAergic inhibitory synaptic function of γ_2-subunit-deficient neurons required transfection of α_2/γ_2-chimeric construct that included the intracellular domain of the γ_2-subunit in addition to TM4. These experiments confirmed that postsynaptic accumulation of

GABA$_A$ Rs under certain conditions can occur independently of gephyrin *(56)* (*see* page 49). The observation that some of the chimeric constructs can target to synapses, yet fail to rescue inhibitory synaptic function is consistent with the existence of an intracellular pool of GABA$_A$ Rs that is closely associated with the postsynaptic density but inaccessible to neurotransmitter. Evidence for an intracellular endocytic pool of GABA$_A$ Rs is based on studies comparing immunohistochemical staining of fixed and live cells under controlled conditions that either allowed or blocked endocytosis of GABA$_A$ Rs *(57)*. Thus, it appears that the γ_2-subunit cytoplasmic-loop region and perhaps recruitment of the gephyrin scaffold is required for recycling of GABA$_A$ Rs from the endocytic-subsynaptic pool to the plasma membrane, thereby enabling synaptic transmission.

Analogous to this approach, Christie et al. *(58)* have used transfection of γ_2/δ-chimeric constructs into wild-type neurons to map γ_2-subunit clustering domains. Consistent with data obtained by Alldred et al. *(56)*, they found that the γ_2-subunit intracellular loop domain was insufficient to induce clustering of GABA$_A$ Rs when tested in chimeric constructs containing the δ-subunit as a backbone. However, unlike in the case of γ_2/α_2-chimeric constructs, substitution of the γ_2-subunit cytoplasmic-loop region with the corresponding region of the δ-subunit completely abolished the clustering function of the γ_2-subunit, suggesting that the γ_2-subunit cytoplasmic-loop region was required for postsynaptic localization. This apparent discrepancy with results obtained from analysis of γ_2/α_2-chimeric subunits is likely owing to differences in the cultures used for analysis (cortical $\gamma_2^{-/-}$ vs hippocampal wild-type neurons). Whereas transfection of γ_2/α_2-chimeric constructs into γ_2-deficient neurons favors assembly of these constructs with endogenous α- and β-subunits, the δ/γ_2-constructs used by Christie et al. *(58)* had to compete with the endogenous γ_2-subunit for assembly with α- and β-subunits. In addition, they were likely competing with endogenous γ_2-subunit for interaction with γ_2-subunit-associated trafficking factors. In addition, the δ-subunit, which in its native form is never found at synapses, might be less permissive for postsynaptic localization than the α_2-subunit, which is invariably concentrated at synapses in vivo.

Using in vitro interaction assays, the γ_2-subunit intracellular domain was recently shown to interact with itself and with intracellular domains of β-subunits and other γ-subunits *(59)*. Interestingly, the γ_2-subunit self-interaction motif was mapped to a 23-amino acid peptide within the major intracellular loop that is identical with the domain that also interacts with the GABA$_A$ R-associated protein (GABARAP). Based on the authors' model, competing interactions of the γ_2-subunit with itself and other subunits on one hand, and GABARAP on the other might regulate the clustering of GABA$_A$ Rs *(59)*. Thus, it is likely that both the major cytoplasmic loop and TM4 region of the γ_2-subunit contribute to clustering and accumulation of $GABA_A$ Rs at synapses.

4. Proteins Associated With the Postsynaptic Cytoskeleton of GABAergic Synapses

As indicated earlier, postsynaptic $GABA_A$ Rs are invariably colocalized with gephyrin, a 93-KDa putative subsynaptic scaffold protein, first identified as a postsynaptic anchoring protein of glycine receptors that interacts directly with a 20-amino acid sequence in the cytoplasmic-loop region of the glycine receptor β-subunit (reviewed in refs. *60,61*). In contrast to glycine receptors, current evidence suggests that $GABA_A$ Rs do not interact directly with gephyrin *(62,63)*. However, loss of postsynaptic localization of $GABA_A$ Rs in γ_2-subunit-deficient mice is associated with a parallel loss of gephyrin from synapses, indicating that the two proteins interact indirectly *(38,44,45,64)*. Conversely, antisense RNA suppression, RNA interference (RNAi), or knockout of gephyrin results in loss of postsynaptic localization of major subsets of $GABA_A$ Rs containing the γ_2-subunit together with the α_2- or α_3-subunit *(38,44,49,65–67)*, along with a reduction in the mean amplitude of miniature inhibitory postsynaptic currents (mIPSCs) *(68)*. However, postsynaptic accumulation of $\alpha_1\beta\gamma_2$-receptors is largely unaffected by loss of gephyrin *(67,68)* (Table 1).

Most of the current information about the composition of the postsynaptic cytoskeleton of inhibitory synapses is based on proteins identified as gephyrin-interacting proteins. Interaction of gephyrin with tubulin and microtubules qualifies gephyrin as a bona fide microtubule interacting protein (MAP) *(69)*. Yeast two-hybrid screens that have used gephyrin as bait have led to the identification of several gephyrin-associated proteins that are invariably concentrated in the postsynaptic specialization of both glycinergic and GABAergic synapses. These gephyrin-interacting proteins include profilin I/IIa and Mena/VASP, which act as regulators and adaptors of the microfilament cytoskeleton, respectively *(70,71)*. However, the functional relevance of interactions between gephyrin and the microfilament or microtubule cytoskeleton remain unclear because clustering of gephyrin and $GABA_A$ Rs in mature neurons is unaffected by drug-induced depolymerization of microtubules and microfilaments *(72)*. A possible effect of similar disruption of the cytoskeleton on the lateral mobility of gephyrin and $GABA_A$ Rs has yet to be analyzed. Other gephyrin-binding proteins that are concentrated at synapses include rapamycin and FKBP12 target (RAFT)-1; also called FKBP12-rapamycin associated protein (FRAP) or mammalian target of rapamycin (mTOR) *(73)* and the dynein light chains (LC)1 and -2 *(74)*. RAFT-1 functions as the in vivo mediator of the effects of the immunosuppressant rapamycin and as an important regulator of messenger

Table 1
Proteins of the Postsynaptic Cytoskeleton That Are Indirectly Associated With Postsynaptic $GABA_A$ Rs

Inhibitory synapse-associated protein	Proposed function	References
Gephyrin	Immobilizes and clusters α_2/γ_2-containing-$GABA_A$ Rs at postsynapses	*38,49,65–67*
Collybistin	Binds to gephyrin and regulates trafficking and clustering of gephyrin	*76,155,156*
Dynein LC1 and -2	Interacts with gephyrin, however not required for transport of gephyrin to synapses	*74*
GRIP-1	Found both pre- and postsynaptically. The protein interacts with GABARAP, which appears to be absent at synapses. Function unknown	*117–119*
Mammalian enabled (Mena)/ vasodilator-stimulated phosphoprotein (VASP) and profilin I/IIa	Mediate the association of gephyrin with microfilaments	*70,71*
Dystrophin–glycoprotein complex	Stabilizes or modulates α_1/α_2-subunit-containing $GABA_A$ Rs at postsynapses	*42,95–97*
NL2	Integral membrane protein that promotes GABAergic synaptogenesis by interaction with presynaptic β-neurexin	*87,89,94*
RAFT1	Colocalized with postsynaptic gephyrin; implicated in subsynaptic regulation of mRNA translation	*73*

RNA translation, allowing speculation on a possible mechanism by which gephyrin might contribute to translational control of postsynaptic protein synthesis. However, the functional relevance of this RAFT-1-gephyrin interaction for the function of GABAergic transmission is not understood. Similarly, the biological significance of interaction between gephyrin and the dynein LCs is unclear because their association is dispensable for localization of gephyrin to synapses *(74)*.

An interesting line of research was initiated by the discovery of collybistin as a gephyrin-binding protein *(75,76)*. Collybistin is a member of the diffuse B-cell lymphoma-like (dbl-like) superfamily of guanine nucleotide exchange factors (GEFs) that catalyze guanosine 5′ triphosphate (GTP)–guanosine 5′ diphosphate

(GDP) exchange on Rho family GTPases *(77)*. As a typical GEF, collybistin has tandem exchange factor (RhoGEF) and pleckstrin homology lipid-binding domains. The protein exists in several alternatively spliced isoforms that are retained by these two domains *(75,76,78)*. The presence of a N-terminal SH3 domain that is present in all of the major collybistin isoforms expressed in neurons appears to negatively regulate the ability of collybistin to translocate gephyrin to submembrane microaggregates in transfected mammalian cells *(75,76,78)*. Thus, protein–protein interactions at the collybistin SH3 domain might relieve this inhibition and promote the deposition of gephyrin to the postsynaptic cytoskeleton. Gephyrin clustering requires collybistin–gephyrin interactions and an intact collybistin pleckstrin homology domain. The vital importance of collybistin for inhibitory synaptogenesis is underlined by the discovery of a mutation (G55A) in exon 2 of the human *collybistin* gene in a patient with clinical symptoms of both hyperekplexia and epilepsy. Overexpression of collybistin^{G55A} (but not native collybistin) in cultured neurons interferes with postsynaptic clustering of gephyrin and α_2-subunit containing $GABA_A$ Rs, suggesting that the clinical manifestations of this collybistin missense mutation result, at least in part, from mislocalization of gephyrin and postsynaptic $GABA_A$ Rs *(78)*.

Using fluorescence recovery after photobleaching (FRAP), Jacob et al. *(49)* compared the mobility of fluorescently-tagged $GABA_A$ Rs at postsynaptic and extrasynaptic plasma membrane sites of live neurons. These experiments revealed significantly greater fluorescence recovery rates at extrasynaptic than at postsynaptic membrane domains, thereby indicating greater mobility of extrasynaptic than postsynaptic $GABA_A$ Rs. Moreover, the fluorescence recovery rate at the periphery of the photobleached area was greater than that at the center, which was interpreted as evidence for lateral diffusion of $GABA_A$ Rs. To assess the role of gephyrin in modulating lateral diffusion, FRAP experiments were combined with RNAi knockdown of gephyrin, a treatment that effectively reduced the expression of gephyrin but did not affect the translocation of $GABA_A$ Rs to the plasma membrane and left at least some $GABA_A$ receptor clusters intact. Interestingly, fluorescently tagged postsynaptic $GABA_A$ Rs in gephyrin-RNAi-treated neurons exhibit a significant increase in the FRAP-recovery rate, indicating that they are more rapidly replaced than under control conditions *(49)*. These experiments suggest that gephyrin acts to restrict the lateral mobility of postsynaptic $GABA_A$ Rs, reminiscent of observations made earlier by video tracking of quantum dot-labeled glycine receptors in cultured spinal neurons *(79)*. Similar to gephyrin-associated glycine receptor clusters, gephyrin-associated $GABA_A$ receptor clusters may therefore be defined as temporary confinement areas in which receptors are concentrated by limited diffusion (Fig. 2).

It is now clear that gephyrin assumes distinct functional roles at glycinergic and GABAergic synapses. Unlike in the case of glycine receptors, gephyrin does not copurify with $GABA_A$ Rs, suggesting an indirect interaction with $GABA_A$ Rs *(62,63)*. At glycinergic synapses, accumulation of gephyrin precedes the clustering of glycine receptors, which is consistent with an instructive function for assembly of the postsynaptic apparatus. In contrast, the appearance of $GABA_A$ receptor clusters in developing spinal cord neurons tends to precede the formation of gephyrin clusters *(80)*. In immature hippocampal cultures, α_2-subunit-containing $GABA_A$ receptor clusters first form extrasynaptically in the apparent absence of gephyrin, whereas later-emerging gephyrin clusters are always colocalized with $GABA_A$ Rs and preferentially postsynaptic *(81)*. Furthermore, transfection of the γ_2-subunit into γ_2-subunit-deficient neurons results in efficient rescue of postsynaptic localization of $GABA_A$ Rs and recruitment of gephyrin to synapses, indicating that $GABA_A$ Rs are able to instruct the proper apposition of pre- and postsynaptic elements of GABAergic synapses *(56)*. Moreover, some γ_2/α_2-chimeric constructs can accumulate at postsynaptic sites of γ_2-deficient neurons, while failing to recruit gephyrin, indicating that clustering of $GABA_A$ Rs and recruitment to synapses can occur independently of gephyrin *(56)*, consistent with gephyrin-independent clustering of α_1-subunit-containing receptors in gephyrin-deficient mice and cultured neurons *(67,68)*.

Gephyrin exists in several structurally distinct variants that are owing to alternative splicing mainly in the N-terminal region *(82–85)*. Analysis of spinal cord neurons that release both glycine and GABA and accumulate glycine receptors and $GABA_A$ Rs in the same postsynaptic densities, suggests that alternative splicing may underlie functional differences of gephyrin at GABAergic and glycinergic synapses *(86)*. The presence of the alternatively spliced exon C5 in gephyrin results in preferential association of the isoform at GABAergic as opposed to glycinergic synapses. Hence, the type of gephyrin isoform expressed in the subsynaptic scaffold contributes to the preferential accumulation of $GABA_A$ or glycine receptors at synapses *(86)*.

Interactions between postsynaptic neuroligins (NL1–4) and presynaptic β-neurexins are believed to play an important role in the formation of glutamatergic and GABAergic synapses *(87–89)* (reviewed in refs. *90–92*). Ectopic expression of NL in heterologous cells that are then cocultured with neurons can induce presynaptic development of glutamatergic and GABAergic synapses *(87,93)*. Conversely, β-neurexin presented on beads or overexpressed in heterologous cells can induce aggregates of separate postsynaptic GABAergic or glutamatergic hemisynapses in cocultured neurons *(87)*. Of particular interest is NL2 because it is thus far the only synaptogenic factor that is found selectively at GABAergic postsynapses *(87,94)*. Thus, neurexin-induced recruitment of postsynaptic NL2 is believed to serve as a critical step in postsynaptic differentiation of GABAergic synapses. The mechanism by which *trans*-synaptic β-neurexin-NL2 complexes induce postsynaptic differentiation is so far not understood; in particular it appears that NL2 does not interact directly with gephyrin or $GABA_A$ Rs *(92)*.

In addition to NL2, gephyrin, and gephyrin-associated proteins, the dystrophin-associated glycoprotein complex is specifically localized to the postsynaptic membrane of a subset of GABAergic synapses *(42,95–97)*. The complex consists of the peripheral membrane protein α-dystroglycan, the integral membrane spanning protein β-dystroglycan and the subsynaptic cytoskeletal component dystrophin *(98)*. The function of this complex is poorly understood as it appears late during synaptogenesis and is dispensable for accumulation of $GABA_A$ Rs and gephyrin in the postsynaptic membrane *(42,97)*. Of note, the postsynaptic accumulation of the dystrophin-associated glycoprotein complex in γ_2-subunit-deficient neurons is unaltered *(42)*, suggesting that this late form of postsynaptic differentiation can occur independently of gephyrin, $GABA_A$ Rs, and synaptic function. The postsynaptic dystroglycan complex also interacts with presynaptic β-neurexin *(99)*, but unlike NL-2 does not mediate β-neurexin-induced postsynaptic accumulation of $GABA_A$ Rs and gephyrin *(87)*.

5. Proteins Implicated in Regulating Exocytosis of $GABA_A$ Rs to the Plasma Membrane

Yeast two-hybrid screens using the γ_2-subunit intracellular-loop domain resulted in isolation of GABARAP as the first $GABA_A$ receptor associated protein that selectively interacts with γ_{1-3} subunits *(100,101)* (Table 2). GABARAP is a member of a small family of ubiquitin-like homologous proteins that includes Golgi-associated transport enhancer of 16 kDa (GATE16), Apg8L, as well as the LC3 of MAP 1A and -B. Interestingly, the four members of this protein family are subject to a post-translational lipid-conjugation mechanism, which involves activating, conjugating, and deconjugating enzymes reminiscent of the ubiquitin-conjugation system *(102,103)*. Accordingly, GABARAP family proteins are believed to be involved in intra-Golgi and endoplasmic reticulum (ER) to Golgi transport, as well as in autophagic processes involved in bulk degradation of proteins and organelles (reviewed in refs. *104,105*). GABARAP itself is implicated in $GABA_A$ receptor trafficking by its binding to $GABA_A$ Rs and microtubules both in vitro and in brain extracts, and by its colocalization with intracellular $GABA_A$ Rs in cultured cortical neurons *(100,106,107)*.

In addition, indirect association of GABARAP with microfilaments in vitro has been proposed to link $GABA_A$ Rs to the actin cytoskeleton *(107)*. Heterologous coexpression of GABARAP and $GABA_A$ Rs promotes the clustering of $GABA_A$ Rs in a γ_2-subunit- and tubulin-binding motif-dependent manner *(108)*. However, whereas GABARAP was initially proposed to act as a postsynaptic targeting protein of $GABA_A$ Rs *(100)*, its colocalization with clustered

$GABA_A$ Rs in the plasma membrane could not be confirmed. Immunoelectron microscopic analysis of brain sections revealed that GABARAP is absent at synapses and instead concentrated in the Golgi complex and other intracellular membrane systems, reminiscent of the cellular distribution of its homolog GATE-16 *(106,109,110)*. Similar to GATE-16, GABARAP also interacts with the vesicular transport factor *N*-ethylmaleimide-sensitive factor (NSF), consistent with a possible role of GABARAP in vesicular trafficking *(106,110)*. Transient overexpression of GABARAP in cultured neurons increases the surface expression of $GABA_A$ Rs, suggesting that GABARAP facilitates exocytosis of $GABA_A$ Rs *(111)* (Fig. 2). However, recent evidence from analyses of GABARAP knockout mice has shown that GABARAP is functionally dispensable, possibly because of functional redundancy with other members of its family *(112)*. GABARAP-deficient mice lack an overt behavioral phenotype and exhibit normal expression and punctuate distribution of γ_2-subunit-containing $GABA_A$ Rs. More conclusive analyses of GABARAP function therefore will have to address the extent by which this entire protein family is involved in trafficking of $GABA_A$ Rs in vivo.

Evidence in support of a function of GABARAP as a trafficking factor of $GABA_A$ Rs includes the identification of phospholipase C-related inactive protein (PRIP-1, p130) as both a GABARAP- and a $GABA_A$ receptor-interacting protein *(113,114)*. A role for PRIP-1 in regulating the function or trafficking of $GABA_A$ Rs is further suggested by the phenotype of PRIP-1 knockout mice, which show impairments in $GABA_A$ receptor modulation by Zn^{2+} and benzodiazepines, as well as altered benzodiazepine-induced behavior *(113)*. PRIP-1 interacts with the β-subunits of $GABA_A$ Rs and functions as an adaptor protein for protein phosphatase (PP)1α, keeping this enzyme in an inactive state *(114,115)*. Phosphorylation of PP1α at threonine 94 leads to dissociation of the PRIP-1/PP1α complex, thereby providing a rationale for phosphorylation-induced dephosphorylation of $GABA_A$ receptor β-subunits *(114)*. Together with recent data summarized below on β_3-subunit phosphorylation-dependent and clathrin-mediated endocytosis of $GABA_A$ Rs *(116)*, the data suggest that the PRIP-1 regulated phosphorylation state of $GABA_A$ Rs modulates the surface stability of $GABA_A$ Rs. The glutamate receptor trafficking factor GRIP1 has been isolated as a GABARAP-binding protein *(117,118)* and shown to localize to pre- and postsynaptic sites of GABAergic synapses *(117–119)*. However, although GRIP-1 is well established as a trafficking factor of α-amino-3-hydroxy-5-methylisoxazole-4-propionate (AMPA)-type glutamate receptors that is important for plasticity of glutamatergic synapses *(120,121)*, its role at GABAergic synapses is not understood.

The NSF is a homohexameric ATPase that acts as an essential chaperone involved in all kinds of localized and regulated membrane fusion events *(122)*. The protein was first implicated in trafficking of $GABA_A$ Rs through its interactions with GABARAP and its homolog GATE-16 *(106,110)*. However, recent

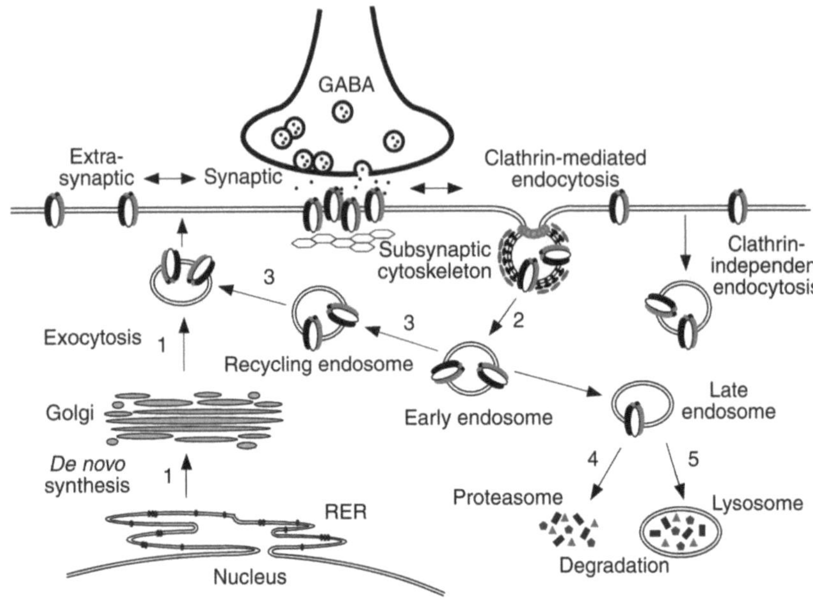

Fig. 2. Major routes of trafficking of postsynaptic $GABA_A$ Rs. $GABA_A$ Rs exiting from the ER traffic through the Golgi complex to the plasma membrane. The available evidence suggests that they are inserted into the extrasynaptic plasma membrane from where they reach the postsynaptic membrane by lateral diffusion. Exit of receptors from the ER may be regulated by interactions between β-subunits and BIG2. Translocation through the Golgi complex appears to be regulated by interactions of the γ_2-subunit with GABARAP family members and NSF and by GODZ-mediated palmitoylation of the γ_2-subunit. The postsynaptic gephyrin-associated cytoskeleton acts as a γ_2-subunit-dependent diffusion trap for receptors that have reached the plasma membrane, thereby leading to the accumulation of receptors in the membrane across from GABA-release sites. $GABA_A$ Rs exit from the plasma membrane through clathrin-dependent and perhaps clathrin-independent mechanisms. Association of $GABA_A$ Rs with clathrin-coated pits is regulated by a phosphorylation-sensitive interaction between $GABA_A$ receptor β-subunits and the clathrin adaptor AP2. This step may also be regulated by the phosphatase adaptor and β-subunit interacting protein PRIP-1. Recycling of $GABA_A$ Rs from early endosomes to recycling endosomes and back to the surface membrane is positively regulated by interactions between β-subunits and HAP1 and perhaps Plic-1. Alternatively, Plic-1 might inhibit degradation of *de novo*-synthesized $GABA_A$ Rs exiting from the ER.

evidence indicates that NSF colocalizes with intracellular neuronal $GABA_A$ Rs and interacts directly with the intracellular loop region of receptor β-subunits in vitro *(123)*. On heterologous overexpression of NSF and $GABA_A$ Rs in Cos7 cells or on overexpression in cultured neurons, NSF reduces the steady state cell surface number of $GABA_A$ Rs. Because the rate of endocytosis of $GABA_A$ Rs remains unaffected by NSF, it appears that NSF inhibits the exocytotic pathway of $GABA_A$ Rs *(123)*. However, the biological significance of such regulation remains unclear, given that the observation involved massive overexpression of a general factor that is essential for all types of membrane fusion events.

Ubiquitination is an evolutionarily conserved post-translational modification of diverse neurotransmitter receptors and other signaling proteins, many of which are concentrated in the postsynaptic density of glutamatergic synapses (reviewed in refs. *124–126*). The process involves addition of single or multiple copies of a 76-amino acid ubiquitin moiety to lysine residues of target proteins. Whereas polyubiquitination generally acts as a signal for degradation by the 26S proteasome, the addition of single ubiquitin moieties is reversible and serves as an active signal in diverse intracellular trafficking pathways, including as a trigger for endocytosis. Evidence that ubiquitination might be involved in regulated trafficking of $GABA_A$ Rs, stems from the isolation of the ubiquitin-like protein, Plic-1, as an interactor of the intracellular loop domains of $GABA_A$ receptor α- and β-subunits *(127,128)*.

Plic-1 and -2 are closely related proteins characterized by an ubiquitin-like N-terminus and an ubiquitin-associated domain. They are further known to interact with the ubiquitin ligase E6-AP and with proteasomes in large complexes and believed to interfere with the in vivo degradation of ubiquitin-dependent proteasome substrates *(129–131)*. Overexpression of Plic-1 increases the half-life and surface expression of cotransfected $GABA_A$ Rs in heterologous cells or in neurons *(127)*. Furthermore, disrupting the interaction between Plic-1 and $GABA_A$ Rs by competing with an α_1-subunit-derived peptide that corresponds to the Plic-1-binding site results in a significant reduction in the cell surface expression of $GABA_A$ Rs, parallel with rapid decay of GABA-evoked whole cell currents, both in transfected heterologous cells and in neurons. The cell surface steady-state expression of $GABA_A$ Rs expressed in heterologous cells is increased following inhibition of the proteasome. This evidence therefore suggests that Plic-1 facilitates exocytosis of $GABA_A$ receptor by inhibiting proteasome-mediated degradation of newly synthesized and/or endocytosed $GABA_A$ Rs *(127)*. Indirectly, the data also suggest that $GABA_A$ Rs might be substrates for ubiquitination, consistent with recently reported preliminary results *(132)*.

Similar to Plic-1, the brefeldin A-inhibited GDP/GTP exchange factor 2 (BIG2) interacts selectively with the intracellular loop region of $GABA_A$ receptor β-subunits *(133)*. BIG2 is a Sec7 domain-containing GEF that catalyzes GDP/GTP exchange at the small G-proteins ADP-ribosylation factor (ARF)1 and -3 *(134,135)*. GEF activation of class I ARF proteins is required for membrane budding of vesicles from the Golgi apparatus, thereby enabling proteins to proceed from the Golgi through the trans-Golgi network toward the plasma membrane *(136)*. Consistent with a role in exocytosis of $GABA_A$ Rs, BIG2 immunoreactivity is concentrated in the trans-Golgi network and has been detected in somatic and dendritic vesicle-like structures, as well as in the postsynaptic density of both inhibitory and excitatory synapses *(133)*. Moreover, coexpression of BIG2 with the β_3-subunit in heterologous cells promotes the translocation of β_3-subunit assemblies to the cell surface. Interestingly, a recent report has further identified BIG2 as a component of recycling endosomes, and provided evidence that BIG2 mediated activation of ARFs contributes to the structural integrity of this trafficking compartment *(137)*. Thus, BIG2, in addition to or instead of exocytosis, might have a role in endocytic recycling of $GABA_A$ Rs.

Palmitoylation is a reversible post-translational modification that involves the addition of the 16-carbon saturated fatty acid palmitate to cysteine residues of a large and diverse class of target proteins, including integral and peripheral membrane proteins, as well as extracellular proteins (reviewed in refs. *138,139*). The transient nature of palmitoylation and the ability of substrate proteins to potentially undergo multiple rounds of palmitoylation and depalmitoylation before degradation suggest that this protein modification is involved in dynamic regulation of diverse cellular processes, reminiscent of phosphorylation. Of particular interest as regulators are palmitoyl acyltransferases, a class of enzymes containing an Asp-His-His-Cys cysteine-rich domain (DHHC-CRD) and first identified in yeast *(140,141)*.

Using a yeast two-hybrid screen that is compatible with isolation of membrane proteins, Keller et al. *(142)* have isolated a Golgi apparatus-specific protein with a DHHC zinc-finger domain (GODZ, also known as DHHC3), as a $GABA_A$ receptor interacting protein and as a first mammalian member of the DHHC-CRD family of palmitoyltransferases. GODZ binds to a 14-amino acid γ_2-subunit cytoplasmic domain containing four Cys residues N-terminal to the GABARAP-binding site. This binding domain is conserved selectively in γ_{1-3}-subunits and essential for GODZ-mediated palmitoylation of $GABA_A$ Rs in transfected 293T cells. Although GODZ is broadly expressed in many tissues, expression of GODZ mRNA and protein in brain is neuron-specific with a regional distribution similar to that of the γ_2-subunit. $GABA_A$ Rs are also subject to palmitoylation in neurons.

Analysis of Cys-Ala substituted γ_2-subunit constructs in transfected Cos7 cells indicates that the γ_2-subunit is palmitoylated at all four cysteines within the GODZ-binding domain *(143)*. Moreover, transfection of the same Cys-Ala substituted γ_2-subunit constructs abolishes the postsynaptic clustering of γ_2-subunit–containing $GABA_A$ Rs in neurons. Drug-induced global inhibition of palmitoylation by Br-palmitate dramatically reduces the expression of $GABA_A$ Rs at the cell surface. Thus, reduced clustering of receptors seen following transfection of neurons with Cys-Ala-substituted γ_2-subunits is largely owing to reduced surface expression of $GABA_A$ Rs *(143)*.

More direct evidence for a role of GODZ-mediated palmitoylation in regulation of GABAergic inhibition was recently obtained by transfection of cultured neurons with a dominant negative GODZ construct and by inhibition of GODZ expression by RNAi *(144)*. Both types of treatments result in a selective reduction of postsynaptic clustering of γ_2-subunit containing GABA$_A$ Rs, which is paralleled by a significant reduction in GABA-evoked whole-cell currents and amplitude of GABAergic mIPSCs. In contrast, the function of AMPA-evoked whole-cell currents and glutamatergic miniature excitatory postsynaptic currents is unaffected by dominant negative GODZ and GODZ-directed RNAi, demonstrating a selective function of GODZ at inhibitory synapses *(144)*. Importantly, GODZ is among several DHHC proteins that can palmitoylate PostSynaptic Density Protein of 95-KDa (PSD-95) and AMPA receptors in vitro and/or in heterologous expression systems *(145,146)*. Lack of an effect of GODZ on glutamatergic currents suggests extensive functional redundancy among DHHC-CRD domain enzymes at glutamatergic synapses, in contrast to inhibitory synapses where GODZ and its close homolog sertoli cell gene with a zinc finger domain-β (SERZ-β)/DHHC7 *(142,147)* are the only DHHC-CRD family members that can use the γ_2-subunit as a substrate *(144)*.

6. Proteins That Regulate Endocytosis and Endocytic Recycling of GABA$_A$ Rs

GABA$_A$ Rs expressed on the cell surface of neurons are subject to constitutive and protein kinase C (PKC)-stimulated endocytosis *(2,57,148,149)*. Therefore, changes in the rate of receptor insertion and endocytosis allow for dynamic physiological adaptation in inhibitory transmission but are also implicated in the etiology of neurological disorders. For example, excessive rates of endocytosis might explain loss of surface expression of GABA$_A$ Rs and pharmacoresistance to benzodiazepine treatment observed in rodents and patients during prolonged status epilepticus *(5,19,20)*. Whereas both clathrin-dependent and -independent endocytosis mechanisms have been described in heterologous cells *(150)*, evidence from brain and cultured neurons suggests that GABA$_A$ Rs are endocytosed through clathrin-coated pits *(2,15,57,148,149,151)*. GABA$_A$ Rs associate directly with the endocytic machinery by binding of receptor-β and γ-subunits to the μ_2-subunit of the clathrin adaptor AP2 *(2,15,116)*. Consistent with this idea, blocking clathrin-dependent endocytosis by a peptide that interferes with binding of amphiphysin to dynamin results in an increased in mIPSC amplitude, reflecting an increase in postsynaptic GABA$_A$ Rs owing to reduced receptor internalization *(2)*.

Approximately 25% of cell surface receptors endocytose in 30 min and 70% of such endocytosed receptors are recycled back to the cell surface within 1 h *(152)*. On a slower time-scale (6 h), a significant fraction (29%) of neuronal

Table 2
$GABA_A$ Receptor-Interacting Proteins and Their Functions in Trafficking of Receptors

$GABA_A$ R-interacting protein and pathway indicated in Fig. 2	Subunit specificity	Proposed function	References
AP2 (pathway 2)	$\beta\gamma$	Recruits $GABA_A$ Rs into clathrin-coated pits and mediates phoshorylation dependent endocytosis of the receptor	*2,116*
BIG2 (pathway 1, 3)	β	Class I ARF-specific GEF, promotes the exit of recombinant $GABA_A$ Rs from the ER of transfected heterologous cells. In HeLa cells, BIG2 is also found in and contributes to their structural integrity of recycling endosomes	*133,137*
GABARAP (pathways 1, 3)	γ_{1-3}	Overexpression facilitates exocytosis of $GABA_A$ Rs	*108,111*
GODZ (pathway 1)	γ	Palmitoyltransferase implicated in exocytosis of γ-subunit-containing receptors	*142*
GRIF-1	β_2	Unknown	*157*
HAP1 (pathway 3, 4)	β	Regulates endocytic sorting by inhibiting $GABA_A$ receptor degradation and facilitating recycling	*152*
NSF (pathway 1, 3)	β	Overexpression negatively regulates exocytosis of $GABA_A$ Rs	*123*
Plic-1 (pathway 1, 3)	α, β	Facilitates exocytosis or recycling by inhibiting proteasome dependent degradation of $GABA_A$ Rs	*127*
PRIP-1 (pathway 2)	β	Adaptor for and negative regulator of protein PP1α regulates phosphorylation state of serine(s) in AP2-binding site of β-subunits, thereby implicated in regulating association of β-subunits with AP2 that contributes to endocytosis of $GABA_A$ Rs	*114,115*

$GABA_A$ Rs is targeted to late endosomes and is then degraded, apparently by the lysosome *(148,152)*. Interestingly, interaction of AP2 with the $GABA_A$ receptor β_3-subunit is negatively regulated by phosphorylation of the major serine phosphorylation site known as an in vitro PKA and PKC substrate and present in all three β-subunits *(116)*. An unphosphorylated β_3-subunit peptide containing the AP2-binding site effectively binds to the μ-subunit of AP2, and on infusion through the patch pipet into neurons results in a sustained increase in the mIPSC amplitude and frequency. In contrast, a phosphorylated form of the same peptide binds to the μ-subunit with sixfold lower affinity and has no such effects *(116)*. The authors argue that the increase in mIPSC frequency reflects increased expression of $GABA_A$ Rs at synapses that under control conditions fail to give rise to detectable miniature currents.

The decision of whether endocytosed $GABA_A$ Rs are recycled or degraded appears to be regulated by interaction of $GABA_A$ receptor β-subunits with huntingtin-associated protein (HAP)1, an endocytic trafficking factor, first identified through its interaction with the Huntington disease protein huntingtin *(152–154)*. Overexpression of HAP1 in cultured cortical neurons interferes with the degradation of $GABA_A$ Rs, while increasing the recycling and surface expression of GABA Rs, concomitant with an increase in the mIPSC amplitude *(152)*. It will be interesting to see whether and how diverse signaling pathways modulate this mechanism and whether and where it is affected under pathological conditions such as in status epilepticus or Huntington's disease.

In summary, recent research has led to major progress in identification of protein–protein interactions and post-translational subunit modifications that contribute to regulated surface expression and postsynaptic targeting of $GABA_A$ Rs. These interactions, with few exceptions, appear to be limited to the major intracellular-loop region of β- and γ-subunits. Future research will need to focus on how these interactions and trafficking routes are regulated by extracellular cues and second messenger pathways during synaptogenesis and at mature synapses, both under physiological and pathological conditions.

References

1. Nusser, Z., Hajos, N., Somogyi, P., and Mody, I. (1998) Increased number of synaptic $GABA_A$ receptors underlies potentiation at hippocampal inhibitory synapses. *Nature* **395,** 172–177.
2. Kittler, J. T., Delmas, P., Jovanovic, J. N., Brown, D. A., Smart, T. G., and Moss, S. J. (2000) Constitutive endocytosis of $GABA_A$ receptors by an association with the adaptin AP2 complex modulates inhibitory synaptic currents in hippocampal neurons. *J. Neurosci.* **20(21),** 7972–7977.
3. Kapur, J., Stringer, J. L., and Lothman, E. W. (1989) Evidence that repetitive seizures in the hippocampus cause a lasting reduction of GABAergic inhibition. *J. Neurophysiol.* **61(2),** 417–426.

4. Kapur, J. and Lothman, E. W. (1989) Loss of inhibition precedes delayed spontaneous seizures in the hippocampus after tetanic electrical stimulation. *J. Neurophysiol.* **61(2),** 427–434.
5. Naylor, D. E. and Wasterlain, C. G. (2005) GABA synapses and the rapid loss of inhibition to dentate gyrus granule cells after brief perforant-path stimulation. *Epilepsia* **46(Suppl. 5),** 142–147.
6. Meier, J., Akyeli, J., Kirischuk, S., and Grantyn, R. (2003) GABA(A) receptor activity and PKC control inhibitory synaptogenesis in CNS tissue slices. *Mol. Cell. Neurosci.* **23(4),** 600–613.
7. Crestani, F., Lorez, M., Baer, K., et al. (1999) Decreased $GABA_A$-receptor clustering results in enhanced anxiety and a bias for threat cues. *Nat. Neurosci.* **2,** 833–839.
8. Brambilla, P., Perez, J., Barale, F., Schettini, G., and Soares, J. C. (2003) GABAergic dysfunction in mood disorders. *Mol. Psychiatry* **8(8),** 721–737.
9. Tunnicliff, G. and Malatynska, E. (2003) Central GABAergic systems and depressive illness. *Neurochem. Res.* **28(6),** 965–976.
10. Yoshikawa, T., Watanabe, A., Ishitsuka, Y., Nakaya, A., and Nakatani, N. (2002) Identification of multiple genetic loci linked to the propensity for "behavioral despair" in mice. *Genome. Res.* **12(3),** 357–366.
11. Yamada, K., Watanabe, A., Iwayama-Shigeno, Y., and Yoshikawa, T. (2003) Evidence of association between gamma-aminobutyric acid type A receptor genes located on 5q34 and female patients with mood disorders. *Neurosci. Lett.* **349(1),** 9–12.
12. N-Wihlback, A. C., Sundstrom-Poromaa, I., and Backstrom, T. (2005) Action by and sensitivity to neuroactive steroids in menstrual cycle related CNS disorders. *Psychopharmacology (Berl.)* 1–14.
13. Sundstrom Poromaa, I., Smith, S., and Gulinello, M. (2003) GABA receptors, progesterone and premenstrual dysphoric disorder. *Arch. Women Ment. Health* **6(1),** 23–41.
14. Grobin, A. C., Matthews, D. B., Devaud, L. L., and Morrow, A. L. (1998) The role of GABA(A) receptors in the acute and chronic effects of ethanol. *Psychopharmacology (Berl).* **139(1–2),** 2–19.
15. Kumar, S., Kralic, J. E., O'Buckley, T. K., Grobin, A. C., and Morrow, A. L. (2003) Chronic ethanol consumption enhances internalization of alpha1 subunit-containing GABA(A) receptors in cerebral cortex. *J. Neurochem.* **86(3),** 700–708.
16. Seil, F. J. and Drake-Baumann, R. (2000) TrkB receptor ligands promote activity-dependent inhibitory synaptogenesis. *J. Neurosci.* **20(14),** 5367–5373.
17. Seil, F. J. (2003) TrkB receptor signaling and activity-dependent inhibitory synaptogenesis. *Histol. Histopathol.* **18(2),** 635–646.
18. Marty, S., Wehrle, R., Fritschy, J. M., and Sotelo, C. (2004) Quantitative effects produced by modifications of neuronal activity on the size of GABA(A) receptor clusters in hippocampal slice cultures. *Eur. J. Neurosci.* **20(2),** 427–440.
19. Goodkin, H. P., Yeh, J. L., and Kapur, J. (2005) Status epilepticus increases the intracellular accumulation of GABA(A) receptors. *J. Neurosci.* **25(23),** 5511–5520.
20. Goodkin, H. P., Liu, X., and Holmes, G. L. (2003) Diazepam terminates brief but not prolonged seizures in young, naive rats. *Epilepsia* **44(8),** 1109–1112.

21. Brunig, I., Penschuck, S., Berninger, B., Benson, J., and Fritschy, J. M. (2001) BDNF reduces miniature inhibitory postsynaptic currents by rapid downregulation of GABA(A) receptor surface expression. *Eur. J. Neurosci.* **13(7),** 1320–1328.
22. Mizoguchi, Y., Kanematsu, T., Hirata, M., and Nabekura, J. (2003) A rapid increase in the total number of cell surface functional GABAA receptors induced by brain-derived neurotrophic factor in rat visual cortex. *J. Biol. Chem.* **278(45),** 44,097–44,102.
23. Wan, G., Xiong, Z. G., Man. H. Y., et al. (1997) Recruitment of functional GABA$_A$ receptors to postsynaptic domains by insulin. *Nature* **388,** 686–690.
24. Stellwagen, D., Beattie, E. C., Seo, J. Y., and Malenka, R. C. Differential regulation of AMPA receptor and GABA receptor trafficking by tumor necrosis factor-alpha. *J. Neurosci.* **25(12),** 3219–3228.
25. Tehrani, M. H. and Barnes, E. M. Jr. (1991) Agonist-dependent internalization of gamma-aminobutyric acidA/benzodiazepine receptors in chick cortical neurons. *J. Neurochem.* **57(4),** 1307–1312.
26. Tehrani, M. H. and Barnes, E. M. Jr. (1997) Sequestration of gamma-aminobutyric acidA receptors on clathrin-coated vesicles during chronic benzodiazepine administration in vivo. *J. Pharmacol. Exp. Ther.* **283(1),** 384–390.
27. Barnard, E. A., Darlison, M. G., Fujita, N., et al. Molecular biology of the GABA(A) receptor. *Adv. Exp. Med. Biol.* **236,** 31–45.
28. Sieghart, W. and Sperk, G. (2002) Subunit composition, distribution and function of GABA(A) receptor subtypes. *Curr. Top. Med. Chem.* **2(8),** 795–816.
29. Mody, I. and Pearce, R. A. (2004) Diversity of inhibitory neurotransmission through GABA(A) receptors. *Trends. Neurosci.* **27(9),** 569–575.
30. Luscher, B. and Keller, C. A. (2004) Regulation of GABA$_A$ receptor trafficking and channel activity in functional plasticity of inhibitory synapses. *Pharmacol. Ther.* **102(3),** 195–221.
31. Ke, Y., Cohen, B. M., Bang, J. Y., Yang, M., and Renshaw, P. F. (2000) Assessment of GABA concentration in human brain using two-dimensional proton magnetic resonance spectroscopy. *Psychiatry. Res.* **100(3),** 169–178.
32. Terpstra, M., Ugurbil, K., and Gruetter, R. (2002) Direct *in vivo* measurement of human cerebral GABA concentration using MEGA-editing at 7 Tesla. *Magn. Reson. Med.* **47(5),** 1009–1012.
33. Brown, N., Kerby, J., Bonnert, T. P., Whiting, P. J., and Wafford, K. A. Pharmacological characterization of a novel cell line expressing human alpha(4)beta(3)delta GABA(A) receptors. *Br. J. Pharmacol.* **136(7),** 965–974.
34. Haas, K. F. and Macdonald, R. L. (1999) GABA(A) receptor subunit gamma2 and delta subtypes confer unique kinetic properties on recombinant GABA(A) receptor currents in mouse fibroblasts. *J. Physiol.* **514(Pt. 1),** 27–45.
35. Saxena, N. C. and Macdonald. R. L. (1996) Properties of putative cerebellar gamma-aminobutyric acid A receptor isoforms. *Mol. Pharmacol.* **49(3),** 567–579.
36. Bianchi, M. T., Haas, K. F., and Macdonald, R. L. (2001) Structural determinants of fast desensitization and desensitization-deactivation coupling in GABAa receptors. *J. Neurosci.* **21(4),** 1127–1136.
37. Perrais, D. and Ropert, N. (1999) Effect of zolpidem on miniature IPSCs and occupancy of postsynaptic GABAA receptors in central synapses. *J. Neurosci.* **19(2),** 578–588.

38. Essrich, C., Lorez, M., Benson, J., Fritschy, J. M., and Luscher, B. (1998) Postsynaptic clustering of major $GABA_A$ receptor subtypes requires the γ2 subunit and gephyrin. *Nat. Neurosci.* **1(7),** 563–571.
39. Rao, A., Cha, E. M., and Craig, A. M. (2000) Mismatched appositions of presynaptic and postsynaptic components in isolated hippocampal neurons. *J. Neurosci.* **20(22),** 8344–8353.
40. Sassoe-Pognetto, M., Panzanelli, P., Sieghart, W., and Fritschy, J. M. (2000) Colocalization of multiple GABA(A) receptor subtypes with gephyrin at postsynaptic sites. *J. Comp. Neurol.* **420(4),** 481–498.
41. Brunig, I., Scotti, E., Sidler, C., and Fritschy, J. M. (2002) Intact sorting, targeting, and clustering of gamma-aminobutyric acid A receptor subtypes in hippocampal neurons *in vitro*. *J. Comp. Neurol.* **443(1),** 43–55.
42. Brunig, I., Suter, A., Knuesel, I., Luscher, B., and Fritschy, J-M. GABAergic terminals are required for postsynaptic clustering of dystrophin but not GABA(A) receptors and gephyrin. *J. Neurosci.* **22(2),** 4805–4813.
43. Christie, S. B., Miralles, C. P., and De Blas, A. L. (2002) GABAergic innervation organizes synaptic and extrasynaptic GABAA receptor clustering in cultured hippocampal neurons. *J. Neurosci.* **22(3),** 684–697.
44. Baer, K., Essrich, C., Benson, J. A., et al. (1999) Postsynaptic clustering of $GABA_A$ receptors by the γ3 subunit in vivo. *Proc. Natl. Acad. Sci. USA* **96(22),** 12,860–12,865.
45. Schweizer, C., Balsiger, S., Bluethmann, H., et al. (2003) The γ2 subunit of $GABA_A$ receptors is required for maintenance of receptors at mature synapses. *Mol. Cell. Neurosci.* **24,** 442–450.
46. Nusser, Z., Sieghart, W., Benke, D., Fritschy, J-M., and Somogyi, P. (1996) Differential synaptic localization of two major γ-aminobutyric acid type A receptor a subunits on hippocampal pyramidal cells. *Proc. Natl. Acad. Sci. USA* **93,** 11,939–11,944.
47. Li, R. W., Yu, W., Christie, S., et al. (2005) Disruption of postsynaptic GABA receptor clusters leads to decreased GABAergic innervation of pyramidal neurons. *J. Neurochem.* **95(3),** 756–770.
48. Thomas, P., Mortensen, M., Hosie, A. M., and Smart, T. G. (2005) Dynamic mobility of functional GABA(A) receptors at inhibitory synapses. *Nat. Neurosci.* **8(7),** 889–897.
49. Jacob, T. C., Bogdanov, Y. D., Magnus, C., et al. (2005) Gephyrin regulates the cell surface dynamics of synaptic GABA(A) receptors. *J. Neurosci.* **25(45),** 10,469–10,478.
50. Fritschy, J. M., Johnson, D. K., Mohler, H., and Rudolph, U. (1998) Independent assembly and subcellular targeting of $GABA_A$-receptor subtypes demonstrated in mouse hippocampal and olfactory neurons in vivo. *Neurosci. Lett.* **249(2–3),** 99–102.
51. Crestani, F., Keist, R., Fritschy, J. M., et al. (2002) Trace fear conditioning involves hippocampal alpha5 GABA(A) receptors. *Proc. Natl. Acad. Sci. USA* **99(13),** 8980–8985.
52. Caraiscos, V. B., Elliott, E. M., You-Ten, K. E., et al. (2004) Tonic inhibition in mouse hippocampal CA1 pyramidal neurons is mediated by alpha5 subunit-containing gamma-aminobutyric acid type A receptors. *Proc. Natl. Acad. Sci. USA* **101(10),** 3662–3667.

53. Christie, S. B. and de Blas, A. L. (2002) alpha5 Subunit-containing GABA(A) receptors form clusters at GABAergic synapses in hippocampal cultures. *Neuroreport* **13(17),** 2355–2358.
54. Devor, A., Fritschy, J. M., and Yarom, Y. (2000) Non-homogeneous distribution of $GABA_A$ receptors in the inferior olivary nucleus revealed by electrophysiology and immunocytochemistry. *Eur. J. Neurosci.* **12(11),** 141.
55. Devor, A., Fritschy, J. M., and Yarom, Y. (2001) Spatial distribution and subunit composition of GABA(A) receptors in the inferior olivary nucleus. *J. Neurophysiol.* **85(4),** 1686–1696.
56. Alldred, M. J., Mulder-Rosi, J., Lingenfelter, S. E., Chen, G., and Luscher, B. (2005) Distinct gamma2 subunit domains mediate clustering and synaptic function of postsynaptic GABA(A) receptors and gephyrin. *J. Neurosci.* **25(3),** 594–603.
57. van Rijnsoever, C., Sidler, C., and Fritschy, J. M. (2005) Internalized GABA-receptor subunits are transferred to an intracellular pool associated with the postsynaptic density. *Eur. J. Neurosci.* **21(2),** 327–338.
58. Christie, S. B., Li, R. W., Miralles, C. P., Yang, B. Y., and De Blas, A. L. (2005) Clustered and non-clustered GABA(A) receptors in cultured hippocampal neurons. *Mol. Cell. Neurosci.* **31(1),** 1–14.
59. Nymann-Andersen, J., Sawyer, G. W., and Olsen, R. W. (2002) Interaction between GABAA receptor subunit intracellular loops: implications for higher order complex formation. *J. Neurochem.* **83(5),** 1164–1171.
60. Sassoè-Pognetto, M. and Fritschy, J. M. (2000) Gephyrin, a major postsynaptic protein of GABAergic synapses. *Eur. J. Neurosci.* **12(7),** 2205–2210.
61. Kneussel, M. and Betz, H. (2000) Receptors, gephyrin and gephyrin-associated proteins: novel insights into the assembly of inhibitory postsynaptic membrane specializations. *J. Physiol.* **525(Pt. 1),** 1–9.
62. Meyer, G., Kirsch, J., Betz, H., and Langosch, D. (1995) Identification of a gephyrin binding motif on the glycine receptor b subunit. *Neuron* **15,** 563–572.
63. Kannenberg, K., Baur, R., and Sigel, E. (1997) Proteins associated with a1-subunit-containing $GABA_A$ receptors from bovine brain. *J. Neurochem.* **68,** 1352–1360.
64. Baer, K., Essrich, C., Balsiger, S., et al. (2000) Rescue of γ2 subunit-deficient mice by transgenic overexpression of the $GABA_A$ receptor γ2S or γ2L subunit isoforms. *Eur. J. Neurosci.* **12(7),** 2639–2643.
65. Kneussel, M., Brandstätter, J. H., Laube, B., Stahl, S., Müller, U., and Betz, H. Loss of postsynaptic $GABA_A$ receptor clustering in gephyrin-deficient mice. *J. Neurosci.* **19(21),** 9289–9297.
66. Fischer, F., Kneussel, M., Tintrup, H., et al. (2000) Reduced synaptic clustering of GABA and glycine receptors in the retina of the gephyrin null mutant mouse. *J. Comp. Neurol.* **427(4),** 634–648.
67. Kneussel, M., Brandstätter, J. H., Gasnier, B., Feng, G., Sanes, J. R., and Betz, H. Gephyrin-independent clustering of postsynaptic $GABA_A$ receptor subtypes. *Mol. Cell. Neurosci.* **17,** 973–982.
68. Levi, S., Logan, S. M., Tovar, K. R., and Craig, A. M. (2004) Gephyrin is critical for glycine receptor clustering but not for the formation of functional GABAergic synapses in hippocampal neurons. *J. Neurosci.* **24(1),** 207–217.

69. Kirsch, J., Langosch, D., Prior, P., Litauer, U. Z., and Betz, H. (1991) The 93-kDa glycine receptor-associated protein binds to tubulin. *J. Biol. Chem.* **266,** 22,242–22,245.
70. Mammoto, A., Sasaki, T., Asakura, T., et al. (1998) Interactions of drebrin and gephyrin with profilin. *Biochem. Biophys. Res. Commun.* **243(1),** 86–89.
71. Giesemann, T., Schwarz, G., Nawrotzki, R., et al. (2003) Complex formation between the postsynaptic scaffolding protein gephyrin, profilin, and Mena: a possible link to the microfilament system. *J. Neurosci.* **23(23),** 8330–8339.
72. Allison, D. W., Chervin, A. S., Gelfand, V. I., and Craig, A. M. (2000) Postsynaptic scaffolds of excitatory and inhibitory synapses in hippocampal neurons: maintenance of core components independent of actin filaments and microtubules. *J. Neurosci.* **20(12),** 4545–4554.
73. Sabatini, D. M., Barrow, R. K., Blackshaw, S., et al. (1999) Interaction of RAFT1 with gephyrin required for rapamycin-sensitive signaling. *Science* **284(5417),** 1161–1164.
74. Fuhrmann, J. C., Kins, S., Rostaing, P., et al. (2002) Gephyrin interacts with dynein light chains 1 and 2, components of motor protein complexes. *J. Neurosci.* **22(13),** 5393–5402.
75. Kins, S., Heinrich Betz, H., and Kirsch, J. (2000) Collybistin, a newly identified brain-specific GEF, induces submembrane clustering of gephyrin. *Nature Neurosci.* **3,** 22–29.
76. Grosskreutz, Y., Hermann, A., Kins, S., Fuhrmann, J. C., Betz, H., and Kneussel, M. (2001) Identification of a gephyrin-binding motif in the GDP/GTP exchange factor collybistin. *Biol. Chem.* **382(10),** 1455–1462.
77. Wherlock, M. and Mellor, H. (2002) The Rho GTPase family: a Racs to Wrchs story. *J. Cell. Sci.* **115(Pt. 2),** 239–240.
78. Harvey, K., Duguid, I. C., Alldred, M. J., et al. (2004) The GDP-GTP exchange factor collybistin: an essential determinant of neuronal gephyrin clustering. *J. Neurosci.* **24(25),** 5816–5826.
79. Dahan, M., Levi, S., Luccardini, C., Rostaing, P., Riveau, B., and Triller, A. (2003) Diffusion dynamics of glycine receptors revealed by single-quantum dot tracking. *Science* **302(5644),** 442–445.
80. Dumoulin, A., Levi, S., Riveau, B., Gasnier, B., and Triller, A. (2000) Formation of mixed glycine and GABAergic synapses in cultured spinal cord neurons. *Eur. J. Neurosci.* **12(11),** 3883–3892.
81. Studler, B., Sidler, C., and Fritschy, J. M. (2005) Differential regulation of GABA(A) receptor and gephyrin postsynaptic clustering in immature hippocampal neuronal cultures. *J. Comp. Neurol.* **484(3),** 344–355.
82. Prior, P., Schmitt, B., Grenningloh, G., et al. (1992) Primary structure and alternative splice variants of gephyrin, a putative glycine receptor-tubulin linker protein. *Neuron* **8,** 1161–1170.
83. Ramming, M., Kins, S., Werner, N., Hermann, A., Betz, H., and Kirsch, J. (2000) Diversity and phylogeny of gephyrin: tissue-specific splice variants, gene structure, and sequence similarities to molybdenum cofactor- synthesizing and cytoskeleton-associated proteins. *Proc. Natl. Acad. Sci. USA* **97(18),** 10,266–10,271.
84. Meier, J., De Chaldee, M., Triller, A., and Vannier, C. (2000) Functional heterogeneity of gephyrins. *Mol. Cell. Neurosci.* **16(5),** 566–577.

85. Hermann, A., Kneussel, M., and Betz, H. (2001) Identification of multiple gephyrin variants in different organs of the adult rat. *Biochem. Biophys. Res. Commun.* **282(1),** 67–70.
86. Meier, J. and Grantyn, R. (2004) A gephyrin-related mechanism restraining glycine receptor anchoring at GABAergic synapses. *J. Neurosci.* **24(6),** 1398–1405.
87. Graf, E. R., Zhang, X. Z., Jin, S-X., Linhoff, M. W., and Craig, A. M. Neurexins induce differentiation of GABA and glutamate postsynaptic specializations via neuroligins. *Cell* **119(7)**, 1013–1026.
88. Prange, O., Wong, T. P., Gerrow, K., Wang, Y. T., and El-Husseini, A. (2004) A balance between excitatory and inhibitory synapses is controlled by PSD-95 and neuroligin. *Proc. Natl. Acad. Sci. USA* **101(38)**, 13,915–13,920.
89. Chih, B., Engelman, H., and Scheiffele, P. (2005) Control of excitatory and inhibitory synapse formation by neuroligins. *Science* **307(5713),** 1324–1328.
90. Levinson, J. N., Chery, N., Huang, K., et al. (2005) Neuroligins mediate excitatory and inhibitory synapse formation: involvement of PSD-95 and neurexin-1beta in neuroligin-induced synaptic specificity. *J. Biol. Chem.* **280(17),** 17,312–17,319.
91. Waites, C. L., Craig, A. M., and Garner, C. C. (2005) Mechanisms of vertebrate synaptogenesis. *Annu. Rev. Neurosci.* **28,** 251–274.
92. Craig, A. M., Graf, E. R., and Linhoff, M. W. (2006) How to build a central synapse: clues from cell culture. *Trends Neurosci.* **29(1),** 8–20.
93. Scheiffele, P., Fan, J., Choih, J., Fetter, R., and Serafini, T. (2000) Neuroligin expressed in nonneuronal cells triggers presynaptic development in contacting axons. *Cell* **101(6),** 657–669.
94. Varoqueaux, F., Jamain, S., and Brose, N. (2004) Neuroligin 2 is exclusively localized to inhibitory synapses. *Eur. J. Cell. Biol.* **83(9),** 449–456.
95. Knuesel, I., Mastrocola, M., Zuellig, R. A., Bornhauser, B., Schaub, M. C, and Fritschy, J. M. (1999) Altered synaptic clustering of $GABA_A$ receptors in mice lacking dystrophin (mdx mice). *Eur. J. Neurosci.* **11(12),** 4457–4462.
96. Knuesel, I., Zuellig, R. A., Schaub, M. C., and Fritschy, J-M. (2000) Alterations in dystrophin and utrophin expression parallel the reorganization of GABAergic synapses in a mouse model of temporal lobe epilepsy. *Eur. J. Neurosci.* **13,** 1113–1124.
97. Levi, S., Grady, R. M., Henry, M. D., Campbell, K. P., Sanes, J. R., and Craig, A. M. (2002) Dystroglycan is selectively associated with inhibitory GABAergic synapses but is dispensable for their differentiation. *J. Neurosci.* **22(11),** 4274–4285.
98. Winder, S. J. (2001) The complexities of dystroglycan. *Trends Biochem. Sci.* **26(2),** 118–124.
99. Sugita, S., Saito, F., Tang, J., Satz, J., Campbell, K., and Sudhof, T. C. (2001) A stoichiometric complex of neurexins and dystroglycan in brain. *J. Cell. Biol.* **154(2),** 435–445.
100. Wang, H., Bedford, F. K., Brandon, N. J., Moss, S. J., and Olsen, R. W. (1999) $GABA_A$-receptor-associated protein links $GABA_A$ receptors and the cytoskeleton. *Nature* **397,** 69–72.
101. Nymann-Andersen, J., Wang, H., Chen, L., Kittler, J. T., Moss, S. J., and Olsen, R. W. (2002) Subunit specificity and interaction domain between GABA(A) receptor-associated protein (GABARAP) and GABA(A) receptors. *J. Neurochem.* **80(5),** 815–823.

102. Hemelaar, J., Lelyveld, V. S., Kessler, B. M., and Ploegh, H. L. (2003) A single protease, Apg4B, is specific for the autophagy-related ubiquitin-like proteins GATE-16, MA1-LC3, GABARAP, and Apg8L. *J. Biol. Chem.* **278(51),** 51,841–51,850.
103. Tanida, I., Komatsu, M., Ueno, T., and Kominami E. (2003) GATE-16 and GABARAP are authentic modifiers mediated by Apg7 and Apg3. *Biochem. Biophys. Res. Commun.* **300(3),** 637–644.
104. Kneussel, M. (2002) Dynamic regulation of $GABA_A$ receptors at synaptic sites. *Brain. Res. Rev.* **39(1),** 74–83.
105. Tanida, I., Ueno, T., and Kominami, E. (2004) LC3 conjugation system in mammalian autophagy. *Int. J. Biochem. Cell. Biol.* **36(12),** 2503–2518.
106. Kittler, J. T., Rostaing, P., and Schiavo, G., et al. (2001) The subcellular distribution of GABARAP and its ability to interact with NSF suggest a role for this protein in the intracellular transport of $GABA_A$ receptors. *Mol. Cell. Neurosci.* **18(1),** 13–25.
107. Wang, H. and Olsen, R. W. (2000) Binding of the GABA(A) receptor-associated protein (GABARAP) to microtubules and microfilaments suggests involvement of the cytoskeleton in GABARAP-GABA(A) receptor interaction. *J. Neurochem.* **75(2),** 644–655.
108. Chen, L., Wang, H. B., Vicini, S., and Olsen, R. W. (2000) The gamma-aminobutyric acid type A ($GABA_A$) receptor-associated protein (GABARAP) promotes $GABA_A$ receptor clustering and modulates the channel kinetics. *Proc. Natl. Acad. Sci. USA* **97,** 11,557–11,562.
109. Kneussel, M., Haverkamp, S., Fuhrmann, J. C., et al. (2000) The γ-aminobutyric acid type A receptor ($GABA_A$ R)-associated protein GABARAP interacts with gephyrin but is not involved in receptor anchoring at the synapse. *Proc. Natl. Acad. Sci. USA.* **97(15),** 8594–8599.
110. Sagiv, Y., Legesse-Miller, A., Porat, A., and Elazar, Z. (2000) GATE-16, a membrane transport modulator, interacts with NSF and the Golgi v-SNARE GOS-28. *EMBO. J.* **19(7),** 1494–1504.
111. Leil, T. A., Chen, Z. W., Chang, C. S., and Olsen, R. W. (2004) GABA(A) receptor-associated protein traffics GABA(A) receptors to the plasma membrane in neurons. *J. Neurosci.* **24(50),** 11,429–11,438.
112. O'Sullivan, G. A., Kneussel, M., Elazar, Z., Betz, H. (2005) GABARAP is not essential for GABA receptor targeting to the synapse. *Eur. J. Neurosci.* **22(10),** 2644–2648.
113. Kanematsu, T., Jang, I. S., Yamaguchi, T., et al. (2002) Role of the PLC-related, catalytically inactive protein p130 in $GABA_A$ receptor function. *EMBO. J.* **21(5),** 1004–1011.
114. Terunuma, M., Jang, I. S., Ha, S. H., et al. (2004) GABAA receptor phospho-dependent modulation is regulated by phospholipase C-related inactive protein type 1, a novel protein phosphatase 1 anchoring protein. *J. Neurosci.* **24(32),** 7074–7084.
115. Yoshimura, K., Takeuchi, H., Sato, O., et al. (2001) Interaction of p130 with, and consequent inhibition of, the catalytic subunit of protein phosphatase 1alpha. *J. Biol. Chem.* **276(21),** 17,908–17,913.

116. Kittler, J. T., Chen, G., Honing, S., et al. (2005) Phospho-dependent binding of the clathrin AP2 adaptor complex to GABA(A) receptors regulates the efficacy of inhibitory synaptic transmission. *Proc. Natl. Acad. Sci. USA* **102(41),** 14,871–14,876.
117. Li, R. W., Serwanski, D. R., Miralles, C. P, et al. (2005) GRIP1 in GABAergic synapses. *J. Comp. Neurol.* **488(1),** 11–27.
118. Kittler, J. T., Arancibia-Carcamo, I. L., and Moss, S. J. (2004) Association of GRIP1 with a GABA(A) receptor associated protein suggests a role for GRIP1 at inhibitory synapses. *Biochem. Pharmacol.* **68(8),** 1649–1654.
119. Charych, E. I., Yu, W., Li, R., et al. (2004) A four PDZ domain-containing solice variant form of GRIP is localized in GABAergic and glutamatergic synapses in brain. *J. Biol. Chem.* **279(37),** 38,978–38,990.
120. Song, I. and Huganir, R. L. (2002) Regulation of AMPA receptors during synaptic plasticity. *Trends Neurosci.* **25(11),** 578–588.
121. Kim, E. and Sheng, M. (2004) PDZ domain proteins of synapses. *Nat. Rev. Neurosci.* **5(10),** 771–781.
122. Morgan, A. and Burgoyne, R. D. (2004) Membrane traffic: controlling membrane fusion by modifying NSF. *Curr. Biol.* **14(22),** R968–R970.
123. Goto, H., Terunuma, M., Kanematsu, T., Misumi, Y., Moss, S. J., and Hirata, M. (2005) Direct interaction of N-ethylmaleimide-sensitive factor with GABA(A) receptor beta subunits. *Mol. Cell. Neurosci.* **30(2),** 197–206.
124. Wojcikiewicz, R. J. (2004) Regulated ubiquitination of proteins in GPCR-initiated signaling pathways. *Trends Pharmacol. Sci.* **25(1),** 35–41.
125. Cremona, O., Collesi, C., and Raiteri, E. (2003) Protein ubiquitylation and synaptic function. *Ann. NY. Acad. Sci.* **998,** 33–40.
126. DiAntonio, A. and Hicke, L. (2004) Ubiquitin-dependent regulation of the synapse. *Annu. Rev. Neurosci.* **27,** 223–246.
127. Bedford, F. K., Kittler, J. T., Muller, E., et al. (2001) $GABA_A$ receptor cell surface number and subunit stability are regulated by the ubiquitin-like protein Plic-1. *Nat. Neurosci.* **4,** 908–916.
128. Luscher, B. and Keller, C. A. (2001) Ubiquitination, proteasomes and GABA(A) receptors. *Nat. Cell. Biol.* **3(10),** E232–E233.
129. Wu, A. L., Wang, J., Zheleznyak, A., and Brown, E. J. (1999) Ubiquitin-related proteins regulate interaction of vimentin intermediate filaments with the plasma membrane. *Mol. Cell.* **4(4),** 619–625.
130. Kleijnen, M. F., Shih, A. H., Zhou, P., et al. (2000) The hPLIC proteins may provide a link between the ubiquitination machinery and the proteasome. *Mol. Cell.* **6(2),** 409–419.
131. Kleijnen, M. F., Alarcon, R. M., and Howley, P. M. (2003) The ubiquitin-associated domain of hPLIC-2 interacts with the proteasome. *Mol. Biol. Cell.* **14(9),** 3868–3875.
132. Saliba, R. S. and Moss, S. J. (2003) Plic-1 regulates ubiquitination of GABA-A receptors, in *13th Neuropharmacology Conference, Posttranslational modifications of protein structure and synaptic function*, New Orleans, LA, Elsevier, pp. 103.

133. Charych, E. I., Yu, W., Miralles, C. P., et al. (2004) The brefeldin A-inhibited GDP/GTP exchange factor 2, a protein involved in vesicular trafficking, interacts with the beta subunits of the GABA receptors. *J. Neurochem.* **90(1),** 173–189.
134. Togawa, A., Morinaga, N., Ogasawara, M., Moss, J., and Vaughan, M. (1999) Purification and cloning of a brefeldin A-inhibited guanine nucleotide-exchange protein for ADP-ribosylation factors. *J. Biol. Chem.* **274(18),** 12,308–12,315.
135. Morinaga, N., Moss, J., and Vaughan, M. (1997) Cloning and expression of a cDNA encoding a bovine brain brefeldin A-sensitive guanine nucleotide-exchange protein for ADP-ribosylation factor. *Proc. Natl. Acad. Sci. USA* **94(24),** 12,926–12,931.
136. Shin, H. W. and Nakayama, K. (2004) Guanine nucleotide-exchange factors for arf GTPases: their diverse functions in membrane traffic. *J. Biochem. (Tokyo).* **136(6),** 761–767.
137. Shin, H. W., Morinaga, N., Noda, M., and Nakayama, K. (2004) BIG2, a guanine nucleotide exchange factor for ADP-ribosylation factors: its localization to recycling endosomes and implication in the endosome integrity. *Mol. Biol. Cell.* **15(12),** 5283–5294.
138. Qanbar, R. and Bouvier, M. (2003) Role of palmitoylation/depalmitoylation reactions in G-protein-coupled receptor function. *Pharmacol. Ther.* **97(1),** 1–33.
139. El-Husseini, A. D. and Bredt, D. S. (2002) Protein palmitoylation: a regulator of neuronal development and function. *Nat. Rev. Neurosci.* **3(10),** 791–802.
140. Putilina, T., Wong, P., and Gentleman, S. (1999) The DHHC domain: a new highly conserved cysteine-rich motif. *Mol. Cell. Biochem.* **195(1–2),** 219–226.
141. Bartels, D. J., Mitchell, D. A., Dong, X., and Deschenes, R. J. (1999) Erf2, a novel gene product that affects the localization and palmitoylation of Ras2 in Saccharomyces cerevisiae. *Mol. Cell. Biol.* **19(10),** 6775–6787.
142. Keller, C. A., Yuan, X., and Panzanelli, P., et al. (2004) The $\gamma 2$ subunit of $GABA_A$ receptors is a substrate for palmitoylation by GODZ. *J. Neurosci.* **24(26),** 5881–5891.
143. Rathenberg, J., Kittler, J. T., and Moss, S. J. (2004) Palmitoylation regulates the clustering and cell surface stability of $GABA_A$ receptors. *Mol. Cell. Neurosci.* **26,** 251–257.
144. Fang, C., Deng, L., Keller, C. A., et al. (2006) GODZ mediated palmitoylation of GABA(A) receptors is required for normal assembly and function of GABAergic inhibitory synapses. *J. Neurosci.* **26(49),** 12,758–12,768.
145. Fukata, M., Fukata, Y., Adesnik, H., Nicoll, R. A., and Bredt D. S. Identification of PSD-95 palmitoylating enzymes. *Neuron* **44(6),** 987–996.
146. Hayashi, T., Rumbaugh, G., and Huganir, R. L. (2005) Differential regulation of AMPA receptor subunit trafficking by palmitoylation of two distinct sites. *Neuron* **47(5),** 709–723.
147. Chaudhary, J. and Skinner, M. K. (2002) Identification of a novel gene product, Sertoli cell gene with a zinc finger domain, that is important for FSH activation of testicular Sertoli cells. *Endocrinology* **143(2),** 426–435.
148. Connolly, C. N., Kittler, J. T., Thomas, P., et al. (1999) Cell surface stability of g-aminobutyric acid type A receptors. Dependence on protein kinase C activity and subunit composition. *J. Biol. Chem.* **274(51),** 36,565–36,572.
149. Herring, D., Huang, R., Singh, M., Dillon, G. H., and Leidenheimer, N. J. PKC modulation of GABA(A) receptor endocytosis and function is inhibited by mutation

of a dileucine motif within the receptor beta 2 subunit. *Neuropharmacology* **48(2),** 181–194.

150. Cinar, H. and Barnes, E. M. Jr. (2001) Clathrin-independent endocytosis of GABA(A) receptors in HEK 293 cells. *Biochemistry* **40(46),** 14,030–14,036.
151. Tehrani, M. H., Baumgartner, B. J., and Barnes, E. M. Jr. (1997) Clathrin-coated vesicles from bovine brain contain uncoupled $GABA_A$ receptors. *Brain. Res.* **776(1-2),** 195–203.
152. Kittler, J. T., Thomas, P., Tretter, V., et al. (2004) Huntingtin-associated protein 1 regulates inhibitory synaptic transmission by modulating gamma-aminobutyric acid type A receptor membrane trafficking. *Proc. Natl. Acad. Sci. USA* **101(34),** 12,736–12,741.
153. Li, X. J., Li, S. H., Sharp, A. H., et al. (1995) A huntingtin-associated protein enriched in brain with implications for pathology. *Nature* **378(6555),** 398–402.
154. Li, Y., Chin, L. S., Levey, A. I., and Li, L. (2002) Huntingtin-associated protein 1 interacts with hepatocyte growth factor-regulated tyrosine kinase substrate and functions in endosomal trafficking. *J. Biol. Chem.* **277(31),** 28,212–28,221.
155. Kins, S., Kuhse, J., Laube, B., Betz, H., and Kirsch, J. (1999) Incorporation of a gephyrin-binding motif targets NMDA receptors to gephyrin-rich domains in HEK 293 cells. *Eur. J. Neurosci.* **11(2),** 740–744.
156. Harvey, R. J., Depner, U. B., Wassle, H., et al. (2004) GlyR alpha3: an essential target for spinal PGE2-mediated inflammatory pain sensitization. *Science* **304(5672),** 884–887.
157. Beck, M., Brickley, K., Wilkinson, H. L., et al. (2002) Identification, molecular cloning, and characterization of a novel GABA(A) receptor-associated protein, GRIF-1. *J. Biol. Chem.* **277,** 30,079–30,090.

4

Subunit Composition and Structure of $GABA_A$-Receptor Subtypes

Werner Sieghart

Summary

γ-Aminobutyric acid ($GABA_A$) receptors are the major inhibitory neurotransmitter receptors in the brain and are the site of action of many clinically important drugs. They are made up of five subunits that can belong to eight different subunit classes. Depending on their subunit composition these receptors exhibit distinct pharmacological and electrophysiological properties. The distinct but overlapping regional and cellular distribution of these subunits gives rise to the formation of a large number of $GABA_A$-receptor subtypes, only a few of which so far have been unequivocally identified in the brain.

Recent homology modeling of the structure of the extracellular and transmembrane domain of $GABA_A$ receptors for the first time provided a glimpse on the three-dimensional organization of this receptor, and placed different protein segments that have been shown to be of functional importance into a "region in space" and into defined neighboring relations. The models obtained could explain experimental observations and propose the location of putative drug-binding sites. They can be now used to design experiments for clarification of pharmacological and structural questions as well as to shed light on conformational changes during binding of agonist, gating, and allosteric modulation of these receptors.

Key Words: Binding sites; composition; distribution; $GABA_A$ receptors; homology models; structure; subunits.

1. Introduction

GABAA receptors are the major inhibitory neurotransmitter receptors in the brain. They are chloride-ion channels that can be opened by GABA and can be

From: *The Receptors: The GABA Receptors, Third Edition*
Edited by: S. J. Enna and H. Möhler © Humana Press Inc., Totowa, NJ

modulated by a variety of pharmacologically and clinically important drugs, such as benzodiazepines (BZ), barbiturates, steroids, anesthetics, and convulsants (1). Based on the pharmacological actions of these drugs it was concluded that GABAA receptors are involved in controlling the excitability of the brain, the modulation of anxiety, of feeding and drinking behavior, circadian rhythms and cognition, vigilance, memory, and learning (2).

2. Multiplicity of $GABA_A$ Receptors

2.1. $GABA_A$-Receptor Subunits

As other members of the superfamily of ligand-gated ion channels that also include the nicotinic acetylcholine (nACh) receptor, the 5-hydroxytryptamine type-3 receptors, and the glycine receptors, $GABA_A$ receptors are made up of five subunits that can belong to different subunit classes (α, β, γ, δ, ε, π, θ, ρ). Within each class the subunits exhibit a 60–70% homology: In between the classes the homology is only 30–40% *(1,3)*. So far, a total of six α-, three β-, three γ-, one δ-, one ε-, one π-, one θ-, and three ρ-subunits of $GABA_A$ receptors as well as alternatively spliced isoforms of several of these subunits have been cloned and sequenced from the mammalian nervous system *(1,3)*. At least for the human brain this subunit set seems to be final. In a recent study, by applying search algorithms designed to recognize sequences of all known $GABA_A$-receptor type subunits in species from man down to nematodes, no new $GABA_A$-receptor subunits were detectable in the human genome *(4)*. However, in nonmammalian species several additional subunit homologs have been identified *(3,5,6)*. A random combination of 19 different subunits theoretically could give rise to an enormous number of receptor subtypes with different subunit composition, arrangement, and pharmacology. The identification of those receptors actually formed in the brain, and their regional, cellular, and subcellular distribution is thus of paramount importance.

2.2. Distribution of $GABA_A$-Receptor Subunits in the Brain

2.2.1. Regional Distribution and Abundance of $GABA_A$-Receptor Subunits

In situ hybridization and immunohistochemical studies have indicated that the individual subunits exhibit a distinct but overlapping regional distribution. Whereas the subunits α_1, β_{1-3}, γ_2, and γ_3 are found throughout the brain, subunits α_{2-6}, γ_1, and δ are more confined to certain brain areas and in some brain regions a complementary distribution of α_2-, α_4-, β_3-, and δ-subunits against α_1-, β_2-, and γ_2-subunits was detected *(7,8)*.

The α_1-subunit is the most abundant $GABA_A$-receptor subunit in the brain. About 70–90% of all $GABA_A$ receptors seem to contain this subunit. The γ_2-and

β_2-subunits are also quite abundant and are present in 50–70% of all GABA$_A$ receptors. The β_1- and β_3-subunits seem to be present in 20–30% and 40–55% of all GABA$_A$ receptors, respectively, whereas the γ_3-subunit seems to be present in only 3–4% of all GABA$_A$ receptors *(7,8)*. The α_{2-5}-, and δ-subunits are present in 35, 14, 6, 7–8, and 11% of all GABA$_A$ receptors, respectively, but are differentially distributed in various forebrain areas and especially enriched in certain brain regions. Thus, in hippocampus 13% of all GABA$_A$ receptors seem to contain α_4, and 31% of receptors seem to contain α_5-subunits. In thalamus, 20% of all receptors seem to contain α_4-subunits, and 16% contain δ-subunits. The α_6- and γ_1-subunits exhibit a quite specific regional distribution. The α_6-subunits are more or less exclusively located in cerebellar granule cells (and in the cochlea) and are present in about 56% of all GABA$_A$ receptors in the cerebellum. The γ_1-subunit is a minor subunit that is present in about 3–7% of all GABA$_A$ receptors and is especially enriched in the central and medial amygdaloid nuclei, in pallidal areas, the substantia nigra pars reticulata, and the inferior olive *(7)*.

The distribution and abundance of the ε-subunit is currently less clear because of significant discrepancies in the published sequences, discrepant results of *in situ* hybridization studies, and the restricted availability of ε-selective antibodies *(7)*. Nevertheless, there is evidence that this subunit is located in major neuronal groups with broad range influence, such as the cholinergic (basal nucleus), dopaminergic (substantia nigra compacta), serotonergic (raphe nuclei), and noradrenergic (locus coeruleus) systems *(9,10)*.

The π-subunit was detected in several peripheral human tissues as well as in the brain (hippocampus and temporal cortex) and was particularly abundant in the uterus *(11)*. So far, no study investigating the detailed regional distribution of the π-subunit in the brain has been published. Similarly, the relative abundance of these subunits in the brain is not known. The θ-subunit seems to be expressed in various brain regions, including the hypothalamus, amygdala, hippocampus, substantia nigra, dorsal raphe, and locus coeruleus *(10)*. θ-subunits showed strikingly overlapping expression patterns with ε-subunits throughout the brain, especially in septum, preoptic areas, various hypothalamic nuclei, amygdala, and thalamus, as well as in monoaminergic cell groups *(10)*. As with the ε-subunit, there were some discrepancies in the cDNA sequence obtained by different groups *(12)*, and again the quantitative importance of θ-containing receptors is currently not known.

The ρ-subunits seem to be preferentially expressed in the retina. So far, no antibodies are available that could distinguish between different ρ-subunits. *In situ* hybridization studies indicated that ρ-subunits are also present in the superior colliculus, dorsal lateral geniculate nucleus, and cerebellar Purkinje cells. In addition, pharmacological effects characteristic of ρ-subunit-containing receptors have been reported in the cerebellum, superior colliculus, amygdala, hippocampus, dorsal geniculate nucleus, and spinal cord *(7)*. This indicates that ρ-subunits may be present in many central nervous system regions and are more prevalent than previously suspected.

2.2.2. Cell-Specific Expression of $GABA_A$-Receptor Subunits

Subunits α_1, β_2, and γ_2, not only seem to be codistributed in many brain regions but studies on their cellular localization indicated that they are colocalized extensively on GABAergic interneurons in hippocampus and other brain regions, suggesting a preferential assembly of this subunit combination in the brain *(8)*. This conclusion was supported by studies in which the colocalization of α_1-, β_2-, and γ_2-subunits on neuronal membranes have been investigated *(7,13,14)*.

In the raphe nuclei the vast majority of serotonergic neurons express strong α_3-subunit-immunoreactivity but are devoid of α_1-subunit staining. Only a small population of serotonergic neurons coexpress these two subunits. In contrast, both the α_1- and α_3-subunit immunoreactivities are present in glutamate decarboxylase-positive neurons *(15)*. These data indicated that serotonergic and GABAergic neurons selectively express distinct patterns of α-subunits, suggesting that they possess distinct subtypes of $GABA_A$ receptors.

Other studies indicated that the cholinergic neurons in the basal forebrain expressed the α_3-, but not the α_1-subunit. In contrast, 45–60% of parvalbumin-positive GABAergic neurons in the various subnuclei of the medial septum-diagonal band of Broca complex expressed both the α_1- and the α_3-subunit, whereas most of the remaining parvalbumin-positive neurons were labeled with only the α_1-subunit antiserum *(16)*. However, the α_3-subunit is not only associated with serotonergic or cholinergic neurons, but also with noradrenergic and dopaminergic neurons in the brainstem. Finally, other data indicating an overlapping distribution of α_3-, θ-, and ε-$GABA_A$-receptor subunits in the dorsal raphe and the locus coeruleus suggest that novel $GABA_A$-receptor subtypes that so far have not been studied in detail, may regulate neuroendocrine and modulatory systems in the brain *(10)*.

2.2.3. Subcellular Localization of $GABA_A$-Receptor Subunits

Other studies indicated that individual subunits exhibit a quite distinct subcellular distribution. For instance, in cerebellar granule cells the $GABA_A$-receptor subunits α_1, α_6, $\beta_{2/3}$, and γ_2 have been found by immunogold localizations to be concentrated in GABAergic Golgi synapses and also are present in the extrasynaptic membrane at a lower concentration. In contrast, δ-subunits could not be immunolocalized in synaptic junctions, although they were abundantly present in the extrasynaptic dendritic and somatic membranes. Receptors containing the δ-subunit also seem to contain α_6- and β-subunits *(17)*. It has been demonstrated that $\alpha_6\beta\delta$-receptors exhibit a 50-fold higher affinity for GABA than the $\alpha_1\beta\gamma_2$-receptors. These receptors also exhibit a smaller single channel conductance, a much longer open time, and do not desensitize on the prolonged presence of GABA *(18)*. Together with the exclusive extrasynaptic localization of these receptors, these properties indicate that tonic inhibition observed in these cells is mediated mainly by the persistent activation of $\alpha_6\beta\delta$-receptors by GABA that is present in the extracellular space of glomeruli. In contrast, phasic inhibition of granule cells is attributable to the transient activation of synaptic $\alpha_6\beta\gamma_2$- and/or $\alpha_1\beta\gamma_2$-receptors that exhibit a lower affinity for GABA, a more pronounced desensitization, and much shorter open times than $\alpha_6\beta\delta$-receptors *(17)*.

Other studies have indicated that two receptor populations with distinct kinetics coexist also in CA1 pyramidal cells and in many other cell types: slow extrasynaptic receptors that dominate the responses of excised patches to exogenous GABA applications and fast synaptic receptors that generate rapid inhibitory postsynaptic currents (IPSCs) *(19)*. The charge carried by the activation of tonically active GABA$_A$ receptors can be more than three times larger than that produced by phasic inhibition, even when the frequency of phasic events is large. It is quite possible that $\alpha_4\beta\delta$-receptors might at least partially be responsible for extrasynaptic tonic inhibition in these cells *(20)*. Experiments indicating that tonic conductances sometimes can also be enhanced by BZ *(21)* might indicate that tonic inhibition can also be produced by γ-subunit-containing receptors. Overall, these data indicate that inputs from different neurons use different receptors that are located in distinct parts of the receptive neurons *(22)*.

2.3. *Subunit Composition of Recombinant GABA$_A$ Receptors*

2.3.1. *Homo-Oligomeric GABA$_A$ Receptors*

Recombinant receptor studies have indicated that depending on the subunits used for transfection of cells, receptors with distinct pharmacological and electrophysiological properties do arise. Some GABA$_A$-receptor subunits, such as the ρ-subunits, robustly form homo-oligomeric receptors *(23)*. Channels formed exhibit properties of the previously characterized GABA$_C$ receptors. Because ρ-subunits are structurally part of the family of GABA$_A$-receptor subunits, it was recommended that ρ-containing receptors should be classified as a specialized set of the GABA$_A$ receptors *(3)*. Formation of homo-oligomeric receptors has also been reported from β_1- or β_3-subunits, but the extent of formation of these receptors seems to be low *(24)*. Data on the formation of homo-oligomeric receptors are contradictory for some other subunits *(2)*, and it thus can be assumed that homo-oligomeric receptors made up of these subunits are physiologically not important. But in the absence of high affinity ligands interacting with homo-oligomeric receptors other than ρ-subunit-containing receptors, and in the absence of a clear pharmacological profile of such receptors, the possible existence of such receptors in the brain could not be investigated.

2.3.2. $GABA_A$ Receptors Made Up of Two Different Subunits

In most heterologous expression systems, channels consisting of two different subunits formed more efficiently than homo-oligomeric channels and could be activated by lower GABA concentrations. In addition, the chloride-ion flux induced in hetero-oligomeric channels was higher than that in homo-oligomeric channels *(1)*. The efficiency of receptor formation, however, again seems to depend on the subunit combination and on the expression system used. Thus, ρ-subunits not only can form homo-oligomeric but also hetero-oligomeric receptors with other ρ-subunits. Whereas originally ρ-subunits were assumed not to coassemble with other classes of $GABA_A$-receptor subunits, recent studies indicated that ρ-subunits can assemble with γ_2-subunits and possibly also with glycine-receptor subunits and form functional receptors with properties found in certain cell types of the retina, brainstem, hippocampus, or other brain regions *(2)*.

Although α- and β-subunits robustly form hetero-oligomeric receptors in heterologous expression systems, in the additional presence of a γ- or a δ-subunit receptors predominantly made up of αβγ- or αβδ-subunits, respectively, are formed *(1,2)*. Thus, it is not yet clear, whether receptors made up of αβ-subunits actually do occur in the brain under normal conditions. However, such receptors have been identified in the brain of several $GABA_A$-receptor subunit knockout mice *(2,7)*. The formation of recombinant receptors made up of αγ- or βγ-subunits is questionable and no information is available on the possible formation of $GABA_A$ receptors made up of αδ-, βδ-, γδ-subunits. In addition, no functional recombinant channels were formed on cotransfection of $\alpha_1\varepsilon$-, $\beta_1\varepsilon$-, $\alpha_1\pi$-, or $\beta_1\pi$-subunit combinantions *(7)*.

2.3.3. $GABA_A$ Receptors Containing Three Different Subunits

Robust $GABA_A$-receptor expression was obtained when α-, β-, and γ-subunits were coexpressed in various heterologous systems, and only α-, β-, and γ-subunits produced $GABA_A$ receptors with a pharmacology resembling that of most native receptors *(1)*. This indicates that the majority of native receptors is made up of αβγ-subunits. In a minority of receptors, the δ-, ε-, and π-subunits seem to be able to replace the γ-subunit in $GABA_A$ receptors, whereas the θ-subunit might be able to replace a β-subunit in these receptors. The subunit composition of receptors containing ε-, π-, or θ-subunits, however, so far has not been extensively investigated.

2.3.4. $GABA_A$ Receptors Containing More Than Three Different Subunits

Several groups have investigated the subunit stoichiometry of αβγ-receptors. Most of these studies agree that these receptors contain two α-, two β-, and one γ-subunit *(7)*. In agreement with this subunit stoichiometry, recombinant receptors studies have indicated that receptors containing two different α-, one β-, and one γ-subunit can assemble and exhibit properties that are distinct from those of receptors containing only a single type of α-subunit. Other studies have indicated that depending on which α-subunit is neighboring the γ-subunit, receptors with distinct properties do arise *(7)*. Similarly, it has been demonstrated that receptors containing two different types of β-subunits together with one type of α- and γ-subunit are able to assemble and to exhibit properties different from receptors that contain only a single β-subunit subtype. Finally, it has been demonstrated that recombinant receptors made up of α_1-, β_1-, the long splice variant of γ_2-, and δ-subunits ($\alpha_1\beta_1\gamma_2$Lδ-receptors) can also be formed and exhibit properties distinct from those of $\alpha_1\beta_1\gamma_2$L- or $\alpha_1\beta_1\delta$-receptors *(7)*. Whether receptors made up of five different subunits can be formed is difficult to investigate because a variety of different receptor subtypes made up of three, four, or five different subunits could have been formed from the five different subunits simultaneously expressed in heterologous expression systems. All these receptors could have contributed to the chloride current measured in these cells. In addition, as many different types of recombinant receptors can be formed and as the efficiency of formation of these receptors depends on the expression system used, it is currently not clear whether all receptors that can be formed from simultaneously and massively overexpressed subunits in heterologous expression systems are actually formed in the brain.

2.4. Subunit Composition of Native $GABA_A$ Receptors

In contrast to heterologous systems expressing recombinant receptors, subunits in the brain are not necessarily expressed at the same time or at the same subcellular location. In addition, assembly in the brain presumably is governed by preferred subunit partnerships that under most conditions lead to receptors with a defined subunit stoichiometry and arrangement. This reduces the heterogeneity of receptors and underlines the importance of determining the subunit composition of receptors actually formed in the brain.

Experiments using antibodies highly selective against individual $GABA_A$-receptor subunits have been used for purifying $GABA_A$-receptor subtypes from brain membrane extracts. Results obtained have indicated an extreme promiscuity of the various subunits. Thus, it has been demonstrated that most, if not all of the other subunits investigated could be copurified with antibodies directed against an individual α- or β-subunit, suggesting that α- and β-subunits can combine with most of the other subunits to form a variety of different receptor subtypes *(7)*.

These studies indicating a copurification of α with other α-subunits, or of β with other β-subunits, of course also indicated that two different α-subunits or two different β-subunits are present in at least some $GABA_A$ receptors in the brain. The presence and abundance of receptors containing two different types of α-subunits was recently determined in mouse brains containing point-mutated α-subunits *(25)*. Interestingly, depending on the α_1-subunit located close to the γ-subunit these receptors exhibit distinct pharmacological properties in agreement with results obtained with recombinant receptors. Whereas the existence in the brain of $GABA_A$ receptors containing two different α- and/or β-subunits is generally accepted, discrepant results were obtained concerning a possible colocalization of different γ-subunits or of γ- and δ-subunits in the same $GABA_A$ receptor. Whereas some studies support a colocalization of these subunits in the same receptors, others do not *(7)*.

Because of the promiscuity of α- and β-subunits it is impossible to purify a single $GABA_A$-receptor subtype by a one step immunoprecipitation. Recently, however, a subtractive purification strategy has been developed that successfully allowed the determination of the subunit composition of solubilized hetero-oligomeric $GABA_A$ receptors *(26)*. Using this method, the subunit composition and quantitative importance of $GABA_A$-receptor subtypes in the cerebellum of mouse and rat have been analyzed *(27)*. Data obtained were consistent with the results from recombinant receptors and indicated that most of the receptors found in the cerebellum consist of two α-, two β-, and one γ- or one δ-subunits. In addition, a significant part of the receptors contain two different α-subunits and/or two different β-subunits, and is thus made up of four or five different subunits. So far, other brain regions have not been investigated using this technique.

Given the promiscuity of subunits discussed earlier and the receptor subtypes so far identified, it was estimated that more than 500 distinct $GABA_A$-receptor subtypes might exist in the brain. The number of receptors that are relatively abundant in the brain ($\alpha_1\beta\gamma_2$-, $\alpha_2\beta\gamma_2$-, $\alpha_3\beta\gamma_2$-, $\alpha_4\beta\gamma_2$-, $\alpha_5\beta\gamma_2$-, $\alpha_6\beta\gamma_2$-, $\alpha_4\beta\delta$-, or $\alpha_6\beta\delta$-receptors), however, is much smaller *(7)*. However, because of the widespread distribution and quantitative importance of the GABAergic system, even minor $GABA_A$-receptor subtypes probably exhibit an abundance comparable with that of major norepinephrine, dopamine, serotonin, or peptide receptors. However, to finally prove the existence of all these receptors in the brain, these receptors have to be functionally identified *in situ* by electrophysiological and pharmacological techniques. Although some progress has been achieved recently in identifying $GABA_A$ receptors in the brain *(28)*, in most cases an unequivocal identification of receptor subtypes *in situ* currently is not possible, owing to the lack of highly selective pharmacological tools.

3. $GABA_A$-Receptor Structure

3.1. Structure of the Extracellular Domain of $GABA_A$ Receptors

As mentioned earlier, most of the $GABA_A$ receptors present in the brain are made up of two α-, two β-, and one γ-subunit. Studies on the subunit arrangement indicated that a total of four alternating α- and β-subunits are connected by a

Fig. 1. Structure, subunit stoichiometry and arrangement of $GABA_A$ receptors. The absolute arrangement of α_1-, β_2-, and γ_2-containing $GABA_A$ receptors is shown as a model structure. The view is from extracellular. The + (plus) and – (minus) sides of the subunits are identified on the inner circumference of the channel. The interfaces at which BZ and GABA bind are labeled. (Adapted from ref. *31* with permission.)

γ-subunit *(29)*. However, these studies could not distinguish between two mirror images of the same subunit arrangement. The absolute subunit arrangement could be determined when a pentameric $GABA_A$ receptor was modeled according to the X-ray crystallographic structure of the acetylcholine-binding protein (AChBP) *(30)*, a remote homolog of the extracellular part of the nACh receptor and $GABA_A$ receptor. Modeling a pentameric receptor's extracellular domain consisting of two α-, two β-, and one γ-subunit results in a single (absolute) subunit arrangement, in which amino acid residues known to contribute to ligand-binding sites and interfaces are correctly positioned in the respective subunits *(31)* (Fig. 1). This absolute subunit arrangement was also independently confirmed using concatenated $GABA_A$-receptor subunits. There were several combinations of concatenated dimers and trimers that resulted in functional $GABA_A$ receptors on expression in *Xenopus* oocytes. All these combinations resulted in an identical pentameric receptor exhibiting a subunit arrangement as suggested by the

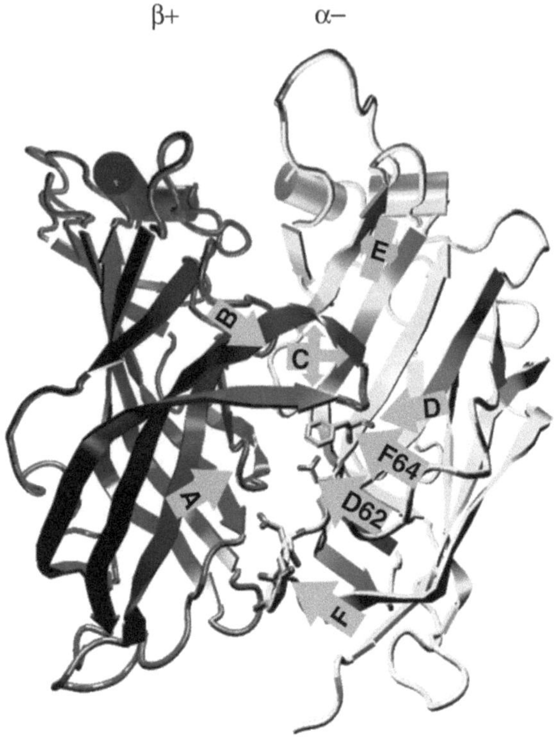

Fig. 2. Model structure of the GABA pocket of the $GABA_A$ receptor extracellular domain in a ribbon representation. The β_2+ and α_1− interface is shown viewed from the side. The membrane would be on the bottom of the figure. So-called binding site "loops" A through F are labeled. Residues implicated in binding at the agonist site are also labeled: The muscimol photolabel target "loop D" α_1F64 and loop D D62, a residue whose accessibility has been shown to change following antagonist binding, but not following GABA binding are shown. Loop F residues α_1 182–184 are shown as well, these residues change their accessibility following agonist and antagonist action. Their homologs in nAChRs participate in interactions with some antagonistic snake toxins. (Adapted from ref. *31* with permission.)

homology model of $GABA_A$ receptors *(32)*. Whether all receptors made up of αβγ-subunits or those made up of αβδ-, αβε-, or αβπ-subunits exhibit the same subunit stoichiometry and subunit arrangement, is not known presently.

The homology models of the $GABA_A$ receptor extracellular domain for the first time provided detailed structural information on part of this important receptor *(31)*. The extracellular domain consists of a variable N-terminus and two β-sheets that form a twisted "sandwich" (Fig. 2). Each subunit has a "principal" (+) and a "complementary" (–) side. Five solvent accessible pockets are found in the extracellular domain at the interfaces between subunits. In $GABA_A$

receptors the agonist-binding sites are located in the pockets at the two β+α– interfaces and the benzodiazepine-binding site is located in the pocket at the α+γ– interface (Fig. 1). So far, no ligands have been unequivocally identified that interact with the pockets at the α+β– or the γ+β– interface. All five pockets have been confirmed to consist of amino acid residues belonging to so-called "loops" A, B, and C from the "principal" part of the binding site, and so-called "loops" D, E, and F from the "complementary" part *(31)* (Fig. 2). It should be noted that this terminology has been established for the ligand-binding segments of pentameric ligand-gated ion channels before the publication of the AChBP crystal structure, and that not all of them turned out to be loops in the structural sense.

Results obtained with the putative active site search package indicated that these "loops" actually confine the solvent accessible space of these pockets *(31)* and this location is consistent with substituted cysteine accessibility mapping data. Observed changes in interface geometry outside of this cavity, as evidenced by accessibility changes following agonist action, probably correspond to the "agonist trapping motion" believed to be responsible for the slow unbinding rates of agonists from desensitized receptors and other allosteric motions.

3.1.1. The GABA-Binding Site of $GABA_A$ Receptors

However, in drawing conclusions from such homology models based on a template with low-sequence homology, it has to be kept in mind that although the overall topology and architecture of receptors are conserved, structural details are less reliable. The model uncertainty to be expected has recently been evaluated *(31,33)*. In addition, further model errors can come from incorrect or ambiguous sequence alignments and from intrinsic limits of the different methods that can be used. Thus, GABA-agonist docking cannot be deduced certainly from present models because of the low homology of the GABA-binding site with the ACh-binding site of the AChBP used as a template for modeling. Also it is not possible to make any definite statement about the access path of GABA to the binding pocket, as the AChBP was crystallized in the agonist-bound conformation *(30)* where part of the access path is likely to be obscured. It is considered that most likely GABA and other ligands enter the cleft from the outside of the channel mouth, not from the luminal side. Large-scale covalent cross-linking experiments, other experimental mapping approaches, or more crystal structures from different members of the superfamily will provide data that can be used to build better models and to map differences between different members of the superfamily.

Competitive antagonists are larger than agonists. The pocket geometry requires that they extend deeper into the membrane-near part of the cleft and thus block allosteric changes that possibly involve motions of loop F on the complementary side and/or loops A and C on the principal side (Fig. 2). Recent work on binding-site mapping in $GABA_A$ receptors *(34)* and on interactions between nACh receptors and snake-toxin antagonists *(35)* suggest that antagonist-binding subsites are conserved within the superfamily and are close to $GABA_A$ receptor α_1D62 or the corresponding homologous positions and loop F (Fig. 2). This indeed supports the notion that antagonistic ligands occupy membrane-near portions of the intersubunit cleft.

3.1.2. The Benzodiazepine-Binding Site of $GABA_A$ Receptors

From the various modulatory sites of $GABA_A$ receptors, so far only the benzodiazepine-binding site has been unequivocally identified. Many residues belonging to "loops" A, B, C, D, and E, at the α+γ– interface have been directly implicated in benzodiazepine-binding, potency, or efficacy. All these are found in solvent accessible parts of the model pockets. Because of the absence of several bulky side chains, the volume of the benzodiazepine pocket is larger than that of the GABA pocket. It extends further toward the membrane and possibly communicates directly with the membrane matrix *(31)*. Computational docking in models of the benzodiazepine site presently is again hampered by model uncertainty, leading to unclear side-chain positions. In addition, out of the same reason it cannot be expected that receptor subtype differences in the binding sites will be modeled properly. In some cases, experimental evidence can possibly discriminate between different ligand positions. For instance, possible dockings of Ro15-4513, an imidazobenzodiazepine used as photoaffinity label were discussed recently *(36,37)*.

3.2. Structure of the $GABA_A$ Receptor Extracellular and Transmembrane Domain

After the release of the AChBP structure, cryo-EM images of the nACh receptor in the open and closed state at modest resolution have been analyzed by fitting the core of the X-ray structure of the AChBP into the two sets of EM-density maps *(38)*. A cryo-EM atomic structure of the transmembrane domain of the nACh receptor consisting of five so-called four α-helix bundles, one bundle per subunit *(39)*, and later on a first model combining the extracellular and transmembrane domains of the nACh receptor have been published *(40)*. Finally, a refined structure of the combined extracellular and transmembrane domain of the nACh receptor in the resting state was released *(41)*. Recently, using these structures as templates, the structure of the $GABA_A$ receptor extracellular and transmembrane domains in the resting state has been calculated in the author's group by comparative modeling *(33)*.

The structural fragments provided by the AChBP for the extracellular domain and by the nAChR transmembrane domain provide a first glimpse on the three-dimensional organization of the superfamily, which places different

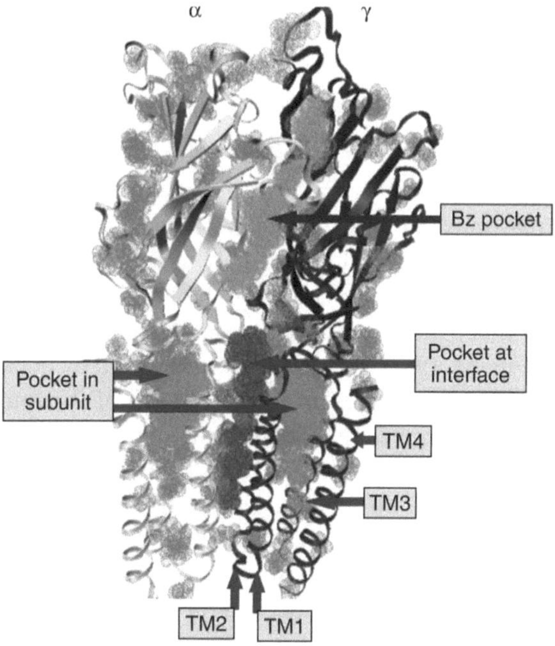

Fig. 3. Solvent accessible space contained in $GABA_A$- receptor models of the extracellular and transmembrane domains. Side view of the αγ-dimer. The proteins are shown in a ribbon representation. The putative pockets are shown in dotted space-filling representation. Clusters of connected solvent accessible volumes that may correspond to drug-binding pockets are labeled, as are TM1–4 of the γ-subunit. (Adapted from ref. *33* with permission.)

protein segments that have been shown to be of functional importance, into a "region in space" and into defined neighboring relations. For example, the four helical segments that pass the membrane, form an "up–down" bundle that is interrupted by the cytoplasmic domain between helices 3 and 4 (Fig. 3). Helix 1 is located in continuation of the minus side of the extracellular part of the subunit, helix 2 is pore forming, helix 3 lies at the plus side, and helix 4 lies at the abluminal side of the subunit *(33)*. The two β-sheets forming the extracellular domain are coupled to this four helix bundle through loop2 and loop7 (cys-loop) that intercalate with the extracellular linker between helix2 and helix3, thus allowing allosteric movement of the extracellular domain being transfered into a conformational change of the helical domain of the receptor *(33)*. The helical pore forming domain's architecture and topology as well as the domain junction is most likely conserved within the superfamily. In addition, the interface between two subunits in the transmembrane domain is now known to be made up not only by the transmembrane domain 2 (TM2) helices of the subunits, but also by the TM3 of one and the TM1 of the other subunit *(33)*. However, which amino acid residues precisely make the intersubunit contacts cannot be determined accurately.

3.2.1. Pockets Within Subunits and at the Subunit Interfaces as Possible Drug-Binding Sites

Interestingly, the transmembrane domain of the nACh-receptor structure is characterized by loose packing, which is likely to be also a conserved feature of this superfamily of receptors *(33)*. Actually, the helical domain of the $GABA_A$ receptors contains much more solvent accessible space than the extracellular β-folded domain (Fig. 3). Two types of cavities are found by pocket finding algorithms: a rather large cavity is present inside of each of the four helix bundles, and in addition, another large cavity exists at the interface between the subunits at and below the domain junction *(33)*. The shape and volume of the pockets varies with different model variants, and in some models, the pockets at the interface of the helical domain seem to communicate with their extracellular counterparts. Thus, the interface between subunits might contain a continuous groove (Fig. 3) that might not only be needed for the movement of helix 2 in gating of the ion channel *(40)*. Because the extracellular pockets between subunits contain the GABA and the benzodiazepine-binding sites of $GABA_A$ receptors, it is quite conceivable that their extension into the junctional and helical domains can also be used by drugs to modulate the function of receptors.

The intrasubunit pockets that are confined by the four helices of each subunit contain a number of amino acid residues in helices 1, 2, and 3, that have been shown to be of key importance in binding or action of different modulatory drugs *(33)*. Thus, residue α_6Ile228 of helix 1 determines the α-subunit selectivity of furosemide action *(42)* and is part of the wall of this intrasubunit pocket. α_1Ser269 of helix 2 and α_1Ala290 of helix 3, which line the wall of the intrasubunit pockets in the respective models, have been proposed to be part of a pocket for volatile anesthetics *(43)*. Similarly, homologous residues in helices 2 and -3 of the β-subunits have been shown to be key components in binding or action of modulatory drugs. Thus, residue β_1Ser265 is located in the wall of the intrasubunit pocket of the β_1-subunit and corresponds to residue α_1Ser269 in α_1-subunits. β_2 and β_3 have an Asn at the homologous position. This Ser/Asn polymorphism of the β-subunits also accounts for the β-subtype selectivity of loreclezole, etomidate, and other related substances *(44)* and might be part of a binding site for these substances. Strong support for a role of this intrasubunit pocket as a possible drug-binding site comes from the fact that in β-subunits helix 3 mutant β_2M286C (homologous to α_1Ala290) is shown to be protected by propofol in a dose-dependent manner against covalent modification by cys-reactive reagents *(45)*. Because functional modulation of $GABA_A$ receptors by furosemide or certain anesthetics can be influenced by amino acid residues in α- and β-subunits, it is tempting to speculate that multiple-binding sites for these compounds are present in the intrasubunit pockets of different GABA$_A$-receptor subunits. Depending on the specific electrostatic and steric interactions of drugs with these pockets, they could stabilize different conformations of the receptors, giving rise to their GABA-modulating, direct gating, or inhibitory action at different drug concentrations.

3.2.2. Pockets Within Subunits and at the Subunit Interfaces are Important for Conformational Changes of the Receptor

The entire helical domain, however, seems also to be crucial for transduction of ligand binding to gating or allosteric modulation. This is indicated by the effects of point mutants in this region on GABA action and benzodiazepine modulation *(46)* and by conformational changes in the helical domain induced by drug binding *(33)*. Thus, it has been demonstrated, that the solvent accessibility of helix 3 residues changes differentially with different drugs, pointing toward a large conformational flexibility of helix 3 and indicating that each drug induces a distinct functional state of the receptor *(33)*. Further mapping of drug-induced changes in solvent accessibility of all four helices as well as of other parts of the receptors using the author's models as a guide will delineate similarities and differences in drug action and drug-induced conformational changes, and thus, provide further information on the molecular function and dynamics of these important receptors.

Thus, in summary, the available homology models of the GABA$_A$ receptor are consistent with most experimental data, could explain experimental observations, and propose the location of putative drug-binding sites. These models can now be used to design new experiments for clarification of pharmacological and structural questions as well as to shed light on conformational changes during binding of agonist, gating, and allosteric modulation of these receptors *(33)*. Overall, these experiments will lead to an improvement in the accuracy of the models and finally pave the way for a structure-based drug design.

Acknowledgments

This work was supported by projects P15165 and P16397 of the Austrian Science Fund.

References

1. Sieghart, W. (1995) Structure and pharmacology of γ-aminobutyric acid$_A$ receptor subtypes. *Pharmacol. Rev.* **47(2),** 181–234.
2. Sieghart, W. and Ernst, M. (2005) Heterogeneity of GABA$_A$ receptors: revived interest in the development of subtype-selective drugs. *Curr. Med. Chem. - Central Nervous System Agents* **5,** 217–242.

3. Barnard, E. A., Skolnick, P., Olsen, R. W., et al. (1998) International Union of Pharmacology. XV. Subtypes of γ-aminobutyric $acid_A$ receptors: classification on the basis of subunit structure and receptor function. *Pharmacol. Rev.* **50(2),** 291–313.
4. Simon, J., Wakimoto, H., Fujita, N., Lalande, M., and Barnard, E. A. (2004) Analysis of the set of $GABA_A$ receptor genes in the human genome. *J. Biol. Chem.* **279(40),** 41,422–41,435.
5. Hosie, A. M., Aronstein, K., Sattelle, D. B., and ffrench-Constant, R. H. (1997) Molecular biology of insect neuronal GABA receptors. *Trends Neurosci.* **20(12),** 578–583.
6. Schuske, K., Beg, A. A., and Jorgensen, E. M. (2004) The GABA nervous system in C. elegans. *Trends Neurosci.* **27(7),** 407–414.
7. Sieghart, W. and Sperk, G. (2002) Subunit composition, distribution and function of $GABA_A$ receptor subtypes. *Curr. Top. Med. Chem.* **2(8),** 795–816.
8. Pirker, S., Schwarzer, C., Wieselthaler, A., Sieghart, W., and Sperk, G. (2000) $GABA_A$ receptors: immunocytochemical distribution of 13 subunits in the adult rat brain. *Neuroscience* **101(4),** 815–850.
9. Moragues, N., Ciofi, P., Lafon, P., Odessa, M. F., Tramu, G., and Garret, M. (2000) cDNA cloning and expression of a gamma-aminobutyric $acid_A$ receptor epsilon-subunit in rat brain. *Eur. J. Neurosci.* **12(12),** 4318–4330.
10. Moragues, N., Ciofi, P., Tramu, G., and Garret, M. (2002) Localisation of $GABA_A$ receptor epsilon-subunit in cholinergic and aminergic neurones and evidence for co-distribution with the theta-subunit in rat brain. *Neuroscience* **111(3),** 657–669.
11. Hedblom, E. and Kirkness, E. F. (1997) A novel class of $GABA_A$ receptor subunit in tissues of the reproductive system. *J. Biol. Chem.* **272(24),** 15,346–15,350.
12. Sinkkonen, S. T., Hanna, M. C., Kirkness, E. F., and Korpi, E. R. (2000) $GABA_A$ receptor epsilon and theta subunits display unusual structural variation between species and are enriched in the rat locus ceruleus. *J. Neurosci.* **20(10),** 3588–3595.
13. Nusser, Z., Roberts, J. D., Baude, A., Richards, J. G., Sieghart, W., and Somogyi, P. (1995) Immunocytochemical localization of the alpha 1 and beta 2/3 subunits of the $GABA_A$ receptor in relation to specific GABAergic synapses in the dentate gyrus. *Eur. J. Neurosci.* **7(4),** 630–646.
14. Somogyi, P., Fritschy, J. M., Benke, D., Roberts, J. D., and Sieghart, W. (1996) The gamma 2 subunit of the $GABA_A$ receptor is concentrated in synaptic junctions containing the alpha 1 and beta 2/3 subunits in hippocampus, cerebellum and globus pallidus. *Neuropharmacology* **35(9–10),** 1425–1444.
15. Gao, B., Fritschy, J. M., Benke, D., and Mohler, H. (1993) Neuron-specific expression of $GABA_A$-receptor subtypes: differential association of the alpha 1- and alpha 3-subunits with serotonergic and GABAergic neurons. *Neuroscience* **54(4),** 881–892.
16. Gao, B., Hornung, J. P., and Fritschy, J. M. (1995) Identification of distinct $GABA_A$-receptor subtypes in cholinergic and parvalbumin-positive neurons of the rat and marmoset medial septum-diagonal band complex. *Neuroscience* **65(1),** 101–117.
17. Nusser, Z., Sieghart, W., and Somogyi, P. (1998) Segregation of different $GABA_A$ receptors to synaptic and extrasynaptic membranes of cerebellar granule cells. *J. Neurosci.* **18(5),** 1693–1703.

18. Saxena, N. C. and Macdonald, R. L. (1994) Assembly of GABA$_A$ receptor subunits: role of the delta subunit. *J. Neurosci.* **14(11 pt. 2),** 7077–7086.
19. Banks, M. I. and Pearce, R. A. (2000) Kinetic differences between synaptic and extrasynaptic GABA$_A$ receptors in CA1 pyramidal cells. *J. Neurosci.* **20(3),** 937–948.
20. Adkins, C. E., Pillai, G. V., Kerby, J., et al. (2001) alpha4beta3delta GABA$_A$ receptors characterized by fluorescence resonance energy transfer-derived measurements of membrane potential. *J. Biol. Chem.* **276(42),** 38,934–38,939.
21. Yeung, J. Y., Canning, K. J., Zhu, G., Pennefather, P., MacDonald, J. F., and Orser, B. A. (2003) Tonically activated GABA$_A$ receptors in hippocampal neurons are high-affinity, low-conductance sensors for extracellular GABA. *Mol. Pharmacol.* **63(1),** 2–8.
22. Klausberger, T., Roberts, J. D., and Somogyi, P. (2002) Cell type- and input-specific differences in the number and subtypes of synaptic GABA$_A$ receptors in the hippocampus. *J. Neurosci.* **22(7),** 2513–2521.
23. Enz, R. and Cutting, G. R. (1998) Molecular composition of GABA$_C$ receptors. *Vision Res.* **38(10),** 1431–1441.
24. Connolly, C. N., Krishek, B. J., McDonald, B. J., Smart, T. G., and Moss, S. J. (1996) Assembly and cell surface expression of heteromeric and homomeric gamma-aminobutyric acid type A receptors. *J. Biol. Chem.* **271(1),** 89–96.
25. Benke, D., Fakitsas, P., Roggenmoser, C., Michel, C., Rudolph, U., and Mohler, H. (2004) Analysis of the presence and abundance of GABA$_A$ receptors containing two different types of alpha subunits in murine brain using point-mutated alpha subunits. *J. Biol. Chem.* **279(42),** 43,654–43,660.
26. Jechlinger, M., Pelz, R., Tretter, V., Klausberger, T., and Sieghart, W. (1998) Subunit composition and quantitative importance of hetero-oligomeric receptors: GABA$_A$ receptors containing alpha6 subunits. *J. Neurosci.* **18(7),** 2449–2457.
27. Poltl, A., Hauer, B., Fuchs, K., Tretter, V., and Sieghart, W. (2003) Subunit composition and quantitative importance of GABA$_A$ receptor subtypes in the cerebellum of mouse and rat. *J. Neurochem.* **87(6),** 1444–1455.
28. Vicini, S. (1999) New perspectives in the functional role of GABA$_A$ channel heterogeneity. *Mol. Neurobiol.* **19(2),** 97–110.
29. Tretter, V., Ehya, N., Fuchs, K., and Sieghart, W. (1997) Stoichiometry and assembly of a recombinant GABA$_A$ receptor subtype. *J. Neurosci.* **17(8),** 2728–2737.
30. Brejc, K., van Dijk, W. J., Klaassen, R. V., et al. (2001) Crystal structure of an ACh-binding protein reveals the ligand-binding domain of nicotinic receptors. *Nature* **411(6835),** 269–276.
31. Ernst, M., Brauchart, D., Boresch, S., and Sieghart, W. (2003) Comparative modeling of GABA$_A$ receptors: limits, insights, future developments. *Neuroscience* **119(4),** 933–943.
32. Baumann, S. W., Baur, R., and Sigel, E. (2002) Forced subunit assembly in alpha1beta2gamma2 GABA$_A$ receptors. Insight into the absolute arrangement. *J. Biol. Chem.* **277(48),** 46,020–46,025.
33. Ernst, M., Bruckner, S., Boresch, S., and Sieghart, W. (2005) Comparative models of GABA$_A$ receptor extracellular and transmembrane domains: important insights in pharmacology and function. *Mol. Pharmacol.* **68,** 1291–1300.

34. Holden, J. H. and Czajkowski, C. (2002) Different residues in the $GABA_A$ receptor alpha 1T60-alpha 1K70 region mediate GABA and SR-95531 actions. *J. Biol. Chem.* **277(21),** 18,785–18,792.
35. Fruchart-Gaillard, C., Gilquin, B., Antil-Delbeke, S., et al. (2002) Experimentally based model of a complex between a snake toxin and the alpha 7 nicotinic receptor. *PNAS* **99(5),** 3216–3221.
36. Sawyer, G. W., Chiara, D. C., Olsen, R. W., and Cohen, J. B. (2002) Identification of the bovine γ-aminobutyric acid type A receptor alpha subunit residues photolabeled by the imidazobenzodiazepine [^{3}H]Ro15-4513. *J. Biol. Chem.* **277(51),** 50,036–50,045.
37. Kucken, A. M., Teissere, J. A., Seffinga-Clark, J., Wagner, D. A., and Czajkowski, C. (2003) Structural requirements for imidazobenzodiazepine binding to $GABA_A$ receptors. *Mol. Pharmacol.* **63(2),** 289–296.
38. Unwin, N., Miyazawa, A., Li, J., and Fujiyoshi, Y. (2002) Activation of the nicotinic acetylcholine receptor involves a switch in conformation of the [alpha] subunits. *J. Mol. Biol.* **319(5),** 1165–1176.
39. Miyazawa, A., Fujiyoshi, Y., and Unwin, N. (2003) Structure and gating mechanism of the acetylcholine receptor pore. *Nature* **423(6943),** 949–955.
40. Unwin, N. (2003) Structure and action of the nicotinic acetylcholine receptor explored by electron microscopy. *FEBS Lett.* **555(1),** 91–95.
41. Unwin, N. (2005) Refined structure of the nicotinic acetylcholine receptor at 4A resolution. *J. Mol. Biol.* **346(4),** 967–989.
42. Thompson, S. A., Arden, S. A., Marshall, G., Wingrove, P. B., Whiting, P. J., and Wafford, K. A. (1999) Residues in transmembrane domains I and II determine gamma-aminobutyric acid type A receptor subtype-selective antagonism by furosemide. *Mol. Pharmacol.* **55(6),** 993–999.
43. Jenkins, A., Greenblatt, E. P., Faulkner, H. J., et al. (2001) Evidence for a common binding cavity for three general anesthetics within the $GABA_A$ receptor. *J. Neurosci.* **21(6),** RC136.
44. Belelli, D., Muntoni, A. L., Merrywest, S. D., et al. (2003) The in vitro and in vivo enantioselectivity of etomidate implicates the $GABA_A$ receptor in general anaesthesia. *Neuropharmacology* **45(1),** 57–71.
45. Bali, M. and Akabas, M. H. (2004) Defining the propofol binding site location on the $GABA_A$ receptor. *Mol. Pharmacol.* **65(1),** 68–76.
46. Boileau, A. J. and Czajkowski, C. (1999) Identification of transduction elements for benzodiazepine modulation of the $GABA_A$ receptor: three residues are required for allosteric coupling. *J. Neurosci.* **19(23),** 10,213–10,220.

5

Differential Activation of $GABA_A$-Receptor Subtypes

Mark Farrant

Summary

γ-Aminobutyric acid A ($GABA_A$) receptor subunit multiplicity results in the generation of diverse receptor subtypes with distinct biophysical and pharmacological properties. These receptor subtypes show different subcellular targeting and can be activated by GABA in different ways, with different functional outcomes. Synaptic receptors underlie brief postsynaptic currents, whereas high-affinity, largely extrasynaptic, receptors generate a "tonic" conductance. These differences have implications for our understanding of normal brain function and the rational design of therapeutic agents.

Key Words: $GABA_A$ receptors; subunit composition; inhibitory postsynaptic current; synaptic; extrasynaptic; tonic inhibition.

1. Introduction

γ-Aminobutyric acid (GABA) is ubiquitous in the mammalian central nervous system (CNS). Ionotropic $GABA_A$ receptors ($GABA_A$Rs) mediate fast inhibitory neurotransmission in all brain regions, and are important for the appropriate regulation of neuronal excitability. This is amply illustrated by the fact that impaired receptor function (resulting from pharmacological intervention or genetic mutation) promotes seizures, whereas augmented receptor function underlies the action of drugs that have beneficial anticonvulsant, anxiolytic, and anesthetic properties. Conventionally, this inhibitory action of GABA is thought of as taking place at synapses, where $GABA_A$Rs are activated following exposure to a high concentration

From: *The Receptors: The GABA Receptors, Third Edition*
Edited by: S. J. Enna and H. Möhler © Humana Press Inc., Totowa, NJ

of GABA released from presynaptic vesicles. This results in a short-lived increase in the anion permeability of the postsynaptic membrane and a transient or "phasic" event in the form of a hyperpolarizing inhibitory postsynaptic potential (IPSP). This statement, though essentially true, ignores many subtle—and not so subtle—features of GABAergic signaling, appreciation of which is surely important for the understanding of the function and diversity of $GABA_A$Rs in the CNS, as well as the therapeutic potential of drugs that act on them. In this chapter, transient point-to-point GABAergic transmission is contrasted with a form of signaling in which a low concentration of GABA in the extracellular space causes the persistent "tonic" activation of $GABA_A$Rs.

2. When is GABA Inhibitory?

Before discussing the properties of specific $GABA_A$Rs and the manner of their activation, it is important to consider, in a general sense, the outcome of such activation. $GABA_A$Rs gate channels that are permeable to chloride (Cl^-) and bicarbonate (HCO_3^-) ions *(1,2)*. Of course, the receptors *per se* are neither inhibitory nor excitatory—by increasing membrane permeability to Cl^- and HCO_3^- they simply act to dissipate any pre-existing transmembrane gradients of these anions established by ion pumps and -carriers. Thus, the immediate response to a change in membrane conductance caused by $GABA_A$R activation depends on the distribution of Cl^- and HCO_3^- and on the membrane potential of the cell. These are clearly dynamic entities, making a "universal" hyperpolarizing IPSP unlikely. Nevertheless, in many mature neurons of the CNS the potassium-chloride cotransporter KCC2 *(3,4)*, by extruding Cl^-, results in a Cl^- equilibrium potential that is more negative than the resting membrane potential. The equilibrium potential for HCO_3^- is positive to the resting membrane potential, but HCO_3^- is about fivefold less permeable than Cl^- *(1,2)*. In this case, $GABA_A$R activation leads to net entry of anions and a hyperpolarizing IPSP. This action is "inhibitory" because both the transient conductance increase (by "shunting" excitatory inputs) and the longer-lived hyperpolarization (by summing with, and thus reducing, any depolarization) reduce the probability of spike firing.

Depolarizing actions of GABA have been known for many years, most notably in peripheral sensory neurons such as dorsal root ganglion cells and in the phenomenon of primary afferent depolarization (*see* refs. *5,6*). Here, the cells accumulate chloride through the sodium- and potassium-coupled cotransporter (NKCC1) *(7)*, resulting in equilibrium potential for GABA that is more depolarized than the resting membrane potential. The same process takes place in immature central neurons, which initially lack KCC2 *(3,4)*, and also in a number of more mature central neurons *(8–13)*. GABA-mediated depolarization does not necessarily equate with "excitation." The depolarization may be modest, but even if it is not, as long as the equilibrium potential of the GABAergic input is below the threshold for spike firing then the immediate response will be inhibitory owing to the inevitable shunting effect of the increase in conductance *(13–15)*. However, subthreshold "IPSPs" may facilitate action potential firing if their location and timing relative to subthreshold "excitatory" inputs allows summation of the voltage responses at a time following decay of the GABA-mediated conductance *(10,16)*.

Along with the recognized changes in Cl^- handling that occur during development, differences exist in Cl^- regulation and GABAergic responses of different cell types in the same brain region *(8,17)*, among different cells of the same type *(12)*, for the same cell type at different times *(18,19)*, and for different subcellular regions of the same cell *(20,21)*. Finally, neuronal activity can change the reversal potential for GABA through transient changes in intracellular Cl^- concentration *(22)*, or persistent Ca^{2+}-induced downregulation of KCC2 *(23–25)*.

3. Two Contrasting Modes of $GABA_A$-Receptor Activation

3.1. Phasic-Receptor Activation

At synapses containing $GABA_A$Rs, the rapid and precise communication of presynaptic activity into a postsynaptic signal is achieved by the release of a high concentration of GABA onto postsynaptic receptors of moderate- to low affinity. In response to each release event, a small number of receptors clustered opposite the release site (from ten to a few hundred *[26–29]*) experience a rapid but transient increase in the concentration of GABA that peaks in the millimolar range *(27)* and decays with a time constant of about 100–500 μs *(30,31)*. Such fleeting exposure (of low-affinity receptors) to GABA is tailored to allow repetitive signaling by discrete synaptic events. With a few exceptions (as noted in Subheading 3.2), $GABA_A$Rs allow ion flux only after they have been "occupied" by agonist, efficient gating of the channel requiring the binding of two molecules of GABA *(32)*. The peak concentration of GABA reached in the synaptic cleft may be sufficient to produce a maximal receptor activation at equilibrium, but because the binding rate of GABA is slow relative to diffusion *(33)*, the short exposure means that not necessarily all receptors are occupied. The degree of receptor occupancy varies between synapses on different neurons and even between those on a single cell *(28,30,31,34,35)*. Because they can affect the time-course of the GABA concentration transient, variations parameters such as vesicle size and content, the nature of vesicle fusion, the geometry of the synaptic cleft, and the number and spatial arrangement of transporters and postsynaptic receptors in relation to the site of GABA release, will all affect the level of receptor occupancy.

For receptors that bind GABA, a conformational change is elicited that may ultimately lead to channel opening. The behavior of the receptors can be envisaged as a series of transitions through various closed, open (ion conducting), and desensitized (relatively long-lived agonist-bound closed) states that can be described in kinetic schemes of microscopic gating *(36–43)*. The time spent in each of the various states, and thus the time-course of the postsynaptic current, is determined by the properties of the receptors and the profile of GABA exposure. Under optimal recording conditions, GABAergic miniature inhibitory postsynaptic currents (mIPSCs) resulting from the release of a single vesicle, have a rapid onset with rise times of only a few hundred microseconds *(28,29,44)*. This reflects the proximity of the receptors to the site of GABA release and the speed of the transition between closed and open states *(38,41,45,46)*. As the GABA concentration transient is brief, the decay of the IPSC reflects the closure of channels following removal of ligand (deactivation). The speed of this process is governed by various transitions of the receptor, notably entry into and exit from agonist-bound closed (desensitized) states that can be viewed as effectively trapping GABA on the receptor before the final unbinding *(38,40,47,48)*. Because the rates of these transitions differ for $GABA_A$Rs of different subunit composition, the expression of different receptor subtypes contributes to differences in IPSC decay observed at different stages of development *(49,50)* and in different cell types *(51–53)*.

3.2. *Tonic-Receptor Activation*

Persistent or "tonic" activation of $GABA_A$Rs is seen in embryonic neurons before synapse formation *(54–57)* as well as in progenitor cells, neuroblasts, and new neurons formed during postnatal neurogenesis *(58–60)*. Here, GABA is suggested to play a role in modulation of neuronal proliferation, migration, growth, and synaptogenesis *(61–63)*. In neonatal hippocampal pyramidal neurons of the CA3 region, the depolarization provided by tonic receptor activity acts to promote voltage-dependent intrinsic bursting *(64,65)*, which in turn gives rise to spontaneous network events termed "giant depolarizing potentials" (GDPs; or extracellularly recorded field GDPs) *(64)*, the proposed counterpart of hippocampal sharp waves *(66)*. Preventing the facilitatory depolarizing action of GABA with bumetanide (an inhibitor of NKCC1), blocks GDPs *(67)* and inhibits sharp waves in the neonatal hippocampus in vivo *(68)*.

Diffuse or paracrine GABAergic signaling also occurs between mature neurons *(69–75)*. In such cells, a tonic mode $GABA_A$R activation, clearly distinct from the current generated by the superimposition of IPSCs *(76–78)*, was first identified in recordings from granule cells in slices of rat cerebellar cortex *(79)*. As well as blocking IPSCs, $GABA_A$R antagonists were shown to decrease the current required to voltage-clamp the cells to a given membrane potential. This decrease in membrane conductance was accompanied by a reduction in current variance, consistent with a decrease in the number of open $GABA_AR$ channels *(79–82)*. Subsequent studies, using acute brain slices, have suggested the existence of GABA-mediated tonic conductances in granule cells of the dentate gyrus *(83–86)*; pyramidal cells and inhibitory interneurons in the CA1 region of the hippocampus *(87–89)*; pyramidal neurons and interneurons in the somatosensory cortex *(90,91)*, and thalamocortical relay neurons of the ventral basal complex and dorsal lateral geniculate nucleus *(92–95)*.

In considering the origin of this current, it should be noted that certain recombinant *(48,96–100)* and native *(101,102)* $GABA_ARs$ can open spontaneously with low probability in the absence of agonists. However, as described earlier, $GABA_ARs$ require the binding of GABA to promote channel opening (*see also* refs. *103,104*). Therefore, the existence of a tonic conductance (blocked by competitive $GABA_AR$ antagonists) suggests that GABA must persist in the extracellular space at a sufficiently high concentration to cause continued receptor activation *(105)*. In accord with this, estimates of the concentration of ambient GABA (in vivo and in vitro) range between tens of nanomolar and a few micromolar *(30,72,106–111)*.

The source of GABA giving rise to the tonic conductance has been investigated most extensively for cerebellar granule cells. In this case, action-potential dependent vesicular release contributes to the generation of the conductance in juvenile animals and may also play a role in mature animals *(112,113)*. One factor underlying this is the apposition of many GABA-releasing Golgi cell axon terminals to intraglomerular dendrites of granule cells *(79,80,82)*. For mature granule cells, a nonvesicular source has also been suggested, though not identified *(82,114)*. Although various mechanisms of nonvesicular GABA release have been characterized in astrocytes *(115–117)* and embryonic neurons *(57,116,118,119)*, it is unclear to what extent they might contribute to ambient GABA levels in the mature nervous system. Importantly, given the uncertainties and potential variations between in vitro experiments, tonic activation of $GABA_ARs$ in granule cells has also been demonstrated to occur in vivo *(120)*.

The concentration of GABA in the extracellular space is set by the number and nature of GABA-releasing elements and by the action of Na^+- and Cl^--dependent GABA uptake carriers. Although these transporters are often viewed as simply "removing" GABA from the extracellular space, they can have an important and dynamic influence on ambient GABA *(121)*. Because their activity is not fixed; the extracellular concentration of GABA at which they are at equilibrium and thus the magnitude of GABA flux, will depend on the membrane potential and the transmembrane gradients for the substrates (GABA, Na^+, Cl^-), each of which can vary under physiological and pathological conditions or during exposure to drugs *(109,121–123)*. Because the transporters can operate in the reverse direction, they could, under certain circumstances, act as a source of GABA *(124)*. However, following pharmacological blockade of transport *(82,83,87,88,91,114)*, and in transporter-deficient mice *(125)*, the magnitude of the tonic current increases, suggesting that reversed uptake does not ordinarily act as a nonvesicular source of GABA.

4. Different $GABA_A$-Receptor Subtypes Mediate Different Forms of Signaling

The molecular basis of $GABA_AR$ diversity has been reviewed extensively *(126–130)*. Briefly, the receptors are heteropentameric assemblies of subunits forming a central ion channel, 19 $GABA_A$-receptor subunits have been cloned from the mammalian CNS (α_{1-6}, β_{1-3}, γ_{1-3}, δ, ε, θ, π, and ρ_{1-3}), with additional variation coming from alternative splicing, notably for the γ_2-subunit *(131)*. Although this diversity predicts enormous heterogeneity of receptor types, basic "rules" of assembly *(132,133)* and the differential distribution of subunit types among brain regions and neuronal populations *(134–136)* restricts the overall number of receptor subtypes that exist. The most abundant receptor type is formed from α_1-, β_2-, and γ_2-subunits *(128,137,138)*. The likely stoichiometry is two α-, two β-, and one γ-subunit *(139–143)*. Common receptors containing three different subunits can also form from other α, β, and γ combinations. In less widely expressed receptors the γ-subunit may be replaced by δ-, ε-, or π-subunits, and the π- and θ-subunits may be capable of coassembling with α-, β-, and γ-subunits to form receptors containing subunits from four families *(144–146)*. Additional variability comes from the fact that individual receptors may contain two different α- or β-subunit isoforms. Finally, although the ρ_1-subunits form homomeric receptors that share properties with those sometimes classified as $GABA_C$ receptors *(147,148)*, they may also form receptors with γ_2-subunits *(149)* or with both α_1- and γ_2-subunits *(150)*.

This molecular heterogeneity has important functional consequences, as subunit composition dictates not only the biophysical properties of the receptors, but also their cell surface distribution and dynamic regulation *(75,128,133,151)*. Using monoclonal antibodies against $GABA_AR$ subunits, high-resolution light-microscopic immunofluorescence, and electron-microscopic immunogold methods revealed an enrichment (up to 200-fold) of the α_{1-3}, α_6, $\beta_{2/3}$, and γ_2-subunits in GABAergic synapses *(152–158)*. Importantly, each subunit type was also found in extrasynaptic membranes, where the total number of receptors can exceed that of receptors at synaptic junctions *(153)*. An unexpected twist to this situation is the apparent exclusion from synaptic junctions of some $GABA_A$-receptor subtypes. Using postembedding electron-microscopic immunogold localization, the δ-subunit was shown to be present exclusively in the extrasynaptic somatic and dendritic membranes of cerebellar granule cells, with gold particles hundreds of nanometres from the edge of the nearest postsynaptic density *(154)*. Later, it was also found to be present both in extrasynaptic membrane and at perisynaptic locations (just outside the postsynaptic density) in hippocampal dentate gyrus granule cells *(159)*.

The δ-subunit associates specifically with the α_6-subunit in cerebellar granule cells ($\alpha_6\beta_{2/3}\delta$ and $\alpha_1\alpha_6\beta_{2/3}\delta$) and with the α_4-subunit ($\alpha_4\beta_x\delta$) in several areas including the thalamus, neostriatum, and dentate gyrus *(126,160)*. As the γ_2-subunit appears essential for clustering of $GABA_ARs$ at synapses (reviewed in refs. *[133,161,162]*) it is the lack of a γ-subunit in each of these receptor subtypes that is likely to be responsible for their failure to be incorporated at the synapse. Although perhaps less clear cut, other receptor subtypes may also be present predominantly outside the synapse. In hippocampal pyramidal cells the α_5-subunit (which most likely forms $\alpha_5\beta_3\gamma_2$-receptors) shows diffuse cell surface labeling that is not clustered with synaptic markers *(158,163)*. In CA1 pyramidal cells, deletion of the α_5-subunit reduces the amplitudes of action-potential dependent and evoked IPSCs *(164)* but not that of mIPSCs *(165)*. This suggests that phasic activation of α_5-containing receptors may require spillover of GABA onto nonsynpatic receptors, as described for α_6- and α_4-subunit-containing receptors in granule cells of the cerebellum *(166)* and dentate gyrus *(159)*. Thus, extrasynaptic α_5-containing receptors may contribute to the generation of slowly rising and decaying IPSCs ($GABA_{A,\,slow}$) seen in CA1 pyramidal neurons *(167)*, and believed to originate at distal dendritic sites *(168)* following activation of receptors remote from the sites of GABA release *(169)*.

Indeed, recent studies of mice carrying a point mutation in either the α_1-, α_2-, or α_3-subunit gene (or combinations of these genes), rendering the respective $GABA_ARs$ insensitive to diazepam, suggest that receptors containing the α_5-subunit (alone or in combination with α_1- and/or α_2-subunits) underlie $GABA_{A,\,slow}$, whereas fast IPSCs from somatic and dendritic sites are mediated by α_2- and α_1-containing $GABA_ARs$, respectively *(170)*. In pyramidal neurons of the subiculum, slow IPSCs are also observed *(170)*, but as α_5-subunits are not present, other subunits may serve similar roles in these neurons. Taken together, these findings suggest that receptors containing a γ_2-subunit in association with α_1-, α_2-, or α_3-subunits ($\alpha_1\beta_{2/3}\gamma_2$, $\alpha_2\beta_{2/3}\gamma_2$, and $\alpha_3\beta_{2/3}\gamma_2$) are the predominant receptor subtypes mediating fast synaptic transmission, whereas receptors containing α_4-, α_5-, or α_6-subunits ($\alpha_6\beta_x\delta$, $\alpha_4\beta_x\delta$, and $\alpha_5\beta_x\gamma_2$, or $\alpha_5\alpha_{1/2}\beta_x\gamma_2$) are predominantly or exclusively extrasynaptic.

A link between α_6- and δ-containing (extrasynaptic) receptors and the tonic GABA-mediated conductance in cerebellar granule cells was proposed *(80,82,154)* when it was recognized that the delayed development of the conductance mirrored the delayed expression of these subunits, and that the receptors have a particularly high affinity for GABA and desensitize more slowly and less extensively than $\alpha\beta\gamma$-receptors (reviewed in ref. *75*). An essential role of extrasynaptically located $\alpha_6\beta_{2/3}\delta$-receptors was confirmed when it was shown that the $GABA_AR$-mediated tonic conductance in cerebellar granule cells was abolished following deletion of the α_6- or δ-subunits *(85,171)*. Similarly, deletion of the δ-subunit (and concomitant loss of α_4 expression *[172]*) reduces tonic-receptor activation in granule cells of the dentate gyrus *(85)*. The δ-subunit, by virtue of its pharmacological signature (*see* Subheading 5.1), is also implicated in combination with α_4- and β_2-subunits, in the generation of a large tonic conductance seen in thalamocortical neurons of the dorsal lateral geniculate and ventral basal thalamus *(93–95)*. Consistent with this, deletion of the δ-subunit reduces the background current variance of ventral basal neurons *(92)*.

Of note, the identification of a key role for δ-containing receptors does not preclude the possibility that under circumstances of elevated extracellular GABA or drug-induced increases in receptor affinity, other extrasynaptic and/or synaptic receptors, even those that exhibit greater desensitization, may also contribute to the generation of a tonic conductance. This is an important consideration when attempting to compare across in vitro studies, some of which, in order to replicate an assumed in vivo condition of sustained extracellular GABA, have measured tonic conductances in the presence of low micromolar concentrations of added GABA *(84,159)*, blockers of GABA uptake *(85,87)*, or blockers of GABA metabolism *(109,165)*. In one of these latter studies, the tonic conductance seen in cultured hippocampal pyramidal neurons treated with the GABA transaminase inhibitor vigabatrin, was greatly reduced by deletion of the α_5-subunit *(165)*. It has recently been suggested that α_5-containing receptors may contribute to the tonic conductance in pyramidal cells only when the concentration of extracellular GABA is elevated, and that under basal conditions the conductance is mediated by δ-containing receptors *(88,170,173)*. Intriguingly, the relative contribution of these two receptor populations, and a third as yet unidentified population, were shown to change after the induction of epilepsy. Such plasticity of receptor subtype expression and/or differences in the concentration of ambient GABA under different recording conditions may explain other apparent discrepancies in the reported drug sensitivity of the tonic current in CA1 pyramidal cells *(87,89)*.

5. Functional Roles of Phasic- and Tonic-Receptor Activation

Phasic $GABA_AR$-mediated signaling involves the short-lived, near synchronous opening of a small number of channels clustered at the synaptic junction. Tonic-receptor activation, on the other hand, is both temporally and spatially dispersed. These different features suggest profound differences in their likely functional impact. Preventing overexcitation of neurons and pathological network activity is an essential task of GABA-mediated inhibition in the adult CNS. However, phasic input from GABAergic interneurons does much more than simply "inhibit" principal cells. It plays a key role in shaping neuronal firing patterns through feedforward and feedback inhibition and participates in the generation of rhythmic activity among large neuronal populations *(13,174–180)*. Two key aspects of phasic signaling dictate the varied actions of interneurons—synapse location and IPSC timing, i.e., precisely where and when the target neuron receives inhibitory input. Thus, numerous examples testify to the significance of spatially segregated phasic GABAergic input (axonic, perisomatic, proximal, or distal dendritic *[181]*) in enabling or regulating important neuronal behaviors, such as backpropagation and regenerative activity in dendrites *(182,183)*, region-specific integration of excitatory inputs *(184)*, or powerful axo-axonic excitation *(21)*. Moreover, in cells that receive spatially segregated IPSCs from multiple sources, phasic inhibitory input may be subject to exquisite modulation, either through changes in the activity of the parent interneurons, or by the regulation of transmitter release from their terminals *(185,186)*.

Unlike phasic $GABA_AR$ activation, tonic activation causes a persistent and distributed increase in a cell's input conductance. This affects the magnitude and duration of the voltage response to an injected current, and increases the decrement of voltage with distance. For any excitatory input, the size and duration of the change in membrane potential change will be reduced, and the temporal and spatial window on which signal integration can occur will be narrowed, making it less likely that an action potential will be generated. This makes cells less excitable and changes the relationship between excitation current and output firing rate *(80,94,120,187,188)*. In cerebellar granule cells, with excitation delivered as random trains of synaptic conductances, "shunting inhibition" shifts the input–output relationship and decreases its slope, corresponding to a change in "gain." This occurs because the slope of the input-output relationship depends on the variability (and thus frequency) of the excitatory conductance *(188–190)*.

In the hippocampus, the differential expression of tonic conductance in interneurons and pyramidal cells is suggested to contribute to a homeostatic regulation of phasic inhibition in the latter *(87,190)*. As the tonic GABA-mediated conductance is small in pyramidal cells compared with that in interneurons *(88)*, pharmacological blockade of the conductance preferentially enhances the excitability of interneurons, which may result in an increase in the frequency of IPSCs in CA1 pyramidal cells. Thus, tonic activation of $GABA_ARs$ can alter network excitability *(87,173)* and modify the way information is processed *(120,187,188)*. Persistent activation of $GABA_ARs$ has also been shown to shift the resting potential of neurons to more hyperpolarized values *(94,120)*. In the case of thalamocortical neurons, block of the tonic conductance causes a depolarization, which tends to promote a shift away from a burst firing toward a tonic firing mode *(94)*.

5.1. Selective Pharmacological Modulation of Tonic-Receptor Activation

The pharmacological properties of $GABA_A$Rs, like their biophysical properties, are determined by their subunit composition (reviewed in refs. *126,130,191*). Thus, differences in subunit composition between synaptic and extra- or perisynaptic receptors are reflected in a differential modulation of phasic- and tonic signaling by various agents of clinical relevance, including benzodiazepine site ligands, endogenous neuroactive steroids, intravenous and inhalation anaesthetics, hypnotics, certain nootropic agents, and somewhat contentiously, alcohol.

In cerebellar granule cells, dentate gyrus granule cells, and thalamocortical cells, where the tonic conductance is mediated by δ-containing receptors, the presence of this subunit imparts unique pharmacological properties. Thus, whereas IPSCs are enhanced by benzodiazepine site ligands (such as diazepam or the α_1-selective imidazopyridine zolpidem) the tonic conductance is generally unaffected *(83,93–95,187)*. Of note, recent experiments *(111)* addressing the effects on tonic conductance in cerebellar granule cells of a naturally occurring single nucleotide polymorphism (R100Q) in the rat α_6 gene *(192)* have shown that $\alpha_6\beta_3\delta$-receptors can in fact be potentiated (albeit modestly) by flunitrazepam at concentrations of GABA that generate incomplete receptor occupancy. This finding may explain the flunitrazepam enhancement of tonic $GABA_A$-receptor activation seen in granule cell cultured from rats *(193)*.

Unlike benzodiazepines, neurosteroids selectively enhance the GABA responsiveness of δ-subunit-containing receptors *(194–197)*. Accordingly, low concentrations of 3α,21-dihydroxy-5α-pregnan-20-one (allotetrahydrodeoxycorticosterone) increase the tonic conductance with no, or only modest, effects on IPSCs *(85,94)*. In this regard, changes in the expression of δ-containing receptors during the ovarian cycle or following short-term steroid administration have been shown to alter the magnitude or nature of the tonic conductance in hippocampal pyramidal and dentate gyrus granule cells associated with alterations in seizure susceptibility and anxiety *(89,198)*.

Selective activation of δ-containing receptors is seen with the hypnotic 4,5,6,7-tetrahydroisoxazolo[5,4-c]-pyridin-3-ol (THIP; Gaboxadol-H, Lundheck A/5). This compound is a partial agonist at $\alpha_4\beta_3\gamma_2$-receptors but behaves as a full- or "super"-agonist at $\alpha_4\beta_3\delta$-receptors, producing a maximum response greater than that produced by GABA *(196,199)*. In thalamocortical neurons, low concentrations of THIP increase the tonic conductance *(93–95)*. This effect is shared by the anaesthetic etomidate *(95)*, suggesting that the hypnotic actions of both agents reflect persistent activation of extrasynaptic receptors and a shift in thalamocortical firing.

A high sensitivity of $\alpha_4\beta_3\delta$-receptors to ethanol *(200,201)* has been suggested to underlie its selective augmentation of tonic inhibition in granule cells of the dentate gyrus *(202)*. Recent evidence also suggests that $\alpha_6\beta_3\delta$-receptors, when formed from α_6-subunits bearing the R100Q mutation found in alcohol nontolerant rats, are particularly sensitive to enhancement by ethanol, and may underlie the motor impairment produced by alcohol consumption *(192)*. Subsequent studies have failed to confirm the unique ethanol sensitivity of the $\alpha_4\beta_3\delta$-receptor *(203)* and have questioned the significance of the R100Q mutation in alcohol nontolerant rats *(204)*. An additional important consideration is that ethanol can alter the local synthesis and release of neurosteroids *(205,206)*. Although conceptually compelling, further studies are required to identify factors controlling the ethanol sensitivity of δ-containing receptors and their significance to its behavioral effects *(207)*.

Lastly, in cultured hippocampal neurons, the amnesic drugs propofol *(208,209)* and isoflurane *(210)* preferentially enhance the $GABA_A$R-mediated tonic conductance, whereas the nootropic α_5-selective inverse agonist L–655,708 preferentially inhibits it *(165)*. This is consistent with a proposed role for extrasynaptic α_5-subunit-containing receptors in regulating learning and memory *(164)*. Although the situation may be more complicated than first assumed, with the contribution of α_5-subunit-containing receptors to the tonic conductance dependent on the ambient GABA concentration *(88)*, this does not preclude a role for these receptors, but may indicate that α_5-selective drugs will have their greatest actions when the hippocampal GABA concentration is high, for example during active physiological states.

6. Conclusions

Phasic activation of synaptic $GABA_A$Rs is surely fundamental to information transfer in the brain, but it is now recognized that less temporally and spatially restricted activation of these ionotropic receptors plays a key role in various brain regions. The receptor subtypes mediating the two forms of signaling display distinct biophysical and pharmacological properties as well as different subcellular locations; and for high-affinity receptors activated tonically, their low occupancy, combined with the low efficacy of GABA, allows their regulation over a large dynamic range. As in most areas of endeavor, our initial assumptions and theories regarding tonic $GABA_A$R activation will no doubt be refined in the light of future findings. Hopefully further insights into the origin and maintenance of ambient GABA concentrations, the dynamic regulation of specific receptor populations, and the increasingly selective pharmacological modulation of neuronal excitability under both physiological and pathological conditions can be anticipated.

Acknowledgments

Supported by the Wellcome Trust. I would like to thank colleagues, past and present, particularly S. G. Brickley (Imperial College London) and Z. Nusser (Institute of Experimental Medicine, Budapest), for valuable discussion over many years.

References

1. Bormann, J., Hamill, O. P., and Sakmann, B. (1987) Mechanism of anion permeation through channels gated by glycine and gamma-aminobutyric acid in mouse cultured spinal neurones. *J. Physiol.* **385,** 243–286.
2. Kaila, K. (1994) Ionic basis of $GABA_A$ receptor channel function in the nervous system. *Prog. Neurobiol.* **42,** 489–537.
3. Payne, J. A., Rivera, C., Voipio, J., and Kaila, K. (2003) Cation-chloride cotransporters in neuronal communication, development and trauma. *Trends Neurosci.* **26,** 199–206.
4. Rivera, C., Voipio, J., and Kaila, K. (2005) Two developmental switches in GABAergic signalling: the K^+-Cl^- cotransporter KCC2 and carbonic anhydrase CAVII. *J. Physiol.* **562,** 27–36.
5. Krnjevic, K. (2004) How does a little acronym become a big transmitter? *Biochem. Pharmacol.* **68,** 1549–1555.
6. Willis, W. D. (2006) John Eccles' studies of spinal cord presynaptic inhibition. *Prog. Neurobiol.* **78,** 189–214.
7. Delpire, E. (2000) Cation-chloride cotransporters in neuronal communication. *News Physiol. Sci.* **15,** 309–312.
8. Martina, M., Royer, S., and Pare, D. (2001) Cell-type-specific GABA responses and chloride homeostasis in the cortex and amygdala. *J. Neurophysiol.* **86,** 2887–2895.
9. DeFazio, R. A., Heger, S., Ojeda, S. R., and Moenter, S. M. (2002) Activation of A-type gamma-aminobutyric acid receptors excites gonadotropin-releasing hormone neurons. *Mol. Endocrinol.* **16,** 2872–2891.
10. Gulledge, A. T. and Stuart, G. J. (2003) Excitatory actions of GABA in the cortex. *Neuron* **37,** 299–309.
11. Stein, V. and Nicoll, R. A. (2003) GABA generates excitement. *Neuron* **37,** 375–378.
12. Chavas, J. and Marty, A. (2003) Coexistence of excitatory and inhibitory GABA synapses in the cerebellar interneuron network. *J. Neurosci.* **23,** 2019–2031.
13. Vida, I., Bartos, M., and Jonas, P. (2006) Shunting inhibition improves robustness of gamma oscillations in hippocampal interneuron networks by homogenizing firing rates. *Neuron* **49,** 107–117.
14. Staley, K. J. and Mody, I. (1992) Shunting of excitatory input to dentate gyrus granule cells by a depolarizing $GABA_A$ receptor-mediated postsynaptic conductance. *J. Neurophysiol.* **68,** 197–212.
15. Gao, X. B., Chen, G., and van den Pol, A. N. (1998) GABA-dependent firing of glutamate-evoked action potentials at AMPA/kainate receptors in developing hypothalamic neurons. *J. Neurophysiol.* **79,** 716–726.

16. Morita, K., Tsumoto, K., and Aihara, K. (2006) Bidirectional modulation of neuronal responses by depolarizing GABAergic inputs. *Biophys. J.* **90,** 1925–1938.
17. Gulacsi, A., Lee, C. R., Sik, A., et al. (2003) Cell type-specific differences in chloride-regulatory mechanisms and $GABA_A$ receptor-mediated inhibition in rat substantia nigra. *J. Neurosci.* **23,** 8237–8246.
18. Wagner, S., Castel, M., Gainer, H., and Yarom, Y. (1997) GABA in the mammalian suprachiasmatic nucleus and its role in diurnal rhythmicity. *Nature* **387,** 598–603.
19. Wagner, S., Sagiv, N., and Yarom, Y. (2001) GABA-induced current and circadian regulation of chloride in neurones of the rat suprachiasmatic nucleus. *J. Physiol.* **537,** 853–869.
20. Gavrikov, K. E., Dmitriev, A. V., Keyser, K. T., and Mangel, S. C. (2003) Cation–chloride cotransporters mediate neural computation in the retina. *Proc. Natl. Acad. Sci. USA* **100,** 16,047–16,052.
21. Szabadics, J., Varga, C., Molnar, G., Olah, S., Barzo, P., and Tamas, G. (2006) Excitatory effect of GABAergic axo-axonic cells in cortical microcircuits. *Science* **311,** 233–235.
22. Voipio, J. and Kaila, K. (2000) GABAergic excitation and K^+-mediated volume transmission in the hippocampus. *Prog. Brain Res.* **125,** 329–338.
23. Woodin, M. A., Ganguly, K., and Poo, M. M. (2003) Coincident pre- and postsynaptic activity modifies GABAergic synapses by postsynaptic changes in Cl^- transporter activity. *Neuron* **39,** 807–820.
24. Rivera, C., Voipio, J., Thomas-Crusells, J., et al. (2004) Mechanism of activity-dependent downregulation of the neuron-specific K-Cl cotransporter KCC2. *J. Neurosci.* **24,** 4683–4691.
25. Fiumelli, H., Cancedda, L., and Poo, M. M. (2005) Modulation of GABAergic transmission by activity via postsynaptic Ca^{2+}-dependent regulation of KCC2 function. *Neuron* **48,** 773–786.
26. Edwards, F. A., Konnerth, A., and Sakmann, B. (1990) Quantal analysis of inhibitory synaptic transmission in the dentate gyrus of rat hippocampal slices: a patch-clamp study. *J. Physiol.* **430,** 213–249.
27. Mody, I., De Koninck, Y., Otis, T. S., and Soltesz, I. (1994) Bridging the cleft at GABA synapses in the brain. *Trends Neurosci.* **17,** 517–525.
28. Nusser, Z., Cull-Candy, S., and Farrant, M. (1997) Differences in synaptic $GABA_A$ receptor number underlie variation in GABA mini amplitude. *Neuron* **19,** 697–709.
29. Brickley, S. G., Cull-Candy, S. G., and Farrant, M. (1999) Single-channel properties of synaptic and extrasynaptic $GABA_A$ receptors suggest differential targeting of receptor subtypes. *J. Neurosci.* **19,** 2960–2973.
30. Overstreet, L. S., Westbrook, G. L., and Jones, M. V. (2002) Measuring and modeling the spatiotemporal profile of GABA at the synapse. In: *Transmembrane Transporters* (Quick, M. W., eds.), Wiley-liss Inc, Hoboken, NJ, 259–275.
31. Mozrzymas, J. W., Zarmowska, E. D., Pytel, M., and Mercik, K. (2003) Modulation of $GABA_A$ receptors by hydrogen ions reveals synaptic GABA transient and a crucial role of the desensitization process. *J. Neurosci.* **23,** 7981–7992.

32. Baumann, S. W., Baur, R., and Sigel, E. (2003) Individual properties of the two functional agonist sites in $GABA_A$ receptors. *J. Neurosci.* **23,** 11,158–11,166.
33. Jones, M. V., Sahara, Y., Dzubay, J. A., and Westbrook, G. L. (1998) Defining affinity with the $GABA_A$ receptor. *J. Neurosci.* **18,** 8590–8604.
34. Perrais, D. and Ropert, N. (1999) Effect of zolpidem on miniature IPSCs and occupancy of postsynaptic $GABA_A$ receptors in central synapses. *J. Neurosci.* **19,** 578–588.
35. Hajos, N., Nusser, Z., Rancz, E. A., Freund, T. F., and Mody, I. (2000) Cell type- and synapse-specific variability in synaptic $GABA_A$ receptor occupancy. *Eur. J. Neurosci.* **12,** 810–818.
36. Weiss, D. S. and Magleby, K. L. (1989) Gating scheme for single GABA-activated Cl^- channels determined from stability plots, dwell-time distributions, and adjacent-interval durations. *J. Neurosci.* **9,** 1314–1324.
37. Twyman, R. E., Rogers, C. J., and Macdonald, R. L. (1990) Intraburst kinetic properties of the $GABA_A$ receptor main conductance state of mouse spinal cord neurones in culture. *J. Physiol.* **423,** 193–220.
38. Jones, M. V. and Westbrook, G. L. (1995) Desensitized states prolong $GABA_A$ channel responses to brief agonist pulses. *Neuron* **15,** 181–191.
39. Jayaraman, V., Thiran, S., and Hess, G. P. (1999) How fast does the γ-aminobutyric acid receptor channel open? Kinetic investigations in the micro-second time region using a laser-pulse photolysis technique. *Biochemistry* **38,** 11372–11378.
40. Haas, K. F. and Macdonald, R. L. (1999) $GABA_A$ receptor subunit γ2 and δ sub-types confer unique kinetic properties on recombinant $GABA_A$ receptor currents in mouse fibroblasts. *J. Physiol. (Lond.)* **514,** 27–45.
41. Burkat, P. M., Yang, J., and Gingrich, K. J. (2001) Dominant gating governing transient $GABA_A$ receptor activity: A first latency and P-o/o analysis. *J. Neurosci.* **21,** 7026–7036.
42. Akk, G., Bracamontes, J. R., Covey, D. F., Evers, A., Dao, T., and Steinbach, J. H. (2004) Neuroactive steroids have multiple actions to potentiate $GABA_A$ receptors. *J. Physiol.* **558,** 59–74.
43. Lema, G. M. and Auerbach, A. (2006) Modes and models of $GABA_A$ receptor gating. *J. Physiol.* **572,** 182–200.
44. Bier, M., Kits, K. S., and Borst, J. G. (1996) Relation between rise times and amplitudes of GABAergic postsynaptic currents. *J. Neurophysiol.* **75,** 1008–1012.
45. Maconochie, D. J., Zempel, J. M., and Steinbach, J. H. (1994) How quickly can $GABA_A$ receptors open? *Neuron* **12,** 61–71.
46. McClellan, A. M. and Twyman, R. E. (1999) Receptor system response kinetics reveal functional subtypes of native murine and recombinant human $GABA_A$ receptors. *J. Physiol.* **515,** 711–727.
47. Chang, Y. and Weiss, D. S. (1999) Channel opening locks agonist onto the GABAC receptor. *Nat. Neurosci.* **2,** 219–225.
48. Bianchi, M. T. and Macdonald, R. L. (2001) Agonist trapping by $GABA_A$ receptor channels. *J. Neurosci.* **21,** 9083–9091.
49. Okada, M., Onodera, K., Van Renterghem, C., Sieghart, W., and Takahashi, T. (2000) Functional correlation of $GABA_A$ receptor α subunits expression with the properties of IPSCs in the developing thalamus. *J. Neurosci.* **20,** 2202–2208.

50. Vicini, S., Ferguson, C., Prybylowski, K., Kralic, J., Morrow, A. L., and Homanics, G. E. (2001) $GABA_A$ receptor $\alpha 1$ subunit deletion prevents developmental changes of inhibitory synaptic currents in cerebellar neurons. *J. Neurosci.* **21,** 3009–3016.
51. Nusser, Z., Sieghart, W., and Mody, I. (1999) Differential regulation of synaptic $GABA_A$ receptors by cAMP- dependent protein kinase in mouse cerebellar and olfactory bulb neurones. *J. Physiol.* **521,** 421–435.
52. Bacci, A., Rudolph, U., Huguenard, J. R., and Prince, D. A. (2003) Major differences in inhibitory synaptic transmission onto two neocortical interneuron subclasses. *J. Neurosci.* **23,** 9664–9674.
53. Ramadan, E., Fu, Z., Losi, G., Homanics, G. E., Neale, J. H., and Vicini, S. (2003) $GABA_A$ receptor beta3 subunit deletion decreases $\alpha 2/3$ subunits and IPSC duration. *J. Neurophysiol.* **89,** 128–134.
54. Valeyev, A. Y., Cruciani, R. A., Lange, G. D., Smallwood, V. S., and Barker, J. L. (1993) Cl- channels are randomly activated by continuous GABA secretion in cultured embryonic rat hippocampal neurons. *Neurosci. Lett.* **155,** 199–203.
55. LoTurco, J. J., Owens, D. F., Heath, M. J., Davis, M. B., and Kriegstein, A. R. (1995) GABA and glutamate depolarize cortical progenitor cells and inhibit DNA synthesis. *Neuron* **15,** 1287–1298.
56. Owens, D. F., Liu, X., and Kriegstein, A. R. (1999) Changing Properties of $GABA_A$ Receptor-mediated signaling during early neocortical development. *J. Neurophysiol.* **82,** 570–583.
57. Demarque, M., Represa, A., Becq, H., Khalilov, I., Ben–Ari, Y., and Aniksztejn, L. (2002) Paracrine intercellular communication by a Ca^{2+}- and SNARE- independent release of GABA and glutamate prior to synapse formation. *Neuron* **36,** 1051–1061.
58. Nguyen, L., Malgrange, B., Breuskin, I., et al. (2003) Autocrine/paracrine activation of the $GABA_A$ receptor inhibits the proliferation of neurogenic polysialylated neural cell adhesion molecule-positive ($PSA-NCAM^+$) precursor cells from postnatal striatum. *J. Neurosci.* **23,** 3278–3294.
59. Liu, X., Wang, Q., Haydar, T. F., and Bordey, A. (2005) Nonsynaptic GABA signaling in postnatal subventricular zone controls proliferation of GFAP-expressing progenitors. *Nat. Neurosci.* **8,** 1179–1187.
60. Ge, S., Goh, E. L., Sailor, K. A., Kitabatake, Y., Ming, G. L., and Song, H. (2006) GABA regulates synaptic integration of newly generated neurons in the adult brain. *Nature* **439,** 589–593.
61. Ben-Ari, Y. (2002) Excitatory actions of GABA during development: The nature of the nurture. *Nat. Rev. Neurosci.* **3,** 728–739.
62. Owens, D. F. and Kriegstein, A. R. (2002) Is there more to GABA than synaptic inhibition? *Nat. Rev. Neurosci.* **3,** 715–727.
63. Lledo, P. M., Alonso, M., and Grubb, M. S. (2006) Adult neurogenesis and functional plasticity in neuronal circuits. *Nat. Rev. Neurosci.* **7,** 179–193.
64. Sipila, S. T., Huttu, K., Soltesz, I., Voipio, J., and Kaila, K. (2005) Depolarizing GABA acts on intrinsically bursting pyramidal neurons to drive giant depolarizing potentials in the immature hippocampus. *J. Neurosci.* **25,** 5280–5289.
65. Sipila, S. T., Huttu, K., Voipio, J., and Kaila, K. (2006) Intrinsic bursting of immature CA3 pyramidal neurons and consequent giant depolarizing potentials are driven by a persistent Na^+ current and terminated by a slow Ca^{2+}-activated K^+ current. *Eur. J. Neurosci.* **23,** 2330–2338.

66. Leinekugel, X., Khazipov, R., Cannon, R., Hirase, H., Ben-Ari, Y., and Buzsaki, G. (2002) Correlated bursts of activity in the neonatal hippocampus in vivo. *Science* **296,** 2049–2052.
67. Dzhala, V. I., Talos, D. M., Sdrulla, D. A., et al. (2005) NKCC1 transporter facilitates seizures in the developing brain. *Nat. Med.* **11,** 1205–1213.
68. Sipila, S. T., Schuchmann, S., Voipio, J., Yamada, J., and Kaila K. (2006) The cation-chloride cotransporter NKCC1 promotes sharp waves in the neonatal rat hippocampus. *J. Physiol.* **573,** 765–773.
69. Mody, I. (2001) Distinguishing between $GABA_A$ receptors responsible for tonic and phasic conductances. *Neurochem. Res.* **26,** 907–913.
70. Mody, I. and Pearce, R. A. (2004) Diversity of inhibitory neurotransmission through $GABA_A$ receptors. *Trends Neurosci.* **27,** 569–575.
71. Semyanov, A., Walker, M. C., Kullmann, D. M., and Silver, R. A. (2004) Tonically active GABA A receptors: modulating gain and maintaining the tone. *Trends Neurosci.* **27,** 262–269.
72. Cavelier, P., Hamann, M., Rossi, D., Mobbs, P., and Attwell, D. (2005) Tonic excitation and inhibition of neurons: ambient transmitter sources and computational consequences. *Prog. Biophys. Mol. Biol.* **87,** 3–16.
73. Kullmann, D. M., Ruiz, A., Rusakov, D. M., Scott, R., Semyanov, A., and Walker, M. C. (2005) Presynaptic, extrasynaptic and axonal $GABA_A$ receptors in the CNS: where and why? *Prog. Biophys. Mol. Biol.* **87,** 33–46.
74. Mody, I. (2005) Aspects of the homeostaic plasticity of $GABA_A$ receptor-mediated inhibition. *J. Physiol.* **562,** 37–46.
75. Farrant, M. and Nusser, Z. (2005) Variations on an inhibitory theme: phasic and tonic activation of $GABA_A$ receptors. *Nat. Rev. Neurosci.* **6,** 215–229.
76. Otis, T. S., Staley, K. J., and Mody, I. (1991) Perpetual inhibitory activity in mammalian brain slices generated by spontaneous GABA release. *Brain Res.* **545,** 142–150.
77. Salin, P. A. and Prince, D. A. (1996) Spontaneous $GABA_A$ receptor-mediated inhibitory currents in adult rat somatosensory cortex. *J. Neurophysiol.* **75,** 1573–1588.
78. Hausser, M. and Clark, B. A. (1997) Tonic synaptic inhibition modulates neuronal output pattern and spatiotemporal synaptic integration. *Neuron* **19,** 665–678.
79. Kaneda, M., Farrant, M., and Cull-Candy, S. G. (1995) Whole-cell and single-channel currents activated by GABA and glycine in granule cells of the rat cerebellum. *J. Physiol.* **485,** 419–435.
80. Brickley, S. G., Cull-Candy, S. G., and Farrant, M. (1996) Development of a tonic form of synaptic inhibition in rat cerebellar granule cells resulting from persistent activation of $GABA_A$ receptors. *J. Physiol.* **497,** 753–759.
81. Tia, S., Wang, J. F., Kotchabhakdi, N., and Vicini, S. (1996) Developmental changes of inhibitory synaptic currents in cerebellar granule neurons: Role of $GABA_A$ receptor $\alpha 6$ subunit. *J. Neurosci.* **16,** 3630–3640.
82. Wall, M. J. and Usowicz, M. M. (1997) Development of action potential-dependent and independent spontaneous $GABA_A$ receptor-mediated currents in granule cells of postnatal rat cerebellum. *Eur. J. Neurosci.* **9,** 533–548.
83. Nusser, Z. and Mody, I. (2002) Selective modulation of tonic and phasic inhibitions in dentate gyrus granule cells. *J. Neurophysiol.* **87,** 2624–2628.

84. Stell, B. M. and Mody, I. (2002) Receptors with different affinities mediate phasic and tonic $GABA_A$ conductances in hippocampal neurons. *J. Neurosci.* **22,** RC223.
85. Stell, B. M., Brickley, S. G., Tang, C. Y., Farrant, M., and Mody, I. (2003) Neuroactive steroids reduce neuronal excitability by selectively enhancing tonic inhibition mediated by δ subunit-containing $GABA_A$ receptors. *Proc. Natl. Acad. Sci. USA* **100,** 14,439–14,444.
86. Mtchedlishvili, Z. and Kapur, J. (2006) High-affinity, slowly desensitizing $GABA_A$ receptors mediate tonic inhibition in hippocampal dentate granule cells. *Mol. Pharmacol.* **69,** 564–575.
87. Semyanov, A., Walker, M. C., and Kullmann, D. M. (2003) GABA uptake regulates cortical excitability via cell type-specific tonic inhibition. *Nat. Neurosci.* **6,** 484–490.
88. Scimemi, A., Semyanov, A., Sperk, G., Kullmann, D. M., and Walker, M. C. (2005) Multiple and plastic receptors mediate tonic $GABA_A$ receptor currents in the hippocampus. *J. Neurosci.* **25,** 10016–10024.
89. Shen, H., Gong, Q. H., Yuan, M., and Smith, S. S. (2005) Short-term steroid treatment increases delta $GABA_A$ receptor subunit expression in rat CA1 hippocampus: pharmacological and behavioral effects. *Neuropharmacology* **49,** 573–586.
90. Yamada, J., Furukawa, T., Ueno, S., Yamamoto, S., and Fukuda, A. (2006) Molecular basis for the $GABA_A$ receptor-mediated tonic inhibition in rat somatosensory cortex. *Cereb. Cortex Advance Access.* Epub ahead of print, Sept. 22, 2006.
91. Keros, S. and Hablitz, J. J. (2005) Subtype-specific GABA transporter antagonists synergistically modulate phasic and tonic $GABA_A$ conductances in rat neocortex. *J. Neurophysiol.* **94,** 2073–2085.
92. Porcello, D. M., Huntsman, M. M., Mihalek, R. M., Homanics, G. E., and Huguenard, J. R. (2003) Intact synaptic GABAergic inhibition and altered neurosteroid modulation of thalamic relay neurons in mice lacking delta subunit. *J. Neurophysiol.* **89,** 1378–1386.
93. Jia, F., Pignataro, L., Schofield, C. M., Yue, M., Harrison, N. L., and Goldstein, P. A. (2005) An extrasynaptic $GABA_A$ receptor mediates tonic inhibition in thalamic VB neurons. *J. Neurophysiol.* **94,** 4491–4501.
94. Cope, D. W., Hughes, S. W., and Crunelli, V. (2005) $GABA_A$ receptor-mediated tonic inhibition in thalamic neurons. *J. Neurosci.* **25,** 11,553–11,563.
95. Belelli, D., Peden, D. R, Rosahl, T. W., Wafford, K. A., and Lambert, J. J. (2005) Extrasynaptic $GABA_A$ receptors of thalamocortical neurons: a molecular target for hypnotics. *J. Neurosci.* **25,** 11,513–11,520.
96. Sigel, E., Baur, R., Malherbe, P., and Mohler, H. (1989) The rat β1-subunit of the $GABA_A$ receptor forms a picrotoxin-sensitive anion channel open in the absence of GABA. *FEBS Lett.* **257,** 377–379.
97. Maksay, G., Thompson, S. A., and Wafford, K. A. (2003) The pharmacology of spontaneously open α1β3ε $GABA_A$ receptor-ionophores. *Neuropharmacology* **44,** 994–1002.
98. Lindquist, C. E., Dalziel, J. E., Cromer, B. A., and Birnir, B. (2004) Penicillin blocks human α1β1 and α1β1γ2s $GABA_A$ channels that open spontaneously. *Eur. J. Pharmacol.* **496,** 23–32.
99. Wagner, D. A., Goldschen-Ohm, M. P., Hales, T. G., and Jones, M. V. (2005) Kinetics and spontaneous open probability conferred by the epsilon subunit of the $GABA_A$ receptor. *J. Neurosci.* **25,** 10,462–10,468.

100. Ranna, M., Sinkkonen, S. T., Moykkynen, T., Uusi-Oukari, M., and Korpi, E. R. (2006) Impact of ε and θ subunits on pharmacological properties of α3β1 $GABA_A$ receptors expressed in Xenopus oocytes. *BMC Pharmacol.* **6,** 1.
101. Birnir, B., Everitt, A. B., Lim, M. S., and Gage, P. W. (2000) Spontaneously opening $GABA_A$ channels in CA1 pyramidal neurones of rat hippocampus. *J. Membr. Biol.* **174,** 21–29.
102. Jones, B. L., Whiting, P. J., and Henderson, L. P. (2006) Mechanisms of anabolic androgenic steroid inhibition of ε-containing $GABA_A$ receptors. *J. Physiol.* **573,** 571–593.
103. Downing, S. S., Lee, Y. T., Farb, D. H., and Gibbs, T. T. (2005) Benzodiazepine modulation of partial agonist efficacy and spontaneously active $GABA_A$ receptors supports an allosteric model of modulation. *Br. J. Pharmacol.* **145,** 894–906.
104. Campo-Soria, C., Chang, Y., and Weiss, D. S. (2006) Mechanism of action of benzodiazepines on $GABA_A$ receptors. *Br. J. Pharmacol.* **148,** 984–990.
105. Wall, M. J. (2002) Furosemide reveals heterogeneous $GABA_A$ receptor expression at adult rat Golgi cell to granule cell synapses. *Neuropharmacology* **43,** 737–749.
106. Lerma, J., Herranz, A. S., Herreras, O., Abraira, V., and Martin del Rio, R. (1986) In vivo determination of extracellular concentration of amino acids in the rat hippocampus. A method based on brain dialysis and computerized analysis. *Brain Res.* **384,** 145–155.
107. Tossman, U., Jonsson, G., and Ungerstedt, U. (1986) Regional distribution and extracellular levels of amino acids in rat central nervous system. *Acta. Physiol. Scand.* **127,** 533–545.
108. Kennedy, R. T., Thompson, J. E., and Vickroy, T. W. (2002) In vivo monitoring of amino acids by direct sampling of brain extracellular fluid at ultralow flow rates and capillary electrophoresis. *J. Neurosci. Methods* **114,** 39–49.
109. Wu, Y., Wang, W., and Richerson, G. B. (2003) Vigabatrin induces tonic inhibition via GABA transporter reversal without increasing vesicular GABA release. *J. Neurophysiol.* **89,** 2021–2034.
110. Xi, Z. X., Ramamoorthy, S., Shen, H., Lake, R., SaWu, Y., and Wang, W. (2003) GABA transmission in the nucleus accumbens is altered after withdrawal from repeated cocaine. *J. Neurosci.* **23,** 3498–3505.
111. Santhakumar, V., Hanchar, H. J., Wallner, M., Olsen, R. W., and Otis, T. S. (2006) Contributions of the $GABA_A$ receptor α6 subunit to phasic and tonic inhibition revealed by a naturally occurring polymorphism in the α6 gene. *J. Neurosci.* **26,** 3357–3364.
112. Brickley, S. G., Cull-Candy, S. G., and Farrant M. (2003) Vesicular release of GABA contributes to both phasic and tonic inhibition of granule cells in the cerebellum of mature mice. *J. Physiol. (Lond.)* **547,** C30.
113. Carta, M., Mameli, M., and Valenzuela, C. F. (2004) Alcohol enhances GABAergic transmission to cerebellar granule cells via an increase in Golgi cell excitability. *J. Neurosci.* **24,** 3746–3751.
114. Rossi, D. J., Hamann, M., and Attwell, D. (2003) Multiple modes of GABAergic inhibition of rat cerebellar granule cells. *J. Physiol. (Lond.)* **548,** 97–110.
115. Kimelberg, H. K., Goderie, S. K., Higman, S., Pang, S., and Waniewski, R. A. (1990) Swelling-induced release of glutamate, aspartate, and taurine from astrocyte cultures. *J. Neurosci.* **10,** 1583–1591.

116. Liu, Q. Y., Schaffner, A. E., Chang, Y. H., Maric, D., and Barker, J. L. (2000) Persistent activation of $GABA_A$ receptor/Cl^- channels by astrocyte-derived GABA in cultured embryonic rat hippocampal neurons. *J. Neurophysiol.* **84,** 1392–1403.
117. Wang, C. M., Chang, Y. Y., Kuo, J. S., and Sun, S. H. (2002) Activation of $P2X_7$ receptors induced [^{3}H]GABA release from the RBA-2 type-2 astrocyte cell line through a Cl^-/HCO_3—dependent mechanism. *Glia* **37,** 8–18.
118. Vautrin, J., Maric, D., Sukhareva, M., Schaffner, A. E., and Barker, J. L. (2000) Surface-accessible GABA supports tonic and quantal synaptic transmission. *Synapse* **37,** 38–55.
119. Valeyev, A. Y., Schaffner, A. E., Skolnick, P., Dunlap, V. S., Wong. G., and Barker, J. L. (1998) Embryonic rat hippocampal neurons and $GABA_A$ receptor subunit-transfected non-neuronal cells release GABA tonically. *J. Membr. Biol.* **164,** 239–251.
120. Chadderton, P., Margrie, T. W., and Hausser, M. (2004) Integration of quanta in cerebellar granule cells during sensory processing. *Nature* **428,** 856–860.
121. Richerson, G. B. and Wu, Y. M. (2003) Dynamic equilibrium of neurotransmitter transporters: Not just for reuptake anymore. *J. Neurophysiol.* **90,** 1363–1374.
122. Overstreet, L. S. and Westbrook, G. L. (2001) Paradoxical Reduction of Synaptic Inhibition by Vigabatrin. *J. Neurophysiol.* **86,** 596–603.
123. Allen, N. J., Rossi, D. J., and Attwell, D. (2004) Sequential release of GABA by exocytosis and reversed uptake leads to neuronal swelling in simulated ischemia of hippocampal slices. *J. Neurosci.* **24,** 3837–3849.
124. Attwell, D., Barbour, B., and Szatkowski, M. (1993) Nonvesicular release of neurotransmitter. *Neuron* **11,** 401–407.
125. Jensen, K., Chiu, C. S., Sokolova, I., Lester, H. A., and Mody, I. (2003) GABA transporter-1 (GAT1)-deficient mice: differential tonic activation of $GABA_A$ versus $GABA_B$ receptors in the hippocampus. *J. Neurophysiol.* **90,** 2690–2701.
126. Barnard, E. A., Skolnick, P., Olsen, R. W., et al. (1998) International union of pharmacology. XV. Subtypes of γ-aminobutyric $acid_A$ receptors: classification on the basis of subunit structure and receptor function. *Pharmacol. Rev.* **50,** 291–313.
127. Korpi, E. R., Grunder, G., and Luddens, H. (2002) Drug interactions at $GABA_A$ receptors. *Prog. Neurobiol.* **67,** 113–159.
128. Sieghart, W. and Sperk, G. (2002) Subunit composition, distribution and function of $GABA_A$ receptor subtypes. *Curr. Top. Med. Chem.* **2,** 795–816.
129. Rudolph, U. and Mohler, H. (2004) Analysis of $GABA_A$ receptor function and dissection of the pharmacology of benzodiazepines and general anesthetics through mouse genetics. *Annu. Rev. Pharmacol. Toxicol.* **44,** 475–498.
130. Sieghart, W. and Ernst, M. (2005) Heterogeneity of $GABA_A$ receptors: revived interest in the development of subtype-selective drugs. *Curr. Med. Chem. Cent. Nerv. Syst. Agents* **5,** 217–242.
131. Simon, J., Wakimoto, H., Fujita, N., Lalande, M., and Barnard, E. A. (2004) Analysis of the set of $GABA_A$ receptor genes in the human genome. *J. Biol. Chem.* **279,** 41,422–41,435.
132. Kittler, J. T., McAinsh, K., and Moss, S. J. (2002) Mechanisms of $GABA_A$ receptor assembly and trafficking - Implications for the modulation of inhibitory neurotransmission. *Mol. Neurobiol.* **26,** 251–268.

133. Luscher, B. and Keller, C. A. (2004) Regulation of $GABA_A$ receptor trafficking, channel activity, and functional plasticity of inhibitory synapses. *Pharmacol. Ther.* **102,** 195–221.
134. Wisden, W., Laurie, D. J., Monyer, H., and Seeburg, P. H. (1992) The distribution of 13 $GABA_A$ receptor subunit mRNAs in the rat brain. I. Telencephalon, diencephalon, mesencephalon. *J. Neurosci.* **12,** 1040–1062.
135. Fritschy, J. M. and Mohler, H. (1995) $GABA_A$-receptor heterogeneity in the adult-rat brain - Differential regional and cellular-distribution of 7 major subunits. *J. Comp. Neurol.* **359,** 154–194.
136. Pirker, S., Schwarzer, C., Wieselthaler, A., Sieghart, W., and Sperk, G. (2000) $GABA_A$ receptors: Immunocytochemical distribution of 13 subunits in the adult rat brain. *Neuroscience* **101,** 815–850.
137. McKernan, R. M. and Whiting, P. J. (1996) Which $GABA_A$-receptor subtypes really occur in the brain? *Trends Neurosci.* **19,** 139–143.
138. Whiting, P. J. (2003) GABA-A receptor subtypes in the brain: a paradigm for CNS drug discovery? *Drug Discov. Today* **8,** 445–450.
139. Tretter, V., Ehya, N., Fuchs, K., and Sieghart, W. (1997) Stoichiometry and assembly of a recombinant $GABA_A$ receptor subtype. *J. Neurosci.* **17,** 2728–2737.
140. Farrar, S. J., Whiting, P. J., Bonnert, T. P., and McKernan, R. M. (1999) Stoichiometry of a ligand-gated ion channel determined by fluorescence energy transfer. *J. Biol. Chem.* **274,** 10,100–10,104.
141. Baumann, S. W., Baur, R., and Sigel, E. (2002) Forced subunit assembly in $\alpha1\beta2\gamma2$ $GABA_A$ receptors. Insight into the absolute arrangement. *J. Biol. Chem.* **277,** 46,020–46,025.
142. Boileau, A. J., Pearce, R. A., and Czajkowski, C. (2005) Tandem subunits effectively constrain $GABA_A$ receptor stoichiometry and recapitulate receptor kinetics but are insensitive to $GABA_A$ receptor-associated protein. *J. Neurosci.* **25,** 11,219–11,230.
143. Baur, R., Minier, F., and Sigel, E. (2006) A $GABA_A$ receptor of defined subunit composition and positioning: concatenation of five subunits. *FEBS Lett.* **580,** 1616–1620.
144. Neelands, T. R., Fisher, J. L., Bianchi, M., and Macdonald, R. L. (1999) Spontaneous and γ-aminobutyric acid (GABA)-activated $GABA_A$ receptor channels formed by ε subunit-containing isoforms. *Mol. Pharmacol.* **55,** 168–178.
145. Neelands, T. R. and Macdonald, R. L. (1999) Incorporation of the π subunit into functional γ-aminobutyric $acid_A$ receptors. *Mol. Pharmacol.* **56,** 598–610.
146. Bonnert, T. P., McKernan, R. M., Farrar, S., et al. (1999) θ, a novel γ-aminobutyric acid type A receptor subunit. *Proc. Natl. Acad. Sci. USA*. **96,** 9891–9896.
147. Bormann, J. (2000) The 'ABC' of GABA receptors. *Trends Pharmacol. Sci* **21,** 16–19.
148. Johnston, G. A. (2002) Medicinal chemistry and molecular pharmacology of $GABA_C$ receptors. *Curr. Top. Med. Chem.* **2,** 903–913.
149. Qian, H. and Ripps, H. (1999) Response kinetics and pharmacological properties of heteromeric receptors formed by coassembly of GABA ρ- and $\gamma2$-subunits. *Proc. R. Soc. Lond. B. Biol. Sci.* **266,** 2419–2425.

150. Milligan, C. J., Buckley, N. J., Garret, M., Deuchars, J., and Deuchars, S. A. (2004) Evidence for inhibition mediated by coassembly of $GABA_A$ and $GABA_C$ receptor subunits in native central neurons. *J. Neurosci.* **24,** 9241–9250.
151. Hevers, W. and Luddens, H. (1998) The diversity of $GABA_A$ receptors. Pharmacological and electrophysiological properties of $GABA_A$ channel subtypes. *Mol. Neurobiol.* **18,** 35–86.
152. Nusser, Z., Roberts, J. D., Baude, A., Richards, J. G., and Somogyi, P. (1995) Relative densities of synaptic and extrasynaptic $GABA_A$ receptors on cerebellar granule cells as determined by a quantitative immunogold method. *J. Neurosci.* **15,** 2948–2960.
153. Nusser, Z., Roberts, J. D., Baude, A., Richards, J. G., Sieghart, W., and Somogyi, P. (1995) Immunocytochemical localization of the $\alpha1$ and $\beta2/3$ subunits of the $GABA_A$ receptor in relation to specific GABAergic synapses in the dentate gyrus. *Eur. J. Neurosci.* **7,** 630–646.
154. Nusser, Z., Sieghart, W., and Somogyi, P. (1998) Segregation of different $GABA_A$ receptors to synaptic and extrasynaptic membranes of cerebellar granule cells. *J. Neurosci.* **18,** 1693–1703.
155. Craig, A. M., Blackstone, C. D., Huganir, R. L, and Banker, G. (1994) Selective clustering of glutamate and γ-aminobutyric acid receptors opposite terminals releasing the corresponding neurotransmitters. *Proc. Natl. Acad. Sci. USA* **91,** 12,373–12,377.
156. Somogyi, P., Fritschy, J. M., Benke, D., Roberts, J. D., and Sieghart, W. (1996) The $\gamma2$ subunit of the $GABA_A$ receptor is concentrated in synaptic junctions containing the $\alpha1$ and $\beta2/3$ subunits in hippocampus, cerebellum and globus pallidus. *Neuropharmacology* **35,** 1425–1444.
157. Fritschy, J. M., Johnson, D. K., Mohler, H., and Rudolph, U. (1998) Independent assembly and subcellular targeting of $GABA_A$-receptor subtypes demonstrated in mouse hippocampal and olfactory neurons *in vivo*. *Neurosci. Lett.* **249,** 99–102.
158. Brunig, I., Scotti, E., Sidler, C., and Fritschy, J. M. (2002) Intact sorting, targeting, and clustering of γ-aminobutyric acid A receptor subtypes in hippocampal neurons *in vitro*. *J. Comp. Neurol.* **443,** 43–55.
159. Wei, W., Zhang, N., Peng, Z., Houser, C. R., and Mody, I. (2003) Perisynaptic localization of δ subunit-containing $GABA_A$ receptors and their activation by GABA spillover in the mouse dentate gyrus. *J. Neurosci.* **23,** 10,650–10,661.
160. Sur, C., Farrar, S. J., Kerby, J., Whiting, P. J., Atack, J. R., and McKernan, R. M. (1999) Preferential coassembly of $\alpha4$ and δ subunits of the γ-aminobutyric acidA receptor in rat thalamus. *Mol. Pharmacol.* **56,** 110–115.
161. Moss, S. J. and Smart, T. G. (2001) Constructing inhibitory synapses. *Nat. Rev. Neurosci.* **2,** 240–250.
162. Fritschy, J. M. and Brunig, I. (2003) Formation and plasticity of GABAergic synapses: physiological mechanisms and pathophysiological implications. *Pharmacol. Ther.* **98,** 299–323.
163. Crestani, F., Keist, R., Fritschy, J. M., et al. (2002) Trace fear conditioning involves hippocampal $\alpha5$ $GABA_A$ receptors. *Proc. Natl. Acad. Sci. USA* **99,** 8980–8985.

164. Collinson, N., Kuenzi, F. M., Jarolimek, W., et al. (2002) Enhanced learning and memory and altered GABAergic synaptic transmission in mice lacking the α5 subunit of the $GABA_A$ receptor. *J. Neurosci.* **22,** 5572–5580.
165. Caraiscos, V. B., Elliott, E. M., You-Ten, K. E., et al. (2004) Tonic inhibition in mouse hippocampal CA1 pyramidal neurons is mediated by α5 subunit-containing γ-aminobutyric acid type A receptors. *Proc. Natl. Acad. Sci. USA* **101,** 3662–3667.
166. Rossi, D. J. and Hamann, M. (1998) Spillover-mediated transmission at inhibitory synapses promoted by high affinity α6 subunit $GABA_A$ receptors and glomerular geometry. *Neuron* **20,** 783–795.
167. Pearce, R. A. (1993) Physiological evidence for two distinct $GABA_A$ responses in rat hippocampus. *Neuron* **10,** 189–200.
168. Banks, M. I., Li, T. B., and Pearce, R. A. (1998) The synaptic basis of $GABA_{A,slow}$. *J. Neurosci.* **18,** 1305–1317.
169. Banks, M. I., White, J. A., and Pearce, R. A. (2000) Interactions between distinct $GABA_A$ circuits in hippocampus. *Neuron* **25,** 449–457.
170. Prenosil, G. A., Schneider Gasser, E. M., Rudolph, U., Keist, R., Fritschy, J. M., and Vogt, K. E. (206) Specific subtypes of $GABA_A$ receptors mediate phasic and tonic forms of inhibition in hippocampal pyramidal neurons. *J. Neurophysiol.* **96,** 846–857.
171. Brickley, S. G., Revilla, V., Cull-Candy, S. G., Wisden, W., and Farrant, M. (2001) Adaptive regulation of neuronal excitability by a voltage-independent potassium conductance. *Nature* **409,** 88–92.
172. Peng, Z., Hauer, B., Mihalek, R. M., et al. (2002) $GABA_A$ receptor changes in δ subunit-deficient mice: altered expression of α4 and γ2 subunits in the forebrain. *J. Comp. Neurol.* **446,** 179–197.
173. Glykys, J. and Mody, I. (2006) Hippocampal network hyperactivity after selective reduction of tonic inhibition in $GABA_A$ receptor α5 subunit-deficient mice. *J. Neurophysiol.* **95,** 2796–2807.
174. Buzsaki, G. and Chrobak, J. J. (1995) Temporal structure in spatially organized neuronal ensembles: a role for interneuronal networks. *Curr. Opin. Neurobiol.* **5,** 504–510.
175. Singer, W. (1996) The changing face of inhibition. *Curr. Biol.* **6,** 395–397.
176. McBain, C. J. and Fisahn, A. (2001) Interneurons unbound. *Nat. Rev. Neurosci.* **2,** 11–23.
177. Buzsaki, G. (2002) Theta oscillations in the hippocampus. *Neuron* **33,** 325-340.
178. Freund, T. F. (2003) Rhythm and mood in perisomatic inhibition. *Trends Neurosci.* **26,** 489–495.
179. Whittington, M. A. and Traub, R. D. (2003) Interneuron diversity series: inhibitory interneurons and network oscillations in vitro. *Trends Neurosci.* **26,** 676–682.
180. Jonas, P., Bischofberger, J., Fricker, D., and Miles, R. (2004) Interneuron Diversity series: Fast in, fast out - temporal and spatial signal processing in hippocampal interneurons. *Trends Neurosci.* **27,** 30–40.
181. Somogyi, P. and Klausberger, T. (2005) Defined types of cortical interneurone structure space and spike timing in the hippocampus. *J. Physiol.* **562,** 9–26.
182. Spruston, N., Schiller, Y., Stuart, G., and Sakmann, B. (1995) Activity-dependent action potential invasion and calcium influx into hippocampal CA1 dendrites. *Science* **268,** 297–300.

183. Miles, R., Toth, K., Gulyas, A. I., Hajos, N., and Freund, T. F. (1996) Differences between somatic and dendritic inhibition in the hippocampus. *Neuron* **16,** 815–823.
184. Pouille, F. and Scanziani, M. (2001) Enforcement of temporal fidelity in pyramidal cells by somatic feed-forward inhibition. *Science* **293,** 1159–1163.
185. Freund, T. F. and Buzsaki, G. (1996) Interneurons of the hippocampus. *Hippocampus* **6,** 347–470.
186. Bacci, A., Huguenard, J. R., and Prince, D. A. (2005) Modulation of neocortical interneurons: extrinsic influences and exercises in self-control. *Trends Neurosci.* **28,** 602–610.
187. Hamann, M., Rossi, D. J., and Attwell, D. (2002) Tonic and spillover inhibition of granule cells control information flow through cerebellar cortex. *Neuron* **33,** 625–633.
188. Mitchell, S. J. and Silver, R. A. (2003) Shunting inhibition modulates neuronal gain during synaptic excitation. *Neuron* **38,** 433–445.
189. Chance, F. S., Abbott, L. F., and Reyes, A. D. (2002) Gain modulation from background synaptic input. *Neuron* **35,** 773–782.
190. Semyanov, A., Walker, M. C., Kullmann, D. M., and Silver, R. A. (2004) Tonically active $GABA_A$ receptors: modulating gain and maintaining the tone. *Trends Neurosci.* **6,** 484–490.
191. Korpi, E. R. and Sinkkonen, S. T. (2006) $GABA_A$ receptor subtypes as targets for neuropsychiatric drug development. *Pharmacol. Ther.* **109,** 12–32.
192. Hanchar, H. J., Dodson, P. D., Olsen, R. W., Otis, T. S., and Wallner, M. (2005) Alcohol-induced motor impairment caused by increased extrasynaptic $GABA_A$ receptor activity. *Nat. Neurosci.* **8,** 339–345.
193. Leao, R. M., Mellor, J. R., and Randall, A. D. (2000) Tonic benzodiazepine-sensitive GABAergic inhibition in cultured rodent cerebellar granule cells. *Neuropharmacology* **39,** 990–1003.
194. Adkins, C. E., Pillai, G. V., Kerby, J., et al. (2001) α4_3_ $GABA_A$ receptors characterised by fluorescence resonance energy transfer-derived measurements of membrane potential. *J. Biol. Chem.* **276,** 38,934–38,939.
195. Belelli, D., Casula, A., Ling, A., and Lambert, J. J. (2002) The influence of subunit composition on the interaction of neurosteroids with $GABA_A$ receptors. *Neuropharmacology* **43,** 651–661.
196. Brown, N., Kerby, J., Bonnert, T. P., Whiting, P. J., and Wafford, K. A. (2002) Pharmacological characterization of a novel cell line expressing human α4β3δ $GABA_A$ receptors. *Br. J. Pharmacol.* **136,** 965–974.
197. Wohlfarth, K. M., Bianchi, M. T., and Macdonald, R. W. (2002) Enhanced neurosteroid potentiation of ternary $GABA_A$ receptors containing the δ subunit. *J. Neurosci.* **22,** 1541–1549.
198. Maguire, J. L., Stell, B. M., Rafizadeh, M., and Mody, I. (2005) Ovarian cycle-linked changes in $GABA_A$ receptors mediating tonic inhibition alter seizure susceptibility and anxiety. *Nat. Neurosci.* **8,** 797–804.
199. Wafford, K. A. and Ebert, B. (2006) Gaboxadol—a new awakening in sleep. *Curr. Opin. Pharmacol.* **6**, 30–36.
200. Sundstrom-Poromaa, I., Smith, D. H., Gong, Q. H., et al. (2002) Hormonally regulated α4β2δ $GABA_A$ receptors are a target for alcohol. *Nat. Neurosci.* **5,** 721–722.

201. Wallner, M., Hanchar, H. J., and Olsen, R. W. (2003) Ethanol enhances α4β3δ and α6β3δ γ-aminobutyric acid type A receptors at low concentrations known to affect humans. *Proc. Natl. Acad. Sci. USA* **100,** 15,218–15,223.
202. Wei, W., Faria, L. C., and Mody, I. (2004) Low ethanol concentrations selectively augment the tonic inhibition mediated by δ subunit-containing $GABA_A$ receptors in hippocampal neurons. *J. Neurosci.* **24,** 8379–8382.
203. Borghese, C. M., Storustovu, S., Ebert, B., et al. (2006) The δ subunit of γ-aminobutyric acid type A receptors does not confer sensitivity to low concentrations of ethanol. *J. Pharmacol. Exp. Ther.* **316,** 1360–1368.
204. Valenzuela, C. F., Mameli, M., and Carta, M. (2005) Single-amino-acid difference in the sequence of α6 subunit dramatically increases the ethanol sensitivity of recombinant $GABA_A$ receptors. *Alcohol Clin. Exp. Res.* **29(7),** 1356–1357, author reply 1358.
205. Sanna, E., Talani, G., Busonero, F., et al. (2004) Brain steroidogenesis mediates ethanol modulation of $GABA_A$ receptor activity in rat hippocampus. *J. Neurosci.* **24,** 6521–6530.
206. Belelli, D. and Lambert, J. J. (2005) Neurosteroids: endogenous regulators of the $GABA_A$ receptor. *Nat. Rev. Neurosci.* **6,** 565–575.
207. Weiner, J. L. and Valenzuela, C. F. (2006) Ethanol modulation of GABAergic transmission: The view from the slice. *Pharmacol. Ther.* **111,** 533–554.
208. Bai, D. L., Zhu, G. Y., Pennefather, P., Jackson, M. F., Macdonald, J. F., and Orser, B. A. (2001) Distinct functional and pharmacological properties of tonic and quantal inhibitory postsynaptic currents mediated by γ-aminobutyric $acid_A$ receptors in hippocampal neurons. *Mol. Pharmacol.* **59,** 814–824.
209. Yeung, J. Y. T., Canning, K. J., Zhu, G., Pennefather, P., MacDonald, J. F., and Orser, B. A. (2003) Tonically activated $GABA_A$ receptors in hippocampal neurons are high-affinity, low-conductance sensors for extracellular GABA. *Mol. Pharmacol.* **63,** 2–8.
210. Caraiscos, V. B., Newell, J. G., You-Ten, K. E., et al. (2004) Selective enhancement of tonic GABAergic inhibition in murine hippocampal neurons by low concentrations of the volatile anesthetic isoflurane. *J. Neurosci.* **24,** 8454–8458.

6

GABA$_A$-Receptor Mutations Associated With Idiopathic Generalized Epilepsies and Febrile Seizures

Robert L. Macdonald, Jing-Qiong Kang, Martin J. Gallagher, and Hua-Jun Feng

Summary

Idiopathic generalized epilepsies (IGEs) are characterized by absence, myoclonic, and/or primary generalized tonic–clonic seizures in the absence of structural brain abnormalities and are believed to have a genetic basis. Mutations in γ-aminobutyric acid (GABA$_A$) receptor γ_2-, α_1-, and δ-subunits have been associated with different IGE syndromes, and all of the γ_2-subunit mutations are also associated with febrile seizures. Here, mutations in GABA$_A$ receptor γ_2-, α_1-, and δ-subunits that have been associated with different IGE syndromes will be reviewed and the basis for the defect in GABAergic inhibition associated with febrile seizures will be explored. These mutations have been shown to alter GABA$_A$ receptor gating, expression, and/or trafficking of the receptor to the cell surface, all pathophysiological mechanisms that result in reduced GABA-evoked currents that, in turn, would cause neuronal disinhibition in neurons and thus predispose affected patients to manifest afebrile and febrile seizures.

Key Words: Electrophysiology; epilepsy; febrile seizures; GABA$_A$ receptor; mutations; myoclonus; review; trafficking.

From: *The Receptors: The GABA Receptors, Third Edition*
Edited by: S. J. Enna and H. Möhler © Humana Press Inc., Totowa, NJ

1. Introduction

GABA$_A$ receptors are the primary mediators of fast inhibitory synaptic transmission in the central nervous system, and are formed by pentameric assembly of multiple subunit subtypes (α_{1-6}, β_{1-3}, γ_{1-3}, δ, ε, $\pi\iota$ θ, and ρ_{1-3}) that form chloride-ion channels. The most common GABA$_A$ receptors contain two α-subunits, two β-subunits, and a γ- or δ-subunit *(1–3)*. GABA$_A$ receptors mediate both phasic, inhibitory synaptic transmission and tonic, perisynaptic inhibition, and GABA$_A$ receptor currents can be modulated by a number of positive and negative allosteric regulators, including barbiturates, benzodiazepines, neurosteroids, bicuculline, picrotoxin, and zinc. Recently, mutations associated with idiopathic generalized epilepsies (IGEs) and febrile seizures have been identified in human GABA$_A$ receptor genes *(4,5)*. This review will focus on these recently described human GABA$_A$ receptor channel epilepsy mutations.

GABA$_A$ receptor mutations associated with IGEs or variants associated with complex epilepsies have been reported in γ_2-, δ-, and α_1-subunits (Fig. 1). The mutations all have autosomal dominant inheritance and have been associated with epilepsy syndromes made up of pure febrile seizures, mixed febrile, and afebrile seizures (childhood absence epilepsy [CAE]), and generalized epilepsy with febrile seizures plus [GEFS$^+$]), or afebrile seizures (juvenile myoclonic epilepsy [JME]). The γ_2-subunit mutations include γ_2(R43Q) associated with febrile seizures and CAE *(6)*, γ_2(R139G) associated with febrile seizures *(6a)*, γ_2(K289M) associated with GEFS$^+$ *(7)*, γ_2(Q351X) associated with GEFS$^+$ *(8)*, γ_2(Q1X) associated with severe myoclonic epilepsy of infancy *(9)*, and γ_2(IVS6 + 2T$\rightarrow$G) associated with GEFS$^+$ *(10)*. The α_1-subunit mutation, α_1(A322D), is associated with JME *(11)*. Two δ-subunit variants, δ(R220H) and δ(E177A), were identified as susceptibility genes associated with GEFS$^+$ and JME *(12)*.

The basis for afebrile and febrile seizures associated with these IGEs has received considerable investigation during the last 5 yr. In general, it has been found that mutations associated with afebrile seizures produce specific functional deficits in either GABA$_A$ receptor trafficking, receptor biophysical, or pharmacological properties, or both for all of the mutations, whereas mutations associated with febrile seizures produce a reduction in the density of surface GABA$_A$ receptors owing to a temperature-dependent trafficking deficiency. These alterations in GABA$_A$ receptor function and or trafficking will be reviewed in this chapter.

2. Afebrile Seizures and GABA$_A$ Receptor Epilepsy Genes

2.1. γ_2-Subunit Missense Mutations

2.1.1. GABA$_A$ Receptor γ_2(K289M)-Subunit Mutation

A mutation in the GABA$_A$ receptor γ_2-subunit, γ_2(K289M), was reported in affected individuals of a family with an autosomal dominant generalized epilepsy

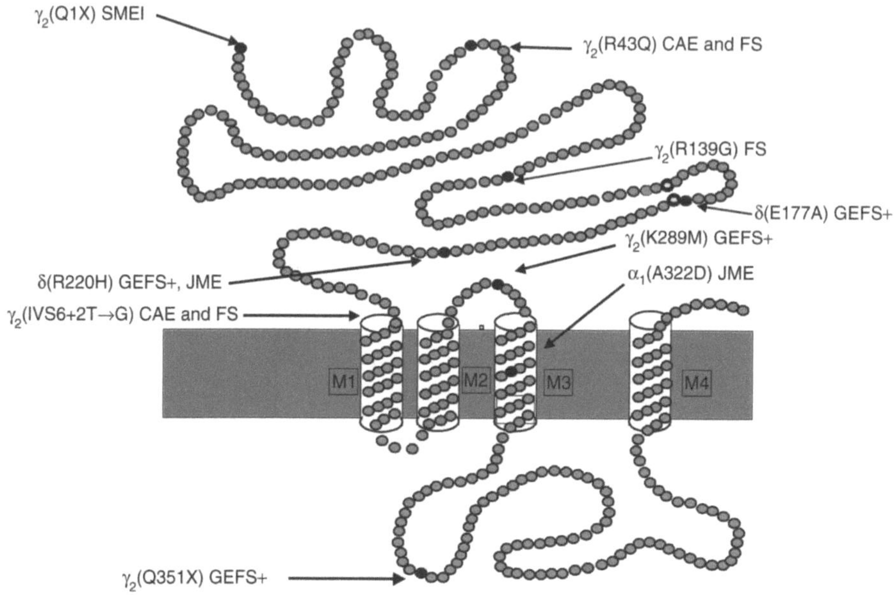

Fig. 1. GABA$_A$-receptor subunit topology and the location of epilepsy mutations.

similar to GEFS$^+$ *(7)*. The mutation is located in the extracellular loop between transmembrane domains M2 and M3 (M2–M3 loop) (Fig. 1), a region implicated in the gating of ligand-gated ion channels *(13–16)*. Recordings from oocytes expressing homozygous $\alpha_1\beta_2\gamma_2$(K289M)-receptors revealed smaller amplitude currents relative to wild-type (WT) receptor current amplitudes *(7)*. The effects of this mutation were also studied in transfected HEK 293T cells using a rapid application, concentration jump technique (open tip application rise time ~400 μs) to apply GABA *(17)* for long (400 ms or 6 s) or brief (2–5 ms) durations and the excised outside-out patch clamp recording technique to determine the effects of these mutations on the pharmacological and biophysical properties of WT and mutant GABA$_A$ receptor currents *(18)* (Fig. 2). Homozygous $\alpha_1\beta_3\gamma_2$L(K289M)-currents evoked by 400 msec applications of 1 m*M* GABA had unchanged current amplitude, rate of activation (Fig. 2D), and rate of desensitization (Fig. 2B,C), but had faster deactivation (~100 msec) (Fig. 2F), relative to WT currents (Fig. 2A–F). Currents evoked by brief applications of GABA (1 m*M*) had reduced weighted current deactivation rates (Fig. 3A1,A2,B) (WT, K→M). Single channel currents from homozygous $\alpha_1\beta_3\gamma_2$(K289M)-receptors had reduced mean open times, consistent with its faster whole-cell current deactivation time. Brief, rapid GABA applications to excised macropatches evoked currents that are similar to inhibitory post synaptic currents (IPSCs) *(19,20)*, and thus reduced duration of rapid GABA-evoked current by the γ_2L(K289M)-mutation,

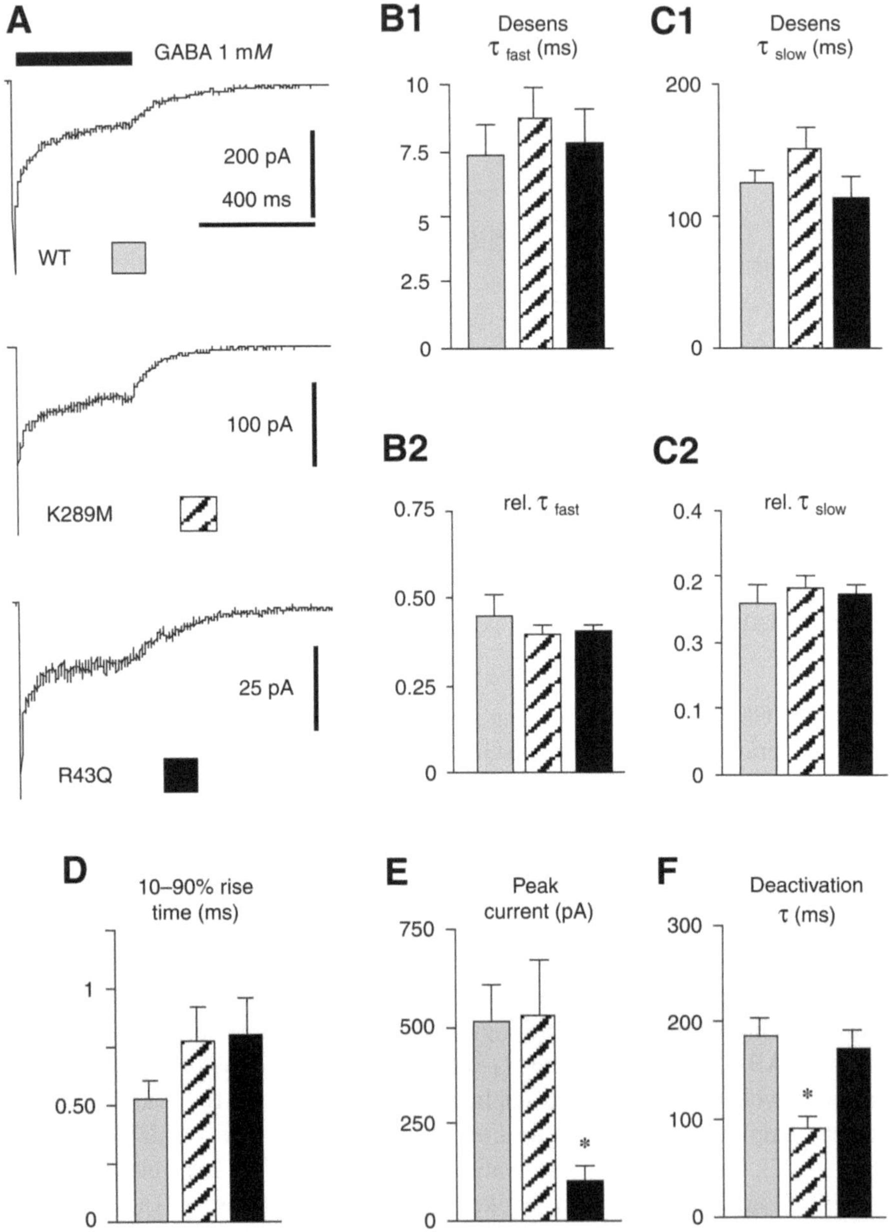

Fig. 2. Wild-type $\alpha_1\beta_3\gamma_2$L and mutant $\alpha_1\beta_3\gamma_2$L(K289M) and $\alpha_1\beta_3\gamma_2$L(R43Q)-GABA$_A$ receptor macroscopic current kinetic properties **(A)** Representative currents evoked by 400 ms jumps into 1 m*M* GABA from wild-type $\alpha_1\beta_3\gamma_2$L- or mutant $\alpha_1\beta_3\gamma_2$L(K289M)- and $\alpha_1\beta_3\gamma_2$L(R43Q)-receptors. The time-scale (top) trace applies to all three traces. **(B,C)** The γ_2L(K289M)-subunit mutation did not alter either the fast

suggesting that the mutation would produce reduced IPSC durations, and thus disinhibition leading to epilepsy.

The structural basis for the mutation-induced acceleration of deactivation and reduced single channel mean open times is unclear, but it has been suggested that the N-terminal domain of cys-loop receptors interacts with M2 and the M2–M3 loop during channel gating *(21,22)*, and that GABA$_A$ receptor α-subunit loops 2 and -7 interact with a lysine residue in the M2–M3 loop to couple GABA binding to gating *(14)*. Although the γ_2L-subunit does not appear to be directly in the binding-gating transduction pathway, it may modify other properties of the receptor channel such as deactivation or desensitization.

2.1.2. GABA$_A$ Receptor γ2(R43Q)-Subunit Mutation

A missense mutation in the N-terminal extracellular domain of the γ_2-subunit, γ_2(R43Q), was reported in affected individuals of a large family having both CAE and febrile seizures *(6)* (Fig. 1). In *Xenopus* oocytes $\alpha_1\beta_2\gamma_2$(R43Q)-GABA$_A$ receptors had unchanged GABA EC_{50}, current amplitude, and desensitization, but were insensitive to enhancement by the benzodiazepine diazepam. Subsequently, it was reported that the γ_2(R43Q)-mutation did not alter benzodiazepine sensitivity, rate of activation, desensitization, or deactivation, but did reduce peak current amplitude of homozygous and heterozygous $\alpha_1\beta_3\gamma_2$(R43Q)-receptor currents (Fig. 2A–F) (18). Importantly, when currents were evoked by brief applications of GABA (1 m*M*), the weighted rate of current deactivation was unchanged but the current amplitude was reduced (Fig. 3A1,A3,B). However, using similar techniques , another study reported that the mutation increased the rate of desensitization and slowed deactivation and slightly reduced benzodiazepine sensitivity *(23)*. The basis for these different effects of the γ_2(R43Q)-mutation are unclear.

The basis for the effect of the γ_2(R43Q)-mutation on peak current amplitude was investigated using cotransfection of individual subunit complementary DNAs (cDNAs) (free assembly) and of tethered constructs consisting of a single cDNA encoding a β_2-subunit connected through a polyglutamine linker to an α_1-subunit *(24)* (forced assembly). With the tethered construct, functional receptors are only formed when the construct

Fig. 2. *(Continued)* **(B1)** nor slow **(C1)** time constants of desensitization, or their relative contributions **(B2,C2)**. **(D)** The current rise times (10–90%) were not significantly altered by the mutations. **(E)** $\alpha_1\beta_3\gamma_2$L(R43Q), but not $\alpha_1\beta_3\gamma_2$L(K289M)-receptors had significantly smaller peak current amplitudes (asterisk, $p < 0.01$). **(F)** The current deactivation rate was significantly faster for $\alpha_1\beta_3\gamma_2$L(K289M), but not $\alpha_1\beta_3\gamma_2$L(R43Q)-receptors (asterisk, $p < 0.001$). (Figure 2 from ref. *18* with permission.)

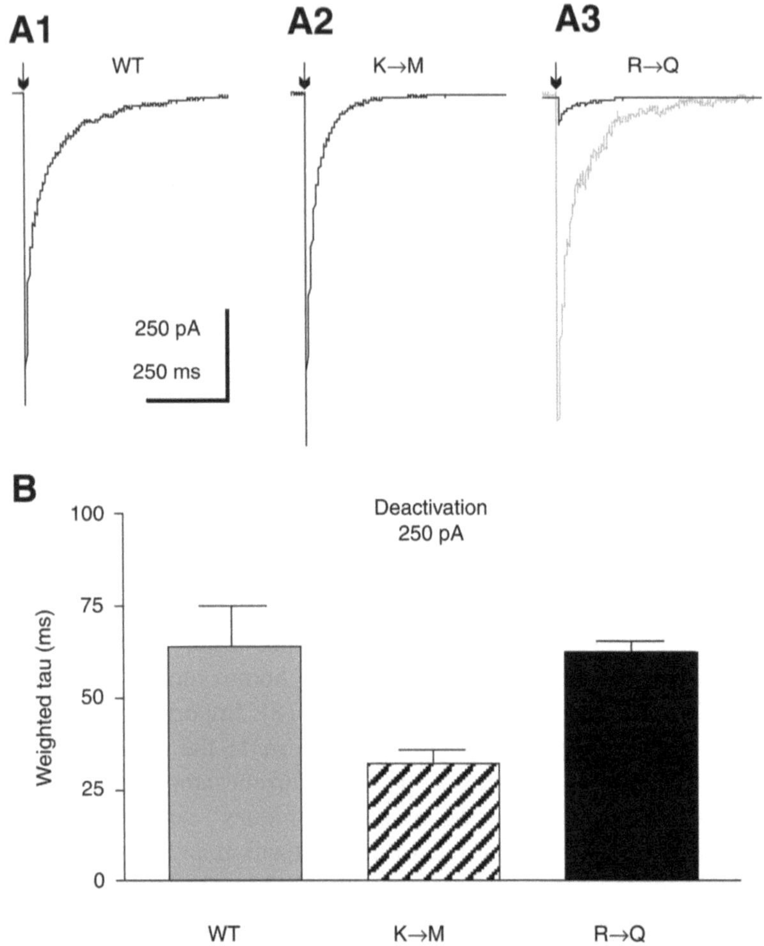

Fig. 3. Brief wild-type-$\alpha_1\beta_3\gamma_2$L and mutant-$\alpha_1\beta_3\gamma_2$L(K289M) and mutant $\alpha_1\beta_3\gamma_2$L(R43Q)-$GABA_A$ receptor currents **(A)** Brief (<5 ms) pulses of GABA (1 mM) were applied to excised macropatches containing $\alpha_1\beta_3\gamma_2$L (A1)-, $\alpha_1\beta_3\gamma_2$L(K289M) (A2)-, or $\alpha_1\beta_3\gamma_2$L(R43Q) (A3)-receptors. Scale bars are applicable to all three black current traces. The black trace in A3 was expanded 10-fold (gray trace) for comparison of deactivation current time-courses. **(B)** Weighted deactivation time constants are for wt and mutated channel currents and were significantly faster for $\alpha_1\beta_3\gamma_2$L(K289M)-receptor currents (hatched bar; asterisk, $p < 0.05$) than wt currents. The $\alpha_1\beta_3\gamma_2$L (R43Q)-receptor deactivation rate (solid bar) was not different than that of wt receptors deactivation rate (gray bar). (Figure 3 from ref. *18* with permission.)

is cotransfected with a β-, γ-, or δ-subunit *(1,2)*. With free or forced assembly, heterozygous and homozygous $\alpha_1\beta_2\gamma_2$S(R43Q) receptors had similar reductions in current amplitude (Fig. 4A). Heterozygous

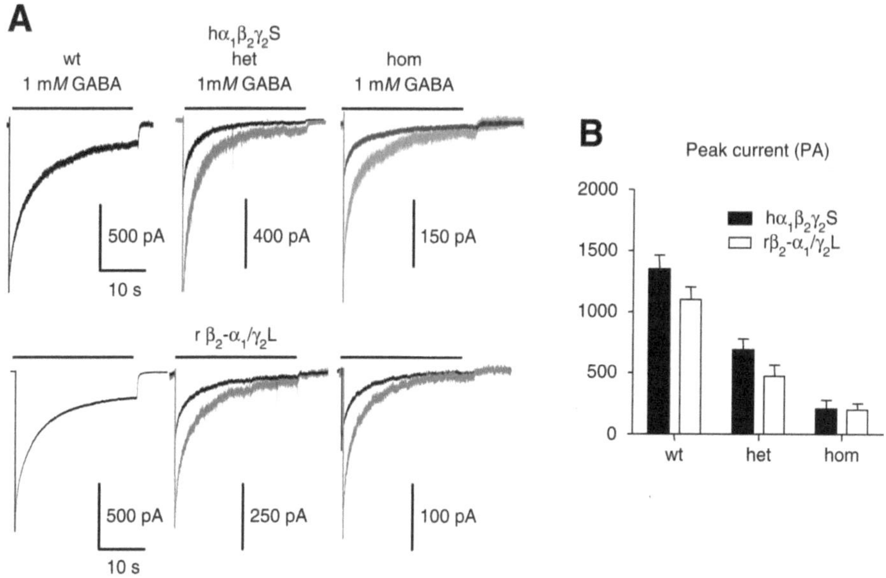

Fig. 4. Heterozygous and homozygous $\alpha_1\beta_2\gamma_2$S(R43Q)-receptors had reduced current amplitudes. **(A)** Representative traces of human $\alpha_1\beta_2\gamma_2$S (free assembly) or rat β_2–$\alpha_1\gamma_2$L (forced assembly) (wt), heterozygous (het) and homozygous (hom) receptor currents evoked by 28-s GABA (1 mM) applications were obtained from transfected HEK 293T cells voltage-clamped at –50 mV. The time-scale for the first trace applies to all traces. Currents are shown to scale (dark traces) and normalize to wt currents (gray traces). **(B)** Peak amplitudes of het and hom $\alpha_1\beta_2\gamma_2$S(R43Q)- and β_2–$\alpha_1\gamma_2$L(R43Q)-receptor currents were significantly reduced with both free and forced assembly. (Modified from ref. *24*.)

$\alpha_1\beta_2\gamma_2$S(R43Q)- and β_2-$\alpha_1\gamma_2$L(R43Q)-receptor peak current amplitudes were significantly smaller than those of WT but were significantly larger than those of homozygous peak current amplitudes (Fig. 4B).

The reduction of GABA peak currents of γ_2(R43Q)-subunit-containing receptors results from reduced surface expression of receptor protein (Fig. 5). Fusion proteins with enhanced yellow fluorescent protein (YFP) inserted between amino acids four and five of the mature γ_2-subunit (γ_2-YFP) were constructed. HEK 293T cells were transfected with WT, heterozygous, and homozygous $\alpha_1\beta_2\gamma_2$S- or $\alpha_1\beta_2\gamma_2$S-YFP receptors, and the amounts of total and surface membrane bound α_1 and γ_2S/γ_2-YFP subunit proteins were determined. With both heterozygous and homozygous receptor expression, the mutation reduced surface expression of the α_1-, γ_2S-, and γ_2-YFP subunit proteins but did not alter the amount of total cellular expression of any of the subunits (Fig. 5).

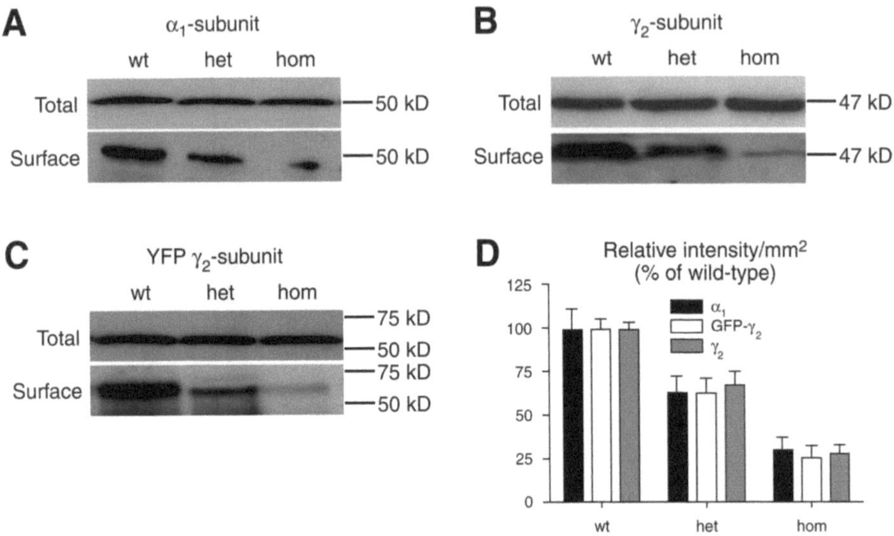

Fig. 5. Heterozygous and homozygous h$\alpha_1\beta_2\gamma_2$S(R43Q)-receptors had reduced surface expression. **(A–C)** Cells were transfected with wild-type (wt), heterozygous (het), or homozygous (hom) $\alpha_1\beta_2\gamma_2$S(R43Q)-receptors **(A,C)** or $\alpha_1\beta_2\gamma_2$S(R43Q)-YFP receptors **(B)** and were biotinylated and immunoblotted with antibodies against the α_1-subunit **(A)**, GFP **(B)**, or the γ_2-subunit **(C)**. In **A–C**, het and hom $\alpha_1\beta_2\gamma_2$S(R43Q)-receptors revealed similar whole-cell protein expression (total) but reduced subunit protein expression on the cell surface (surface). **(D)** The optical absorbance of the Western blots was quantified. Expression of het or hom $\alpha_1\beta_2\gamma_2$S(R43Q)-receptors resulted in similar levels of total whole-cell protein expression (not shown), but reduced cell surface protein expression (surface) compared with wt receptors. In each group, het protein intensities were lower than wt intensities but higher than hom intensities. (Modified from ref. *24.*)

The cellular distribution of WT and mutant $\alpha_1\beta_2\gamma_2$S(R42Q)-YFP receptors was determined using confocal microscopy (Fig. 6A). The endoplasmic reticulum (ER) was identified using the ER-marker, pECFP-ER, and the membrane was identified using the dye FM4-64 to mark the plasma membrane. In live COS-7 cells, homozygous and heterozygous $\alpha_1\beta_2\gamma_2$S(R43Q)-YFP cell surface fluorescence was reduced relative to WT receptor fluorescence (Fig. 6B), but intracellular fluorescence colocalizing with the ER increased. The basis for the ER retention is likely impaired receptor assembly. The R43 residue is part of a sequence, YxxxxRP, that is highly conserved in the cys-loop family of ligand-gated ion channels. The conserved R43 residue was shown to be critical for forming intersubunit contacts between the plus face of the γ_2-subunit and the

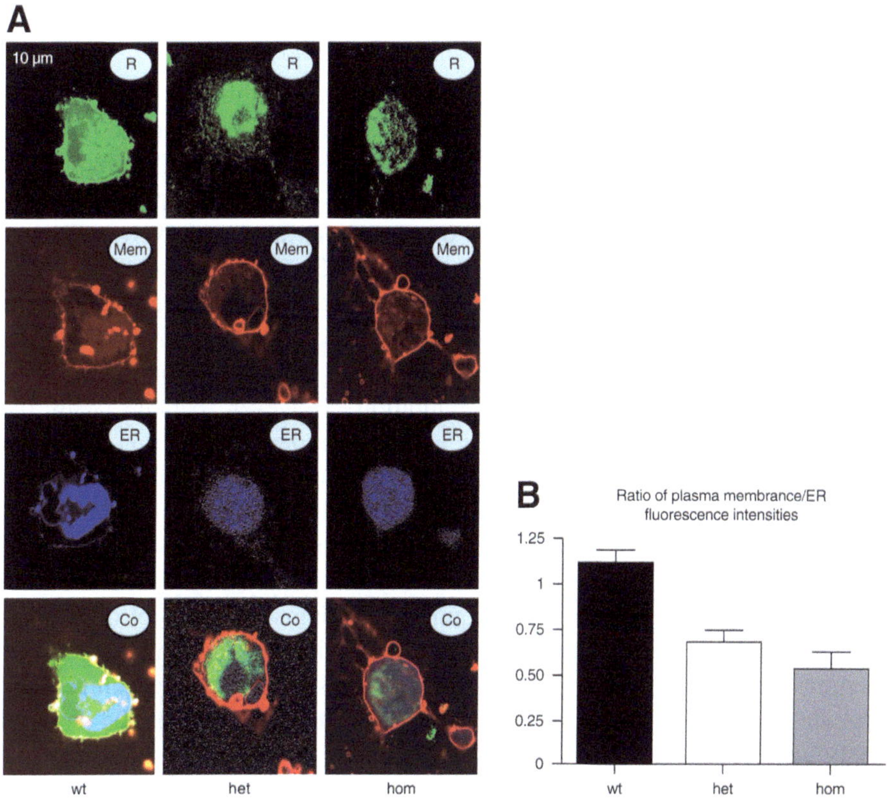

Fig. 6. Heterozygous and homozygous $\alpha_1\beta_2\gamma_2$S(R43Q)-receptors were retained in the ER. **(A)** COS-7 cells were cotransfected with human $\alpha_1\beta_2\gamma_2$S-YFP wild-type (wt) or $\alpha_1\beta_2\gamma_2$S(R43Q)-YFP heterozygous (het) or homozygous (hom) receptors with enhanced CFP-tagged ER marker and labeled with FM4-64 membrane marker. R stands for receptor, Mem for membrane, ER for endoplasmic reticulum and Co for colocalized image. Wt $\alpha_1\beta_2\gamma_2$S-YFP receptors were primarily localized to the cell membrane. Het $\alpha_1\beta_2\gamma_2$S(R43Q)-YFP receptors were found in both the membrane and in intracellular compartments. Hom $\alpha_1\beta_2\gamma_2$S(R43Q)-YFP receptors were primarily found in intracellular compartments with minimal cell surface localization. Both het- and hom receptors had fluorescence patterns that were similar to that of the ER fluorescence pattern. **(B)** The relative membrane/ER fluorescence intensity ratios for het and hom receptors were significantly reduced compared with that for wt receptors. (Modified from ref. *24*.)

minus face of the β_2-subunit *(25)*, thus preventing incorporation of the γ_2-subunit into the receptor pentamer. Thus, the reduced surface expression of mutant $\alpha_1\beta_2\gamma_2$S(R43Q)-receptors was a consequence of receptor ER retention owing to impaired subunit assembly.

2.2. γ_2-Subunit Truncation Mutations

2.2.1. GABA$_A$ Receptor γ_2(Q351X)-Subunit Mutation

A γ_2-subunit mutation, Q351X, localized in the intracellular loop between M3 and M4, which introduced a premature translation-termination codon (PTC) at residue Q351, was identified in a family with GEFS$^+$ (Fig. 1). With homozygous expression in oocytes, sensitivity to GABA was absent, suggesting that the GABA$_A$ receptors were not expressed on the cell surface, or if expressed, were nonfunctional. Using a green fluorescent protein (GFP)-tagged γ_2-subunit *(8)*, it was shown that although possibly assembled, the receptor did not exhibit surface expression and was retained in the ER. Thus, the γ_2(Q351X)-mutation would be expected to reduce surface expression of functional GABA$_A$ receptor complexes, leading to decreased GABAergic inhibition. However, patients with the γ_2(Q351X)-mutation are heterozygous, and not homozygous, for this mutation.

Cellular messenger RNA (mRNA) quality control mechanisms may affect γ_2(Q351X)-subunit expression. Mutations producing PTCs can result in C-terminally truncated proteins that can produce dominant negative inhibition of full length proteins, thus potentially harming the cells. Quality control checkpoints during gene expression are required to maintain a low level of aberrant gene products and prevent them from interfering with the functioning of the cell. Nonsense-mediated decay (NMD), or mRNA surveillance, recognizes and degrades aberrant mRNAs in which the open reading frame is truncated owing to a PTC *(26,27)*. The position of the PTC relative to the position of a down stream intron is the primary determinant of whether the PTC elicits NMD. PTCs not followed by an intron that is located more than 50–55 nucleotides downstream generally elicit NMD. Thus, it would be predicted that NMD would not affect expression of γ_2(Q351X)-subunits, as the PTC is in the terminal exon of the *GABRG2* gene. Therefore, the γ_2(Q351X)-mutation likely produces a truncated γ_2-subunit protein in patients. However, the cellular fate of the truncated protein is unknown. The net functional cell surface expression of receptors with heterozygous expression of γ_2(Q351X)-subunits might not be predicted from simple summation of the results obtained with homozygous and WT expression. It is possible that the truncated γ_2-subunit is simply subject to proteasomal degradation owing to ER quality control. Alternatively, the truncated protein may be able to assemble with the α- and β-subunits, thus causing a dominant negative action to retain α- and β-subunits in the ER and impair assembly of WT receptors.

2.2.2. GABA$_A$ Receptor γ_2(Q1X)-Subunit Truncation Mutation

A γ_2-subunit mutation that introduced a PTC, Q1X, between the signal peptide and mature peptide was identified in a family with severe myoclonic epilepsy of infancy (9). Whereas the functional consequence of the mutation is unknown, NMD is likely to be involved and affect the expression of this mutant subunit. The mutation should prevent production of full length protein and should also trigger NMD, thus preventing production of the signal peptide. Thus, the γ_2(Q1X)-mutation would be predicted to produce haploinsufficiency, which occurs when an individual who is heterozygous for a certain gene mutation or hemizygous at a particular locus, often as a result of deletion of the corresponding allele, is clinically affected because a single copy of the normal gene is incapable of providing sufficient protein for normal function.

However, assuming that mRNA transcription for each gene is the same owing to the $2\alpha_2\beta_1\gamma_2$S-pentameric stoichiometry of $GABA_A$ receptors, there should be an extra copy of the γ_2S-subunit. If the γ_2(Q1X)-mutation provokes NMD and if the mutant γ_2S-subunit does not interfere with transcription of the WT γ_2S-subunit at the mRNA level, there should be sufficient WT γ_2S-subunit to form a normal complement of functional WT $\alpha_1\beta_2\gamma_2$S-receptors and no disinhibition should be produced. To determine the effect of this mutation on expression of functional surface receptor, conventional study of WT and mutant receptor expression with intronless cDNA constructs would not trigger activation of NMD, and thus this method of study would be nonphysiological. Future studies in which heterologous cells are transfected with cDNA constructs that contain the γ_2S(Q1X)-subunit cDNA, interrupted with the cDNA encoding the appropriate intron (a minigene construct) or with knockin animals are required to elucidate the underlying molecular pathology.

2.2.3. $GABA_A$ Receptor γ_2(IVS6 + 2T→G)-Subunit Truncation Mutation

A splice-site mutation in the sixth intron of the *GABRG2* gene, IVS6 + 2T→G, has been identified in a family with CAE and febrile seizures *(10)* (Fig. 1). The effect of this mutation on γ_2-subunit expression and $GABA_A$ receptor function is unknown but was predicted to lead to a nonfunctional, truncated, γ_2-subunit protein through exon skipping. Electrophysiological and expression experiments to study the effects of this mutation have not been reported. Because of the site of the putative truncation (just upstream of M1), it is doubtful that if expressed, the truncated subunit would be incorporated into a functional $GABA_A$ receptor. In addition, the predicted exon skipping would result in a new PTC at the fifth- and seventh exon junction site, and thus it is likely that the PTC would also trigger NMD, thus eliminating the expression of mutant protein at the mRNA level. Therefore, the underlying mechanism for this splice donor site mutation may also be owing to haplo-insufficiency.

2.3. α_1-Subunit Mutations

2.3.1. $GABA_A$ Receptor α_1(A322D)-Subunit Mutation

A mutation in the third transmembrane domain (M3) of the α_1-subunit, α_1(A322D), is associated with an autosomal dominant JME *(11)*. Because this mutation adds a negatively-charged aspartate residue into the M3 transmembrane helix, it might be predicted to destabilize transmembrane helix formation and/or alter subunit folding. When coexpressed with WT β_2- and γ_2-subunits, mutant α_1(A322D)-subunits reduced both surface and total α_1-subunit expression *(28)*. Consistent with its reduction in α_1-subunit protein expression, the α_1(A322D)-subunit mutation reduced peak GABA-evoked currents by 94% in α_1(A322D)$\beta_2\gamma_2$-receptors *(11,28,29)* and by 50% in "heterozygous" $\alpha_1\alpha_1$(A322D)$\beta_2\gamma_2$-receptors *(28)*. Thus, the α_1(A322D)-mutation altered α_1-subunit trafficking differently than the γ_2(R43Q)-mutation altered γ_2-subunit trafficking; γ_2(R43Q) reduced surface but not total γ_2-subunit expression.

Three lines of evidence suggest that the α_1(A322D)-mutation reduced α_1-subunit protein expression through ER-associated degradation (ERAD) *(30)*. First, the α_1(A322D)-mutation reduced total α_1-subunit protein expression by approximately similar amounts either when expressed alone, or when coexpressed with β_2- and γ_2-subunits (Fig. 7A,B); thus α_1(A322D)-reduced α_1-subunit expression before subunit oligomerization. Second, deglycosylation assays demonstrated sequestration and trafficking arrest of mutant α_1(A322D)$\beta_2\gamma_2$-receptors in the ER, WT α_1-subunits contained mature, low-mannose N-linked glycosylation, but α_1(A322D)$\beta_2\gamma_2$-receptors contained only immature, high-mannose ER-associated glycosylation (30) (Fig. 7C). Third, confocal microscopy studies of YFP-tagged α_1- and α_1(A322D)-subunits demonstrated that WT α_1-YFP$\beta_2\gamma_2$-receptors localized to the cell surface but that mutant α_1(A322D)-YFP$\beta_2\gamma_2$-receptors were localized to the ER (Fig. 8) *(30,31)*. Therefore, α_1(A322D)-subunit expression was reduced before subunit oligomerization, but after biosynthesis, residual α_1(A322D)-subunit protein was ER-associated, implying that a substantial fraction of α_1(A322D)-subunits was eliminated by ERAD, similar to another misfolded, disease-associated protein, the cystic fibrosis transporter regulator *(32)*.

It was of interest to determine the cellular fate of α_1(A322D)-subunit-containing $GABA_A$ receptors that were not eliminated by ERAD. Most $GABA_A$ receptors contain one α-subunit positioned between two β-subunits ($\alpha_{\beta\alpha\beta}$) and one α-subunit is positioned between a β- and a γ-subunit (Fig. 9A) *(1,2)* Thus, the autosomal dominant, α_1(A322D)-mutation could theoretically produce four different $GABA_A$ receptor pentameric assemblies: $\alpha_1\beta_2\alpha_1\beta_2\gamma_2$ (WT), $\alpha_1\beta_2\alpha_1$(A322D)$\beta_2\gamma_2$ (Het$_{\beta\alpha\beta}$), α_1(A322D)$\beta_2\alpha_1\beta_2\gamma_2$ (Het$_{\beta\alpha\gamma}$), and α_1(A322D)$\beta_2\alpha_1$(A322D)$\beta_2\gamma_2$ (homozygous). Het$_{\beta\alpha\beta}$- or Het$_{\beta\alpha\gamma}$-receptors were selectively formed by using tethered concatamers consisting of a single cDNA that encoded either the β_2-subunit linked to the α_1-subunit or a cDNA that encoded a γ_2-subunit linked to the β_2-subunit, which was linked to the α_1-subunit. Mutating the β_2–α_1 concatamer (β_2–α_1[A322D]) and coexpressing it with the WT γ_2–β_2–α_1-concatamer, produced Het$_{\beta\alpha\gamma}$-receptors, and mutating the γ_2–β_2–α_1-concatamer

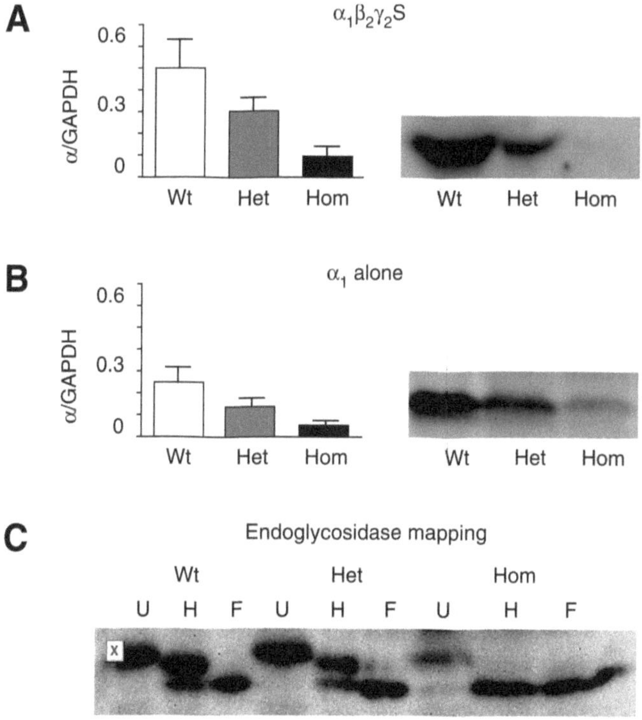

Fig. 7. The JME α_1(A322D)-mutation reduced α_1-subunit expression before receptor oligomerization and prevented Golgi-associated N-linked glycosylation. **(A,B)** Cells were transfected with α_1-subunits (2 μg) and either 2 μg of both β_2- and γ_2S-subunits **(A)** or 4 μg of empty pcDNA3.1 vector. **(B)** Whole-cell lysates were fractionated on 10% SDS-PAGE gels and Western blots were probed with antibodies directed to the α_1-subunit and glyceraldehydes-3-phosphate dehydrogenase (GAPDH), which controlled the amount of protein loaded on the gel. The fraction of the α1-subunit band relative to GAPDH was quantified and depicted to the left of the gel. **(C)** Cells were transfected with wild-type (wt, $\alpha_1\beta_2\gamma_2$), heterozygous (het, $\alpha_1\alpha_1$[A322D]$\beta_2\gamma_2$), or homozygous (hom, α_1[A322D]$\beta_2\gamma_2$)-receptors. Cellular lysates (2.5 mg/mL) were left undigested (U), or digested with endoglycosidase-H (endo-H, H), or peptide-*N*-glycosidase F (PNGaseF, F). The digestion products were fractionated through 12.5% SDS-PAGE, which was stained by Western blot with an antibody directed against the α_1-subunit. Because wt-, het-, and hom receptors differ in α_1-subunit expression **(A)**, the lysates were loaded on the gel in the ratios 8 mg:15 mg:50 mg:wt:het:hom to balance the amount of the α_1-subunit on the gel. The α_1-subunits from undigested lysates from wt-, het-, and hom receptors migrated at 50 kDa, and those digested with PNGaseF migrated at 46 kDa. Following endo-H digestion, wt- and het-subunits migrated in two bands at 48.4 (endo-H resistant) and 46 kDa (endo-H sensitive), but hom α_1-subunits migrated in a single endo-H sensitive band at 46 kDa. (Modified from ref. *30.*)

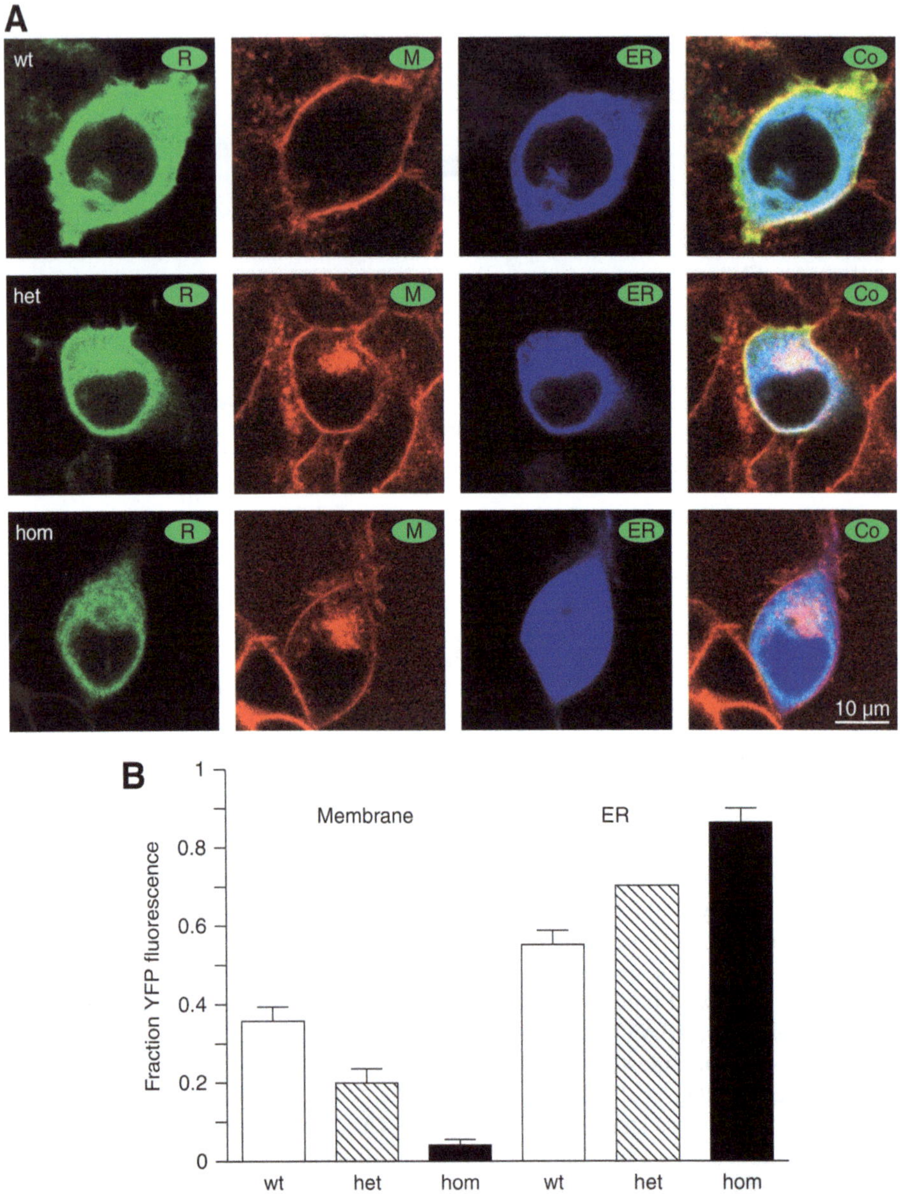

Fig. 8. The JME α_1(A322D)-mutation colocalized to the ER Cells were transfected with the ER marker, CFP-ER (ER, blue) and either wild-type (wt, α_1-YFP$\beta_2\gamma_2$S), heterozygous (het, α_1-YFPα_1[A322D]–YFP$\beta_2\gamma_2$S), or homozygous (hom, α_1[A322D]-YFP$\beta_2\gamma_2$S) receptors (R, green) and stained with the membrane marker FM 4-64 (M, red). The majority of all receptors colocalized (Co) with the ER (cyan). Only wt and het α_1-YFP-subunit colocalized with the plasma membrane marker (yellow). The fraction of total wt, het, and hom α_1-YFP that colocalized to the ER or membrane is graphed. (Modified from ref. *30*.)

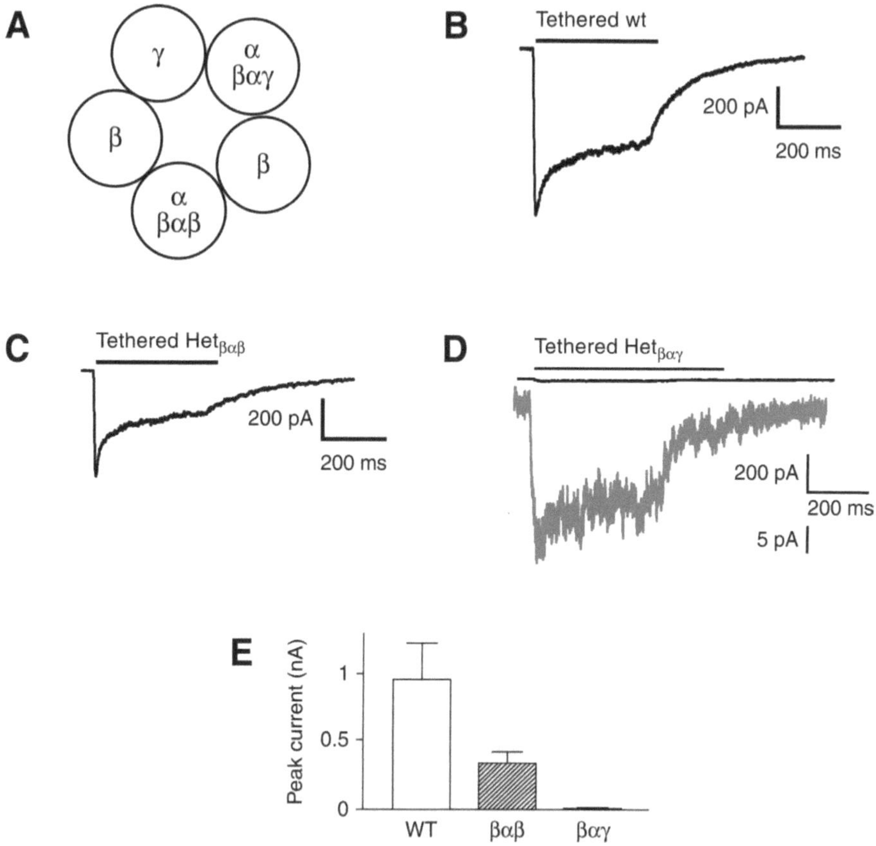

Fig. 9. The JME α_1(A322D)-mutation causes position-dependent reductions in whole-cell GABA-evoked currents. **(A)** αβγ-receptor subunits are arranged (as viewed from the synapse) in the counterclockwise sequence γ-β-α-β-α. **(B–D).** Sample current traces are depicted from **(B)** wild-type (WT), **(C)** Het$_{\beta\alpha\beta}$, or **(D)** Het$_{\beta\alpha\gamma}$ receptors. The gray trace in **(D)** is shown in an expanded scale to display the time-course of the current. **(E)** Mean peak currents of wt concatamer receptors ($n = 20$) were larger than those of Het$_{\beta\alpha\beta}$ ($n = 17$, $p < 0.05$), which were substantially larger than those from Het$_{\beta\alpha\gamma}$ concatameric receptors ($n = 8$, $p < 0.01$). (Modified from ref. *28.*)

(γ_2–β_2–α_1[A322D]), produced Het$_{\beta\alpha\beta}$-receptors. Het$_{\beta\alpha\beta}$-receptor mean peak current amplitude was 65% less than WT, but Het$_{\beta\alpha\gamma}$-receptors had essentially no GABA-evoked currents (Fig. 9B–E) *(28)*. Furthermore, α_1(A322D) asymmetrically reduced β_2–α_1-protein expression; Het$_{\beta\alpha\gamma}$-receptors had 72% of the β_2–α_1-expression as Het$_{\beta\alpha\beta}$-receptors (not shown). These results indicate that not only is most of the α_1(A322D)-mutation eliminated by ERAD before subunit oligomerization, but the α_1(A322D)-subunits that are not degraded are formed into Het$_{\beta\alpha\beta}$- but not Het$_{\beta\alpha\gamma}$-receptors.

2.4. δ-Subunit Susceptibility Variants

Although the monogenic mutations provide useful models for the pathogenesis of epilepsy syndromes, they only account for a small portion of complex epilepsies, suggesting that most of the complex epilepsies are polygenic and require additive actions of a set of susceptibility genes. To date, two susceptibility genes for complex epilepsies have been reported *(33)*. The gene encoding the T-type calcium channel *CACNA1H* is reported to be a susceptibility gene for CAE and IGE *(34,35)*. Several T-type calcium channel mutations resulted in enhanced calcium channel current in a heterologous expression system, which might cause hyperexcitability *(36–38)*. Some of the single nucleotide polymorphisms in *CACNA1H* gene, present in both epileptic patients and general population, exhibited altered channel function, which might contribute to epileptogenesis *(35)*. The first $GABA_A$ receptor susceptibility gene, *GABRD* (the gene encoding the δ-subunit of $GABA_A$ receptors), was reported for complex epilepsies *(12)*. The δ(E177A)-variant is heterozygously associated with $GEFS^+$ patients. A polymorphic allele, δ(R220H), is both heterozygously and homozygously associated with JME patients, but is also found in the general population. Both these variants led to a reduction in hα1β2δ-receptor currents expressed in HEK 293T cells *(12)*.

The mechanistic basis for the current reduction was examined using h$\alpha_4\beta_2\delta$-receptors, which are more physiologically relevant and linked to epileptogenesis *(39)*. The mean $\alpha_4\beta_2\delta$(E177A)- and $\alpha_4\beta_2\delta$(R220H)-receptor GABA EC_{50}s were not significantly different from that of WT receptors. Saturating (1 m*M*) GABA-evoked currents were obtained from WT $\alpha_4\beta_2\delta$, a mixed population of WT and variant, and pure variant receptors. Consistent with the previous report for either variant *(12)*, mean whole-cell current amplitude for mixed and pure variant $\alpha_4\beta_2\delta$(E177A)- or $\alpha_4\beta_2\delta$(R220H)-receptors was significantly smaller than for WT $\alpha_4\beta_2\delta$-receptors (Figs. 10A,B and 11A,B). Macroscopic desensitization and deactivation were also altered for δ(E177A), but not δ(R220H)-variant receptors (Figs. 10A,C,D and 11A,C,D) *(39)*.

Surface expression of WT and variant receptors was investigated using Western blot. Compared with surface expression of WT $\alpha_4\beta_2\delta$-receptors, surface receptor protein was slightly but significantly reduced for pure variant $\alpha_4\beta_2\delta$(E177A)-receptors (Fig. 12A,C), and surface receptor protein was substantially reduced for pure variant $\alpha_4\beta_2\delta$(R220H)-receptors (Fig. 12B,D). Surface receptor protein for mixed $\alpha_4\beta_2\delta$(E177A)- or $\alpha_4\beta_2\delta$(R220H)-receptors was not significantly altered compared with WT receptors (Fig. 12). Total receptor protein (surface + intracellular compartments) was not altered for both mixed and pure variant $\alpha_4\beta_2\delta$(E177A)- or $\alpha_4\beta_2\delta$(R220H)-receptors compared with WT receptors (Fig. 12) *(36)*.

To determine the basis for the reduction of variant current, single channel recordings from WT and variant receptors were obtained (Fig. 13). The single

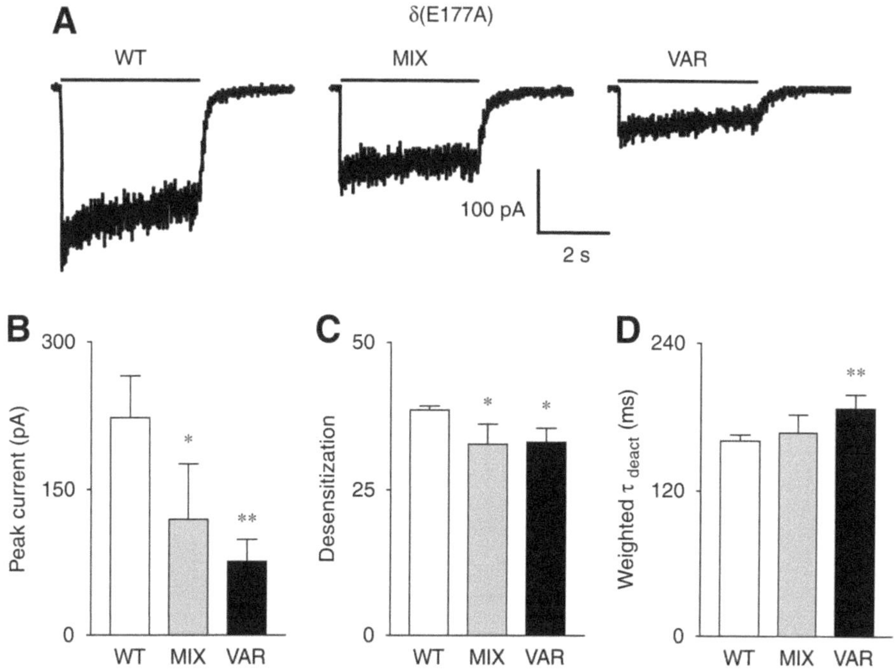

Fig. 10. Peak current amplitudes were reduced and their kinetic properties were altered for $\alpha_4\beta_2\delta$-receptors containing the δ(E177A)-variant. **(A)** Examples of whole-cell current traces evoked by saturating GABA (1 m*M*) from wild-type α4β2δ (wt), mixed wt, and variant (mix), and pure variant $\alpha_4\beta_2\delta$(E177A) (var)-receptors. **(B)** Compared with wt receptors, mean current amplitude was smaller for mix- and var receptors. **(C)** Mean macroscopic desensitization was decreased for mix- and var receptors compared with wt receptors. **(D)** Comparison of the mean time constant of current deactivation among wt-, mix-, and var receptors. The solid bar above each current trace denotes the duration of GABA application (4 s). The blank bars represent the mean current features for wt receptors, the hatched bars represent those for the mix receptors, and the black bars represent those for the var-receptors. The error bars denote the SEM. (Modified from ref. *39*.) *Significantly different from WT at $p < 0.05$, **$p < 0.01$.

channel conductance of pure variant receptors was not altered compared with WT receptors. However, the distributions of the pure variant $\alpha_4\beta_2\delta$(E177A)- and $\alpha_4\beta_2\delta$(R220H)-receptors were shifted toward shorter durations than the distribution of WT $\alpha_4\beta_2\delta$-currents, which resulted in a significant reduction of mean open time for the pure variant receptors *(39)*.

The finding that $\alpha_4\beta_2\delta$-receptors containing either δ(E177A)- or δ(R220H)-variant had reduced GABA$_A$ receptor currents suggests that these susceptibility variants contribute to complex epilepsies by disinhibition. The mechanistic study suggests that the epilepsy in individuals heterozygous for a variant may

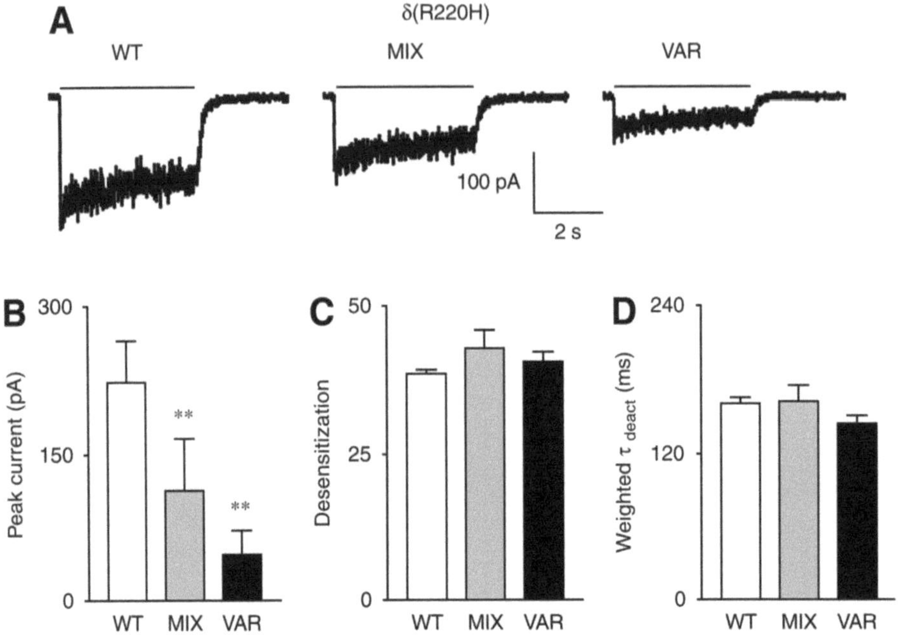

Fig. 11. Peak current amplitudes were reduced but kinetic properties were not altered for $\alpha_4\beta_2\delta$-receptors containing the δ(R220H)-variant. **(A)** Examples of whole-cell current traces evoked by saturating GABA (1 m*M*) from wild-type $\alpha_4\beta_2\delta$ (wt), mixed wt and variant (mix), and pure variant $\alpha_4\beta_2\delta$(R220H)-receptors (var). **(B)** Compared with wt receptors, mean current amplitude was smaller for the mix receptors and for var receptors. **(C)** Mean macroscopic desensitization was not different among wt-, mix-, and var receptors. **(D)** Mean current deactivation time constant for mix- and var receptors was not altered compared with wt receptors. The blank bars represent the mean current features for wt receptors, the hatched bars represent those for mix receptors, and the black bars represent those for var receptors. The error bars denote the SEM. (Modified from ref. *39*.) **Significantly different from WT at $p < 0.01$.

be predominantly owing to a receptor channel gating defect and in patients homozygous for a variant may be owing to both gating and trafficking defects.

3. Febrile Seizures and $GABA_A$ Receptor Epilepsy Gene Mutations

Febrile seizures are the most common convulsions for children under age 6. Although most are self-limited; complex, prolonged febrile seizures have been proposed to lead to hippocampal mesiotemporal sclerosis and complex partial epilepsy, and thus understanding the underlying mechanisms of febrile seizures is of great clinical importance. However, whereas the reports reviewed earlier provide some understanding of the basis for afebrile seizures, the specific cause(s) of febrile seizures and reason(s) why fever evokes convulsions are poorly understood.

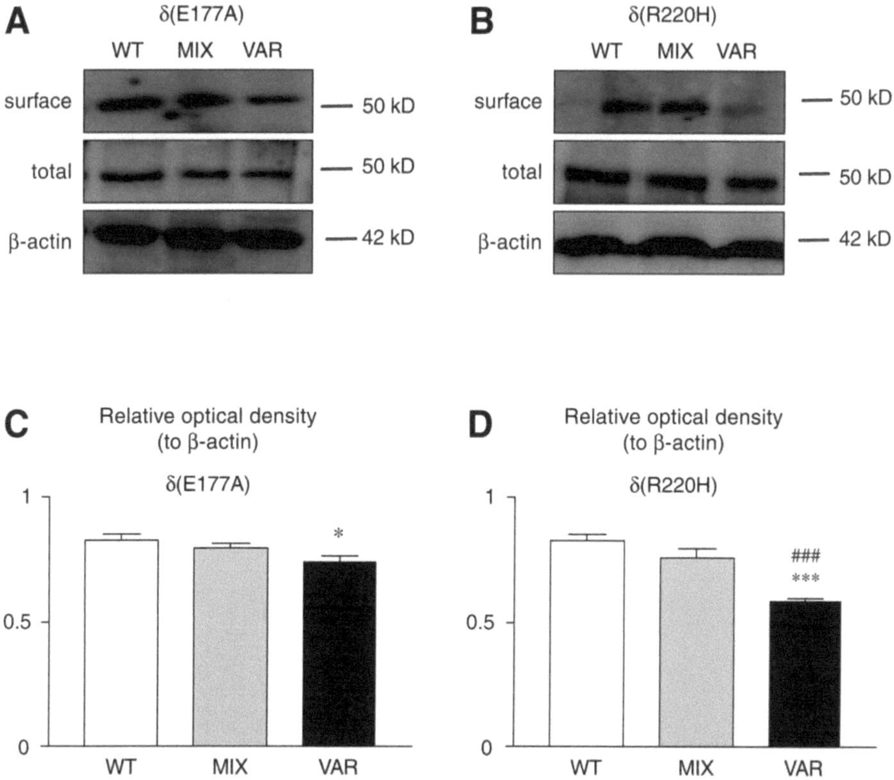

Fig. 12. Surface receptor protein was reduced for $\alpha_4\beta_2\delta$-receptors containing variant δ-subunits. **(A)** Examples of Western blots for wild-type $\alpha_4\beta_2\delta$ (wt), mixed (mix), and variant $\alpha_4\beta_2\delta$(E177A)-receptors. **(B)** Examples of Western blots for wt $\alpha_4\beta_2\delta$-, mix-, and var $\alpha_4\beta_2\delta$(R220H)-receptors. **(C)** Wt and δ(E177A)-var receptor surface receptor proteins were quantified by normalizing to β-actin and expressed as relative optical density. Relative optical density of var receptors was reduced compared with wt receptors. Relative optical density of mix receptors was not different from that of wt- and var receptors. **(D)** Relative optical density of var $\alpha_4\beta_2\delta$(R220H)-receptors was substantially reduced compared with wt receptors. Relative optical density of mix receptors was not different from that of wt receptors but was significantly different from var receptors. A V5 epitope was tagged at the N-terminus of wt or var δ-subunits. Total receptor proteins were analyzed by Western blot using antibody against the V5 epitope. Surface receptor proteins were captured by biotinylation and analyzed by Western blot. A protein band around 50 kDa was detected for the wt $\alpha_4\beta_2\delta$-receptor or the $\alpha_4\beta_2\delta$-receptor containing either of the var subunits. The error bars denote the SEM. In both **A** and **B**, β-actin was applied as internal loading control. (Modified from ref. *39*.) *Significantly different from WT at $p < 0.05$, *** $p < 0.001$, $^{\#\#\#}$, Significantly different from MIX at $p < 0.001$.

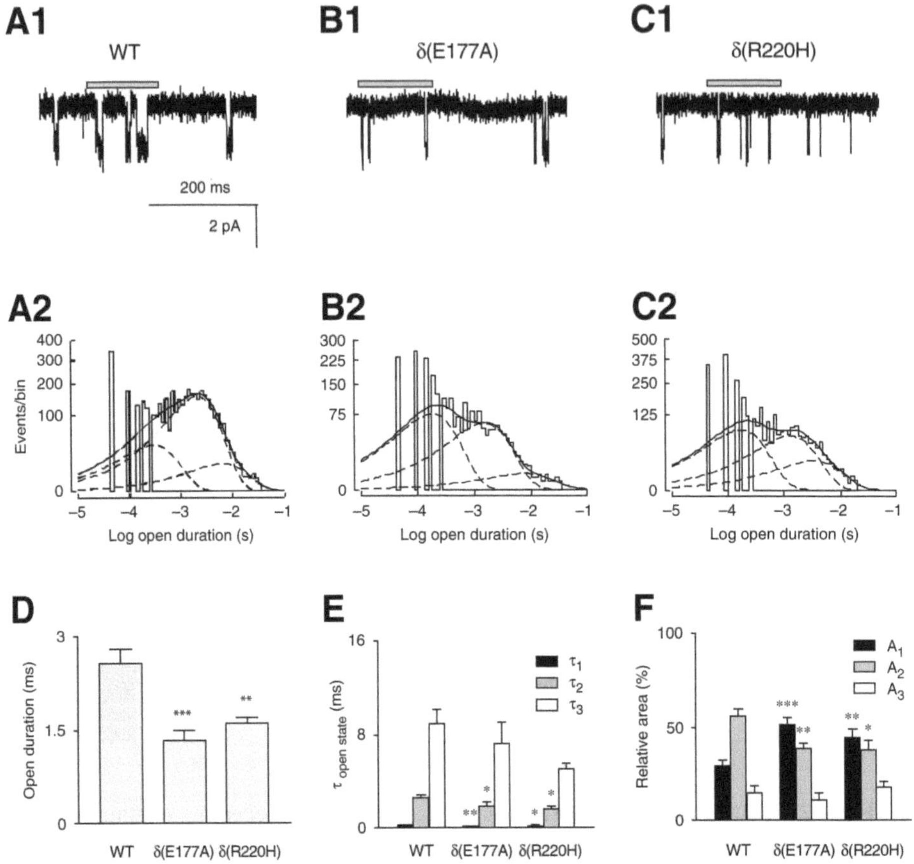

Fig. 13. Mean open duration of single channel currents was reduced for $\alpha_4\beta_2\delta$-receptors containing pure variant δ-subunits. **(A1–C1)** Examples of single channel currents evoked by 1 m*M* GABA from wild-type $\alpha_4\beta_2\delta$ (wt), and variant (var) $\alpha_4\beta_2\delta$(E177A)- and $\alpha_4\beta_2\delta$(R220H)-receptors. Single channel current open durations were longer for wt receptors than for var receptors. **(A2–C2)** Distributions of wt- and var receptor single channel current open states were plotted. **(D)** Single channel current mean open durations were decreased for var $\alpha_4\beta_2\delta$(E177A)- and $\alpha_4\beta_2\delta$(R220H)-receptors compared with wt $\alpha_4\beta_2\delta$-receptors. **(E)** Comparison of the mean value of exponential function time constants, τ_1, τ_2, and τ_3, among wt- and var receptors. **(F)** Comparison of the mean exponential function relative amplitudes, A_1, A_2, and A_3, among wt- and var receptors. The error bars represent the SEM. (Modified from ref. *39*.) *Significantly different from WT at $p < 0.05$, **$p < 0.01$, ***$p < 0.001$.

3.1. Febrile Seizures Have a Genetic Component

There are great variations in the incidence of febrile seizures in different populations and regions. For example, the incidence of febrile seizures in

Caucasians is 4–5%, and in China is about 1%, whereas it is increased to more than 8% in Japan and 14% in Guam *(40)*. These racial variations in the incidence of febrile seizures strongly suggests genetic involvement. For a decade, mutations in sodium channel genes have been associated with $GEFS^+$. Recently, $GABA_A$ receptor γ_2-subunit mutations *(7)* and δ-subunit variants *(41)* have also been associated with febrile seizures. However, in all these pedigrees, mutations (γ_2[R43Q], (γ_2[K289M], (γ_2[Q351X], (γ_2[Q1X], (γ_2[IVS6T-G]), and variants (δ[E177A] or δ[R220H]) are associated with both febrile seizures alone and extended phenotypes. A pedigree associated only with febrile seizures has just been identified and shown to be associated with a γ_2-subunit mutation (R139G) (Audenaert et al., in press).

3.2. A Temperature-Dependent Defect in Trafficking of $GABA_A$ Receptors Containing γ_2-Subunit Epilepsy Mutations

Interestingly, all γ_2-subunit epilepsy mutations that have been identified are associated with febrile seizures. The common association of these mutations with febrile seizures suggests that elevated temperature somehow reduces the function of $GABA_A$ receptor channels, thus causing disinhibition and febrile convulsions; however, the basis for this temperature-dependent effect on inhibition has been unclear. Temperature changes have been shown to affect plasma membrane states *(42)* and synaptic transmission *(43)*. For example, synaptic vesicle recycling has been temperature-dependent. The size of recycling vesicles is twice as large, and the speeds of both endocytosis and exocytosis are faster at physiological temperature than at room temperature *(44)*. Furthermore, it has been demonstrated that inhibitory synaptic strength can be modulated within 10 min by recruiting more functional $GABA_A$ receptors to the synaptic plasma membrane *(45)*.

Previous studies had demonstrated that the γ_2-subunit mutations associated with idiopathic generalized epilepsy (R43Q, K289M, Q351X) impaired receptor channel function either by reducing the number of receptors on the surface owing to receptor ER retention *(46)*, shortening channel mean open time *(18)*, or accumulating receptor in desensitized states *(23)*. These alterations of channel function would lead to disinhibition and thus afebrile seizures. The authors recent study with temperature variations has shed some light on why seizures are provoked by fever in these γ_2-mutation harboring mutants *(47)*.

Expression of surface receptors containing γ_2-subunit epilepsy mutations (R43Q, K289M, Q351X) were reduced compared with WT receptors and were further reduced by elevated temperature (incubation at 40°C for 1 h) (Fig. 14) *(46,47)*. HEK 293T cells transfected with human $\alpha_1\beta_2\gamma_2$S-YFP WT or mutant receptors were incubated at 37° or 40°C, and the cell membrane-bound proteins

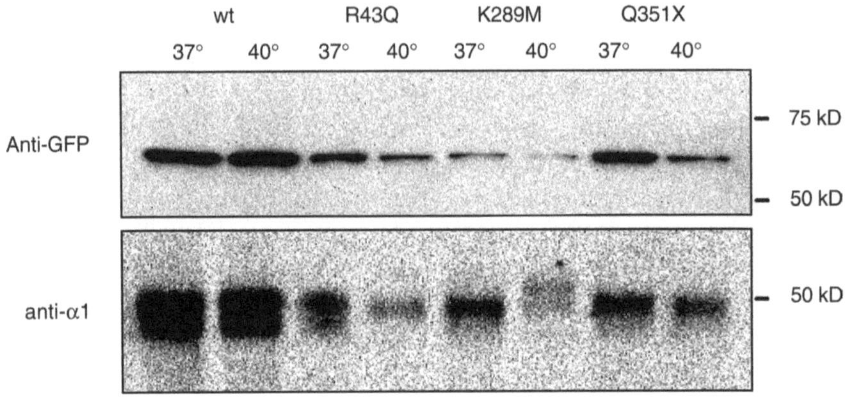

Fig. 14. Mutant γ_2S-subunit-containing $\alpha_1\beta_2\gamma_2$S-enhanced yellow fluorescent protein receptors had reduced surface expression, and temperature elevation further reduced surface protein expression of mutant γ_2S-subunit-containing receptors. Western analysis of biotinylated γ_2S- and α_1-subunit surface proteins in HEK-293T cells expressing wt- and het-mutant $\alpha_1\beta_2\gamma_2$S-EYFP receptors incubated at 37°C or 40°C for 1 h. The cell membranes were biotinylated, and equal amounts of protein were conjugated with beads and loaded and resolved by SDS-PAGE. Membranes were immunoblotted with mouse monoclonal anti-GFP and anti-α_1-antibodies and revealed through conjugation with goat, antimouse secondary antibody. (Modified from ref. *47.*)

were resolved by sodium dodecyl sulfate (SDS-PAGE) and detected with both anti-GFP and anti-α_1-antibodies. Although a previous study suggested that γ_2-subunits may traffic-k to the cell surface alone *(48)*, both YFP-coupled mutant γ_2- and α1-subunits were reduced on the cell surface and further reduced with incubation at 40°C. Interestingly, a JME mutation (α_1[A322D]) did not show further reduction of receptors on the cell surface with high temperature challenge (incubation at 40°C for 1 h). These data suggested that it was the pentameric receptors instead of individual γ_2-subunit mutant proteins that were reduced on the cell surface with elevated temperature.

As both the height of the fever and the rapidity of temperature elevation are involved in provoking convulsions, it was tested how rapidly elevated temperature induced subcellular redistribution of receptors. Consistent with the reduction of receptor surface expression demonstrated using immunoblotting, it was found that after a 30 min incubation at 40°C, heterozygous $\alpha_1\beta_2\gamma_2$S(R43Q)-, $\alpha_1\beta_2\gamma_2$S(K289M)-, and $\alpha_1\beta_2\gamma_2$S(Q351X)-receptors had reduced membrane surface and increased intracellular expression in COS-7 cells (Fig. 15). WT γ_2S-subunits tagged with cyan fluorescent protein (CFP) (blue) and YFP (green) and the plasma membrane marker FM4-64 (red) were primarily colocalized (white) on the membrane surface (Fig. 15, top row) and expressed a small

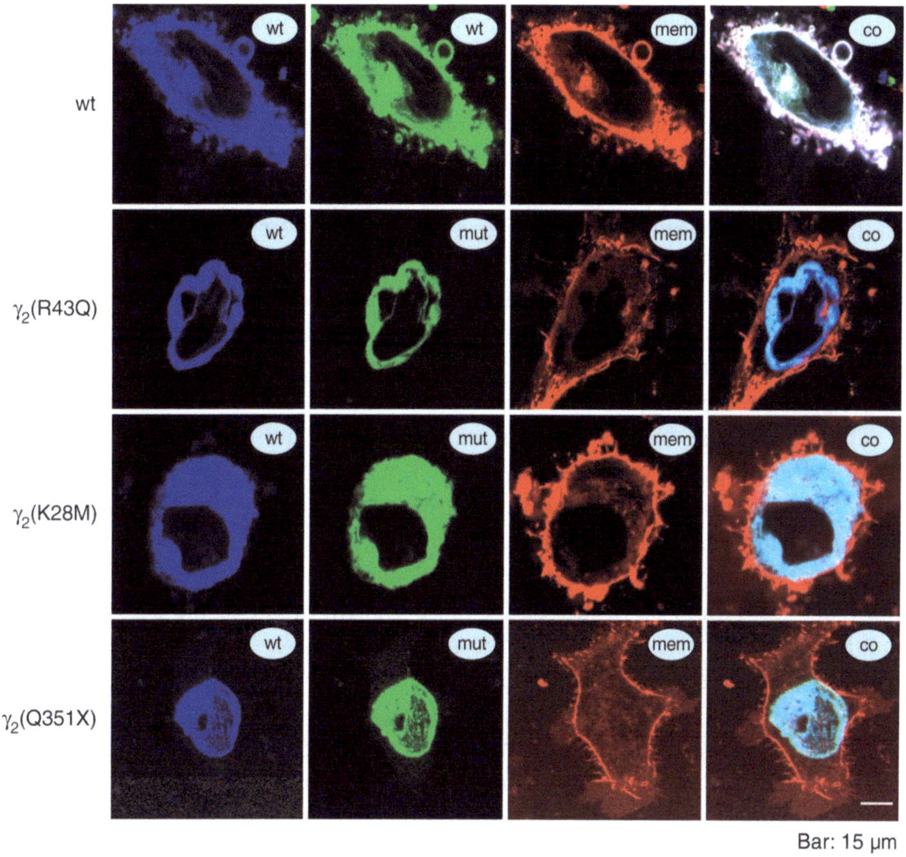

Fig. 15. Elevated temperature rapidly decreased surface receptors and increased intracellular retention of mutant γ_2S-subunit-containing receptors. Confocal microscopic images of fluorescence-tagged wt and mutant γ_2S-subunit-containing $\alpha_1\beta_2\gamma_2$S-receptors in COS-7 cells after a 30 min incubation at 40°C. With het expression, mutant and wt γ_2S(R43Q)-, γ_2S(K289M)-, and γ_2S(Q351X)-subunits were localized intracellularly in compartments that had the morphology of the ER and colocalized with ECFP-ER (not shown). Wild-type receptors containing both ECFP-tagged (blue) and EYFP-tagged (green) γ_2S-subunits were mainly on the cell surface and coregistered with FM 4-64 (merged as white), whereas het mutant receptors containing both ECFP-tagged (blue) wt and EYFP-tagged (green) mutant γ_2S-subunits were coregistered intracellularly and were separated from FM4-64 (red). In the inserts, wt stands for ECFP-tagged wt-subunits, mut for EYFP-tagged mutant subunits, mem for membrane marker FM4-64, and co for colocalization of all ECFP, EYFP, and FM4-64 channels. (Modified from ref. *47.*)

amount of intracellular receptor (aqua) after a 30 min incubation at 40°C (Fig. 15). For heterozygous expression, WT and mutant γ_2S-subunits were tagged with CFP and YFP, respectively. After a 30 min incubation at 40°C, there was reduced membrane surface expression of receptor (loss of white), and both mutant and coexpressed WT γ_2S-subunits were localized intracellularly (cyan) (Fig. 15, middle three rows).

The reduction of receptor surface expression following temperature elevation was very rapid in both heterologous cells and hippocampal neurons. After 5 min on a heated microscope stage, the receptors displayed a merged aqua color both on the surface and intracellularly (Fig. 16A, 5 min, lower panel, red box). When incubated at 40°C for 7 min, the heterozygous $\alpha_1\beta_2\gamma_2$-CYP/γ_2(K289M)-YFP receptors in COS-7 cells were localized more intracellularly. In the next 6 min at 40°C, the intensity of the surface receptors rapidly diminished, and the intensity and amount of intracellular receptors were increased (Fig. 16A, 7 and 11 min, lower panel, red boxes).

The same rapid loss of mutant receptors from the cell surface was also observed in hippocampal neurons. Rat WT γ_2L- and mutant γ_2L(Q351X)-subunits tagged with pHluorin were coexpressed with rat α_1- and β_2-subunits in hippocampal neurons for 6 d. Fluorescence should only be effectively generated by pHluorin when it is on the cell surface (pH > 7.0), and no or minimal fluorescence should be produced by the pHluorin tag at acidic pHs (pH<6.5) that are characteristic for vesicular compartments *(49)*. After 6 d, γ_2L-subunit-coupled fluorescence was visible and appeared as puncta on neurons expressing both WT and mutant receptors (Fig. 16B). As pHluorin only fluoresces at the surface, the puncta on neurons were assumed to be surface receptors. When challenged with high temperature (40°C), neurons transfected with heterozygous mutant receptors had fewer fluorescent puncta compared with WT receptors from the same areas (Fig. 16B), indicating that there were fewer receptors trafficked to the surface in neurons. The fluorescent puncta with WT receptors showed minimal reduction in fluorescence following 40 min at 40°C, whereas fluorescent puncta with heterozygous mutant receptors showed substantial reduction in fluorescence within 20 min at 40°C. After incubation at 40°C for 30 min, there was more loss of puncta expressing mutant than WT receptors.

Whereas high temperature could further compromise surface stability, endocytic recycling, forward trafficking, ER retention or degradation in receptors containing trafficking-deficient γ_2-mutations, and high temperature could also have effects on cellular events in normal neurons. The temperature sensitivity of normal $GABA_A$ receptors has never been addressed. The authors study on WT $GABA_A$ $\alpha_1\beta_2\gamma_2$-receptors suggested that WT $\alpha_1\beta_2\gamma_2$-receptors have impaired trafficking during prolonged temperature elevation compared with the mutant γ_2-subunit-containing $\alpha_1\beta_2\gamma_2$-receptors (Figs. 16B and 17). In hippocampal neurons, WT $GABA_A$ $\alpha_1\beta_2\gamma_2$-pHluorin fluorescence puncta were slightly reduced after incubation at 40°C for 40 min (Fig. 16B $\alpha_1\beta_2\gamma_2$L). In

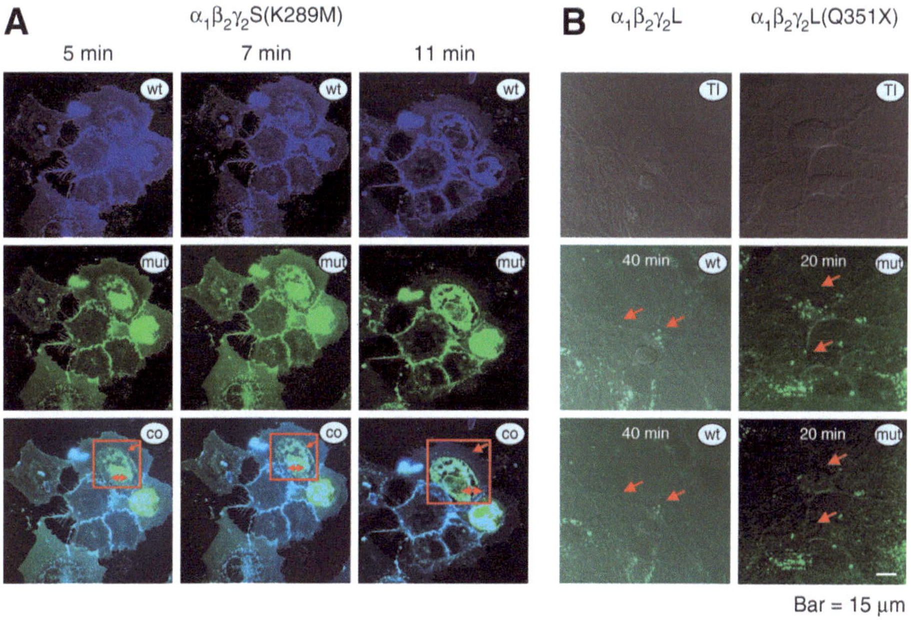

Fig. 16. Reduced surface expression and increased intracellular retention of mutant γ_2-subunit-containing $\alpha_1\beta_2\gamma_2$-receptors with elevated temperature was rapid and dynamic. (**A**) COS-7 cells were cotransfected with het $\alpha_1\beta_2\gamma_2S/\gamma_2S$(K289M)-receptors with the wt γ_2S-subunit tagged with ECFP and the mutant γ_2S-subunit tagged with EYFP. The receptors displayed a merged aqua color both on the surface and intracellularly after incubation at 40°C for 5 min (boxed area). With brief incubation at 40°C, the fluorescence intensity of the surface receptors progressively decreased (7 and 11 min, lower panel, red boxes) with loss of cyan color on the cell surface (changes of cyan color in the red box, red arrow), and the intracellular fluorescence intensity progressively increased with the accumulation of intracellular receptors (changes of cyan color of the coregistered image in the red box, red double arrow). The insert wt stands for ECFP-tagged wt subunits, mut for EYFP tagged mutant subunits, and co for colocalization of all ECFP, EYFP channels. (**B**) Heterozygous $\alpha_1\beta_2\gamma_2$L-pHluorin/$\alpha_1\beta_2\gamma_2$L(Q351X)-pHluorin receptor on the surface of rat hippocampal neurons was reduced rapidly by temperature elevation to 40°C. As illustrated in the left panels, neurons were cotransfected with het $\alpha_1\beta_2\gamma_2$L-pHluorin/$\alpha_1\beta_2\gamma_2$L(Q351X)-pHluorin receptors, and receptors were imaged as puncta on the surface of neurons. With incubation at 40°C, the fluorescent puncta were reduced, with loss or fading of fluorescence on the cell surface (red arrows). In the inserts TI stands for transmitted image, wt stands for wt $\alpha_1\beta_2\gamma_2$L-pHluorin receptors, mut for het mutant $\alpha_1\beta_2\gamma_2$L-pHluorin/$\alpha_1\beta_2\gamma_2$L(Q351X)-pHluorin receptors. (Modified from ref. *47*.)

HEK 293T cells, $\alpha_1\beta_2\gamma_2$S-receptors had a reduced peak current amplitude after incubation for 2.5 h at 40°C (Fig. 17A), and the peak current was increased 30-fold more than 45 min at room temperature and then was stable for the next

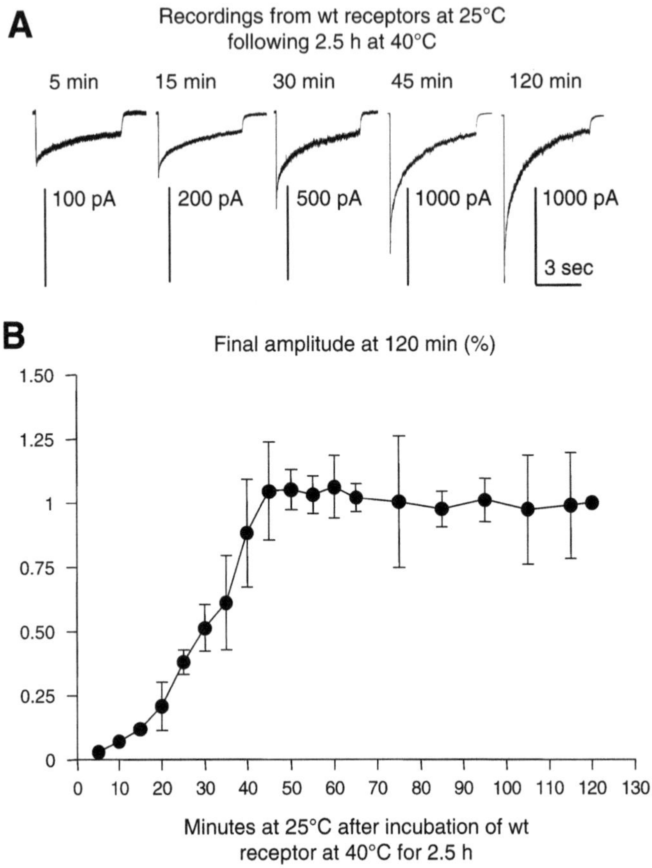

Fig. 17. Wild-type γ_2S-subunit-containing GABA$_A$ receptors had reversible, impaired trafficking with temperature elevation. **(A)** Representative currents are presented from a single cell expressing wt $\alpha_1\beta_2\gamma_2$S-receptors recovering at 25°C from a 2.5-h incubation at 40°C. **(B)** Wild-type receptor peak current amplitudes increased in a 45 min time-course at 25°C and then stabilized following incubation at 40°C for 2.5 h. The ordinate denotes current amplitudes at each time-point over the current amplitude obtained at 120 min after incubation at 25°C. Data were averaged from four cells and all currents were recorded in HEK 293T cells under lifted whole-cell configurations with voltage clamped at –50 mV. (Modified from ref. *47.*)

2 h (Fig. 17B). The data suggest that WT $\alpha_1\beta_2\gamma_2$S-receptors are not spared from the high temperature insult and that this heat-induced receptor trafficking impairment is reversible. This functional impairment of WT $\alpha_1\beta_2\gamma_2$S-receptors by extended exposure at 40°C may explain why children without γ_2-subunit mutations can also manifest seizures when fever spikes.

4. Conclusions

$GABA_A$-receptor subunit mutations associated with IGEs alter $GABA_A$ receptor channel gating, expression, or trafficking. The decreased whole-cell current and/or increased rate of current deactivation of $GABA_A$ receptors containing the γ_2(K289M)-, δ(E177A)-, and δ(R220H)-mutations/variants are explained primarily by reduced mean single channel open times. In contrast, the reduction in whole-cell current for $GABA_A$ receptors containing the α_1(A322D)-, γ_2(R43Q)-, and γ_2(Q351X)-mutant subunits is because of decreased $GABA_A$ receptor cell surface expression. Whereas the "gating" and "trafficking" mutants appear to produce different defects in $GABA_A$ receptor function, the mechanisms are actually overlapping. The γ_2(K289M)-, δ(E177A)-, and δ(R220H)-mutations/variants all produce some defects in receptor trafficking *(39,47)*, and the α_1(A322D)- and γ_2(Q351X)-subunit mutations were associated with altered current kinetic properties *(28)* (Kang and Macdonald, unpublished). Furthermore, the mechanisms by which gating, expression, and/or trafficking are altered are all different. It is likely that the α_1(A322D)-, γ_2(R43Q)-, and γ_2(Q351X)-subunit mutations inhibit proper protein folding and assembly, the α_1(A322D)-mutation disrupts the M3 transmembrane helix, the γ_2(R43Q)-mutation prevents γ_2–β_2 contact interaction during assembly, and the γ_2(Q351X)-mutation truncates the C-terminus before the M4 transmembrane domain. Such misfolded $GABA_A$-receptor subunit proteins do not assemble properly in the ER and are then degraded (for a review *see [50]*).

All the mechanisms described earlier are likely to be related to the development of afebrile seizures, as these defects in expression, trafficking, or function were demonstrated by studying mutant receptors expressed at 37°C. Reasonable hypotheses for the basis for febrile seizures are that the mutations

1. Confer a temperature-dependency to the defect in expression, trafficking, or function; or
2. Confer an independent, temperature-dependent alteration if $GABA_A$ receptor function itself confers susceptibility to fever-induced convulsions.

The later mechanism could be because of enhancement of a normal, temperature-dependent alteration in $GABA_A$ receptor function or a pathological temperature-dependent alteration in $GABA_A$ receptor function. The study discussed in this review provides evidence that febrile seizures may be owing to enhancement of a normal, temperature-dependent alteration of $GABA_A$ receptor *(47)*. This study demonstrated that $GABA_A$ receptor surface expression was somewhat temperature-dependent over a relatively long period (hours). However, all of the studies of temperature-dependent effects were carried out in heterologous expression systems or cultured neurons with transient transfection. The expression variations and the overexpression of the receptor proteins with transient transfection might obscure to some extent the temperature-sensitivity of the receptor trafficking deficiency or alterations in channel function compared with physiological conditions. All the mutations associated with febrile seizures were γ_2-subunit missense mutations or truncations and had increased temperature-dependent reduction of $GABA_A$ receptor surface expression. The authors data suggest that WT γ_2-subunit protein may be somehow very sensitive to temperature changes and any defects during protein folding, conformational stabilization, glycosylation, receptor assembly, forward trafficking, and recycling caused by mutations may render the overall effect of these receptors further compromised, thus further lowering the inhibitory synaptic strength in the brain during rapidly rising fever. Thus, the afebrile and febrile seizures associated with γ_2-subunit missense mutations or truncations have different mechanisms. If these data can be generalized to all or most IGEs with febrile seizures, they also suggest that there might be different strategies for development of therapies for afebrile and febrile seizures. Treatment for afebrile seizures might be directed toward the basic underlying defect associated with the missense mutation or. Based on the authors data, WT $GABA_A$ $\alpha_1\beta_2\gamma_2$-receptor channel function was reversibly temperature-sensitive, suggesting that for febrile seizures, the current therapeutic approach of immediate cooling and reduction of body temperature should be effective in preventing and terminating convulsions. Alternatively, treatments directed toward stabilization of cell surface $GABA_A$ receptors and fast recruitment of receptors to the surface and synapse would be another approach. Finally, validation of the underlying mechanisms of the febrile and afebrile seizures of IGEs will require confirmation of these findings in knockin mice and in neurons in vitro, but studies to date have cast new light on these old and vexing epilepsies.

References

1. Baumann, S. W., Baur, R., and Sigel, E. (2001) Subunit Arrangement of gamma - Aminobutyric Acid Type A Receptors. *J. Biol. Chem.* **276(39),** 36,275–36,280.
2. Baumann, S. W., Baur, R., and Sigel, E. (2002) Forced Subunit Assembly in alpha 1beta 2gamma 2 $GABA_A$ Receptors. Insight Into the absolute arrangement. *J. Biol. Chem.* **277(48),** 46,020–46,025.
3. Chang, Y., Wang, R., Barot, S., and Weiss, D. S. (1996) Stoichiometry of a recombinant $GABA_A$ receptor. *J. Neurosci.* **16(17),** 5415–5424.
4. Hirose, S., Mitsudome, A., Okada, M., and Kaneko, S. (2005) Genetics of idiopathic epilepsies. *Epilepsia* **46(Suppl 1),** 38–43.
5. Mulley, J. C., Scheffer, I. E., Petrou, S., and Berkovic, S. F. (2003) Channelopathies as a genetic cause of epilepsy. *Curr. Opin. Neurol.* **16(2),** 171–176.

6. Wallace, R. H., Marini, C., Petrou, S., et al. (2001) Mutant GABA(A) receptor gamma2-subunit in childhood absence epilepsy and febrile seizures. *Nat. Genet.* **28(1),** 49–52.

6a. Audenaert, D., Schwartz, E., Claeys, K. G., et al. (2006) The GABRG2 gene and febrile seizures: extension of the mutational spectrum. *Neurology.* **67,** 687–690.

7. Baulac, S., Huberfeld, G., Gourfinkel-An, I., et al. (2001) First genetic evidence of GABA(A) receptor dysfunction in epilepsy: a mutation in the gamma2-subunit gene. *Nat. Genet.* **28(1),** 46–48.

8. Harkin, L. A., Bowser, D. N., Dibbens, L. M., et al. (2002) Truncation of the GABA(A)-receptor gamma2 subunit in a family with generalized epilepsy with febrile seizures plus. *Am. J. Hum. Genet.* **70(2),** 530–536.

9. Hirose, S. (2004) Severe myoclonic epilepsy in infancy (SMEI) resulting from SCN1A and GABRG2 mutations inherited from non-SMEI parents. *Am. J. Hum. Genet.*

10. Kananura, C., Haug, K., Sander, T., et al. (2002) A splice-site mutation in GABRG2 associated with childhood absence epilepsy and febrile convulsions. *Arch. Neurol.* **59(7),** 1137–1141.

11. Cossette, P., Liu, L., Brisebois, K., et al. (2002) Mutation of GABRA1 in an autosomal dominant form of juvenile myoclonic epilepsy. *Nat. Genet.* **31(2),** 184–189.

12. Dibbens, L. M., Feng, H. J., Richards, M. C., et al. (2004) GABRD encoding a protein for extra- or peri-synaptic $GABA_A$ receptors is a susceptibility locus for generalized epilepsies. *Hum. Mol. Genet.* **13(13),** 1315–1319.

13. Campos-Caro, A., Sala, S., Ballesta, J. J., Vicente-Agullo, F., Criado, M., and Sala, F. (1996) A single residue in the M2-M3 loop is a major determinant of coupling between binding and gating in neuronal nicotinic receptors. *Proc. Natl. Acad. Sci. USA* **93(12),** 6118–6123.

14. Kash, T. L., Jenkins, A., Kelley, J. C., Trudell, J. R., and Harrison, N. L. (2003) Coupling of agonist binding to channel gating in the GABA(A) receptor. *Nature* **421(6920),** 272–275.

15. Lynch, J. W., Rajendra, S., Pierce, K. D., Handford, C. A., Barry, P. H., and Schofield, P. R. (1997) Identification of intracellular and extracellular domains mediating signal transduction in the inhibitory glycine receptor chloride channel. *EMBO J.* **16(1),** 110–120.

16. Miyazawa, A., Fujiyoshi, Y., and Unwin, N. (2003) Structure and gating mechanism of the acetylcholine receptor pore. *Nature* **424(6943),** 949–955.

17. Hinkle, D. J., Bianchi, M. T., and Macdonald, R. L. (2003) Modifications of a commercial perfusion system for use in ultrafast solution exchange during patch clamp recording. *Biotechniques* **35(3),** 472–474, 476.

18. Bianchi, M. T., Song, L., Zhang, H., and Macdonald, R. L. (2002) Two different mechanisms of disinhibition produced by $GABA_A$ receptor mutations linked to epilepsy in humans. *J. Neurosci.* **22(13),** 5321–5327.

19. Haas, K. F. and Macdonald, R. L. (1999) $GABA_A$ receptor subunit gamma2 and delta subtypes confer unique kinetic properties on recombinant $GABA_A$ receptor currents in mouse fibroblasts. *J. Physiol.* **514(Pt. 1),** 27–45.

20. Jones, M. V. and Westbrook, G. L. (1995) Desensitized states prolong $GABA_A$ channel responses to brief agonist pulses. *Neuron* **15(1),** 181–191.
21. Brejc, K., van Dijk, W. J., Klaassen, R. V., et al. (2001) Crystal structure of an ACh-binding protein reveals the ligand-binding domain of nicotinic receptors. *Nature* **411(6835),** 269–276.
22. Smit, A. B., Syed, N. I., Schaap, D., et al. (2001) A glia-derived acetylcholine-binding protein that modulates synaptic transmission. *Nature* **411(6835),** 261–268.
23. Bowser, D. N., Wagner, D. A., Czajkowski, C., et al. (2002) Altered kinetics and benzodiazepine sensitivity of a $GABA_A$ receptor subunit mutation [gamma 2(R43Q)] found in human epilepsy. *Proc. Natl. Acad. Sci. USA* **99(23),** 15,170–15,175.
24. Kang, J. and Macdonald, R. L. (2004) The $GABA_A$ receptor gamma2 subunit R43Q mutation linked to childhood absence epilepsy and febrile seizures causes retention of alpha1beta2gamma2S receptors in the endoplasmic reticulum. *J. Neurosci.* **24(40),** 8672–8677.
25. Hales, T. G., Tang, H., Bollan, K. A., et al. (2005) The epilepsy mutation, gamma2(R43Q) disrupts a highly conserved inter-subunit contact site, perturbing the biogenesis of $GABA_A$ receptors. *Mol. Cell. Neurosci.* **29(1),** 120–127.
26. Maquat, L. E. (2004) Nonsense-mediated mRNA decay: splicing, translation and mRNP dynamics. *Nat. Rev. Mol. Cell. Biol.* **5(2),** 89–99.
27. Lejeune, F. and Maquat, L. E. (2005) Mechanistic links between nonsense-mediated mRNA decay and pre-mRNA splicing in mammalian cells. *Curr. Opin. Cell. Biol.* **17(3),** 309–315.
28. Gallagher, M. J., Song, L., Arain, F., and Macdonald, R. L. (2004) The juvenile myoclonic epilepsy GABA(A) receptor alpha1 subunit mutation A322D producesasymmetrical, subunit position-dependent reduction. *J. Neurosci.* **24(24),** 5570–5578.
29. Fisher, J. L. (2004) A mutation in the GABA(A) receptor alpha1 subunit linked to human epilepsy affects channel gating properties. *Neuropharmacology* **46(5),** 629–637.
30. Gallagher, M. J., Shen, W., Song, L., and Macdonald, R. L. (2005) Endoplasmic reticulum retention and associated degradation of a $GABA_A$ receptor epilepsy mutation that inserts an aspartate in the M3 transmembrane segment of the alpha 1 subunit. *J. Biol. Chem.* **280(45),** 37,995–38,004.
31. Krampfl, K., Maljevic, S., Cossette, P., et al. (2005) Molecular analysis of the A322D mutation in the GABA receptor alpha-subunit causing juvenile myoclonic epilepsy. *Eur. J. Neurosci.* **22(1),** 10–20.
32. Amaral, M. D. (2004) CFTR and chaperones: processing and degradation. *J. Mol. Neurosci.* **23(1–2),** 41–48.
33. Mulley, J. C., Scheffer, I. E., Harkin, L. A., Berkovic, S. F., and Dibbens, L. M. (2005) Susceptibility genes for complex epilepsy. *Hum. Mol. Genet.* **14(Spec. No. 2),** R243–R249.
34. Chen, Y., Lu, J., Pan, H., et al. (2003) Association between genetic variation of CACNA1H and childhood absence epilepsy. *Ann. Neurol.* **54(2),** 239–243.

35. Heron, S. E., Phillips, H. A., Mulley, J. C., et al. (2004) Genetic variation of CACNA1H in idiopathic generalized epilepsy. *Ann. Neurol.* **55(4),** 595–596.
36. Khosravani, H., Altier, C., Simms, B., et al. (2004) Gating effects of mutations in the Cav3.2 T-type calcium channel associated with childhood absence epilepsy. *J. Biol. Chem.* **279(11),** 9681–9684.
37. Khosravani, H., Bladen, C., Parker, D. B., Snutch, T. P., McRory, J. E., and Zamponi, G. W. (2005) Effects of Cav3.2 channel mutations linked to idiopathic generalized epilepsy. *Ann. Neurol.* **57(5),** 745–749.
38. Vitko, I., Chen, Y., Arias, J. M., Shen, Y., Wu, X. R., and Perez-Reyes, E. (2005) Functional characterization and neuronal modeling of the effects of childhood absence epilepsy variants of CACNA1H, a T-type calcium channel. *J. Neurosci.* **25(19),** 4844–4855.
39. Feng, H. J., Kang, J. Q., Song, L., Dibbens, L., Mulley, J., and Macdonald, R. L. (2006) Delta subunit susceptibility variants E177A and R220H associated with complex epilepsy alter channel gating and surface expression of alpha4beta2delta GABA$_A$ receptors. *J. Neurosci.* **26(5),** 1499–1506.
40. Hauser, W. A. (1994) The prevalence and incidence of convulsive disorders in children. *Epilepsia* **35(Suppl. 2),** S1–S6.
41. Dibbens, L. M., Feng, H. J., Richards, M. C., et al. (2004) GABRD encoding a protein for extra- or peri-synaptic GABA$_A$ receptors is a susceptibility locus for generalized epilepsies. *Hum. Mol. Genet.* **13(13),** 1315–1319.
42. Thompson, S. M., Masukawa, L. M., and Prince, D. A. (1985) Temperature dependence of intrinsic membrane properties and synaptic potentials in hippocampal CA1 neurons in vitro. *J. Neurosci.* **5(3),** 817–824.
43. Volgushev, M., Vidyasagar, T. R., Chistiakova, M., and Eysel, U. T. (2000) Synaptic transmission in the neocortex during reversible cooling. *Neuroscience* **98(1),** 9–22.
44. Micheva, K. D. and Smith, S. J. (2005) Strong effects of subphysiological temperature on the function and plasticity of mammalian presynaptic terminals. *J. Neurosci.* **25(33),** 7481–7488.
45. Wan, Q., Xiong, Z. G., Man, H. Y., et al. (1997) Recruitment of functional GABA(A) receptors to postsynaptic domains by insulin. *Nature* **388(6643),** 686–690.
46. Kang, J. Q. and Macdonald, R. L. (2004) The GABA$_A$ receptor gamma2 subunit R43Q mutation linked to childhood absence epilepsy and febrile seizures causes retention of alpha1beta2gamma2S receptors in the endoplasmic reticulum. *J. Neurosci.* **24(40),** 8672–8677.
47. Kang, J. Q., Shen, W., and Macdonald, R. L. (2006) Why does fever trigger febrile seizures? GABA$_A$ receptor gamma2 subunit mutations associated with idiopathic generalized epilepsies have temperature-dependent trafficking deficiencies. *J. Neurosci.* **26(9),** 2590–2597.
48. Connolly, C. N., Kittler, J. T., Thomas, P., et al. (1999) Cell surface stability of gamma-aminobutyric acid type A receptors. Dependence on protein kinase C activity and subunit composition. *J. Biol. Chem.* **274(51),** 36,565–36,572.

49. Miesenbock, G., De Angelis, D. A., and Rothman, J. E. (1998) Visualizing secretion and synaptic transmission with pH-sensitive green fluorescent proteins. *Nature* **394(6689),** 192–195.
50. Barnes, E. M. Jr. (2000) Intracellular trafficking of GABA(A) receptors. *Life Sci.* **66(12),** 1063–1070.

7

Abuse and Dependence Liability of $GABA_A$-Receptor Modulators

Antecedents and Pharmacology

James K. Rowlett, Angela N. Duke, and Donna M. Platt

Summary

Ligands that act at γ-aminobutyric type A ($GABA_A$) receptors, in particular the benzodiazepines and related drugs, have broad clinical use but also the liability for abuse and dependence. Recent epidemiological data suggest that abuse of benzodiazepine-type drugs may be on the upswing, with a shift from primary misuse of benzodiazepines by people in a therapeutic setting to use by younger people engaging in recreational abuse. Laboratory findings suggest that benzodiazepine-type drugs have reinforcing effects both in human and non-human subjects. However, benzodiazepine-type drugs appear to have lower reinforcing effectiveness compared to other drugs of abuse, such as psychomotor stimulants. Recent research has begun to explore the role of $GABA_A$ receptor subtypes in the reinforcing effects of benzodiazepine-type drugs, and unlike other behavioral effects (e.g., motor coordination deficits); reinforcing effects are not easily attributed to a single receptor subtype. Perhaps the most firm conclusion that can be made at this point is that stimulation of $GABA_A$ receptors containing α1 subunits (α1$GABA_A$ receptors) is not necessary for self-administration of benzodiazepine-type compounds. Benzodiazepine use also is associated with physical dependence, characterized by a withdrawal syndrome. In both human and non-human subjects, this withdrawal syndrome is considered to be intermediate in severity. Preliminary results suggest that compounds with selectivity for α2$GABA_A$, α3$GABA_A$, and/or α5$GABA_A$ receptors do not induce physical

From: *The Receptors: The GABA Receptors, Third Edition*
Edited by: S. J. Enna and H. Möhler

dependence. As with reinforcing effects, systematic studies with selective compounds having relatively high intrinsic efficacy at particular subtypes should shed light on theses important mechanistic issues.

Key Words: Benzodiazepine; $GABA_A$ receptor; abuse; self-administration; physical dependence; history.

1. Introduction

The abuse of substances that exert some or all of their action at γ-aminobutryic acid type A ($GABA_A$) receptors has occurred for centuries. Substances ranging from α-thujone, the active ingredient of absinthe, to ethyl alcohol are documented to alter chloride currents mediated by $GABA_A$ receptors *(1,2)*. At the beginning of the 20th century, the barbiturates were introduced into clinical practice and soon thereafter, reports of abuse began to surface *(3)*. Given the selectivity of barbiturates for $GABA_A$ receptors, it has become clear that activity at these receptors carries with it the risk for abuse and dependence.

The modern history of psychiatric medicine and pharmacotherapy is tied intrinsically to the development of anxiety reducing or "anxiolytic" compounds (for review *see* ref. *4*). Although barbiturates originally were used for anxiety reduction, these compounds proved too toxic for practical use; and the search for safe and effective anxiolytics led to the introduction of meprobamate in 1955 by the pharmaceutical companies Wallace Laboratories and Wyeth *(4)*. Meprobamate (marketed as Miltown [Wallace Laboratories, Cranbury, NJ] and Equanil [Wyeth-Ayerst, Philadelphia, PA]) indeed was safer than any previous compounds, but reports of overuse and illicit diversion gradually negated its clinical usefulness. Interestingly, the mechanism of action of meprobamate is not yet understood fully, although a primary site of action for this drug is the $GABA_A$ receptor *(1)*.

Today, meprobamate and the other early anxiolytics have been replaced largely by the benzodiazepines.* As discussed in the following section, benzodiazepine-type drugs are similar to older anxiolytic drugs in that they act at the $GABA_A$ receptor. Unfortunately, benzodiazepine-type drugs also retain the liability for abuse and dependence associated with the earlier anxiolytics *(5,6)*. Thus, action at $GABA_A$ receptors likely plays a key role in both the therapeutic as well as abuse-related

*"Benzodiazepine" typically refers to a compound with the 1,4-benzodiazepine structure, e.g., diazepam, chlordiazepoxide, or related drugs, such as triazolobenzodiazepines, e.g., alprazolam. However, other compounds with nonbenzodiazepine structures bind at the same receptor as benzodiazepines. Throughout this review, both types of compounds will be referred to as "benzodiazepine-type drugs" for clinically available agents and "benzodiazepine-type compounds" for experimental agents. "Benzodiazepines" will refer exclusively to drugs with benzodiazepine-based structures.

effects of this important class of drugs, and the extent to which therapeutic efficacy and abuse potential can be dissociated is not yet fully understood.

This chapter will explore issues related to the abuse and dependence potential of benzodiazepine-type drugs and the role that GABA$_A$ receptors play in this phenomenon. The addictive properties of benzodiazepine-type drugs have been, and continue to be, subject to lively debate *(5,7)*. Rather than emphasize any particular aspect of this debate, the authors intend to review the issues related to benzodiazepine abuse potential and provide historical context. Additionally, we will provide an overview of GABA$_A$ receptor pharmacology relevant to understanding benzodiazepine abuse and dependence. A final aim of this chapter is to identify gaps in the knowledge of the mechanisms of action underlying these side-effects that have historically plagued the development and use of anxiolytic drugs.

2. Benzodiazepines and GABA$_A$ Receptor Heterogeneity

As reviewed elsewhere in this book, GABA$_A$ receptors have received considerable attention as the site of action for drugs that act not only as anxiolytics, but also as sedatives, anticonvulsants, and muscle relaxants. All of these clinically beneficial effects are exhibited by the benzodiazepine-type drugs, which act by allosterically binding to GABA$_A$ receptors and enhancing the ability of GABA to increase chloride conductance. Many studies over the past decades have revealed the existence of multiple subtypes of the GABA$_A$ receptor (for e.g., *see* ref. *8*), and research with transgenic mice and subtype-selective ligands has postulated that the diverse behavioral effects of benzodiazepine-type drugs may reflect action at different subtypes of GABA$_A$ receptors *(9–12)*.

The GABA$_A$ receptors in the central nervous system are pentamers constituted from structurally distinct proteins, with each protein family consisting of different subunits. The majority of GABA$_A$ receptors consist of α-, β-, and γ-subunit families, and benzodiazepine action appears to be determined by the presence of particular α-subunits. Benzodiazepine-type drugs bind predominantly to a site on the native GABA$_A$ receptor that occurs at the interface of the γ_2-subunit with either α_1-, α_2-, α_3-, or α_5-subunits, whereas these drugs are inactive at corresponding α_4- and α_6-subunit containing receptors. More than 90% of the GABA$_A$ receptors in the brain contain α_1-, α_2-, and α_3-subunits *(13)*, and GABA$_A$ receptors containing α_1-subunits (α_1GABA$_A$ receptors) have been recently implicated in the sedative effects of benzodiazepine-type drugs, whereas GABA$_A$ receptors containing α_2- and α_3-subunits (α_2GABA$_A$- and α_3GABA$_A$ receptors) have been implicated in the anxiolytic effects of benzodiazepine-type drugs *(9,10)*. Receptors containing α_5-subunits (α_5GABA$_A$ receptors), in contrast, are a relatively minor population that may play a role in memory processes, but likely not anxiolysis or motor effects *(14,15)*.

To the extent that the different behavioral effects of benzodiazepines are attributable to different receptor subtypes, it is possible that a subset of receptors is responsible for the abuse-related effects of these drugs. Consequently, the heterogeneity of $GABA_A$ receptors sparks the hope that compounds lacking abuse liability can be found. However, as will be reviewed in later sections of this chapter, a complex picture is emerging with respect to abuse of benzodiazepine-type drugs and the role of different $GABA_A$-receptor subtypes.

3. The Emergence of "Mother's Little Helper"

The introduction of meprobamate raised the hopes that anxiolytics could be developed that were safer than barbiturates. These drugs would, in turn, begin to meet the needs of the then-recognized large portion of the population suffering from anxiety disorders. Meprobamate was by all accounts a hit, spurring the interest in the development of novel anxiolytic drugs *(4)*. The apex of anxiolytic drug development was reached in 1960 with the introduction of chlordiazepoxide (Librium®) and with diazepam in 1963 (Valium®), by the Swiss company Hoffman-La Roche (Basel, Switzerland). The popularity of diazepam has been well documented, with this compound becoming the most widely prescribed drug irrespective of therapeutic use in the US and Europe between 1968 and 1987 *(16)*. But by as early as 1967, reports in the popular media were warning against the potential for illicit use and abuse, particularly by youth and the counterculture *(4,16)*. Use of benzodiazepines in fact entered the popular culture, as typified by The Rolling Stone's song "Mother's Little Helper," referring to a street name associated with the perceived widespread use of diazepam by middle-class housewives (similarly, "Executive Exedrin" was a street name associated with the popular use of benzodiazepines and related anxiolytics by business executives). By the 1970s, the lurid accounts of abuse and dependence had escalated (for e.g., *see* ref. *17*) and in 1979, the US Congress initiated hearings on the "Valium scare" *(18)*.

As a consequence of the 1979 Congressional hearings, benzodiazepines are regulated in the US by the Drug Enforcement Agency (DEA). Benzodiazepine-type drugs are categorized as Schedule IV,* which in regulatory terms reflects the idea that the abuse potential of these drugs is considered low relative to other drugs of abuse *(19)*. The consequence of this scheduling in the US and similar regulations worldwide primarily is restrictions on the period of time for which a benzodiazepine-type drug can be prescribed.

---

*The DEA categorizes drugs into five schedules based on medical use and relative abuse liability. Schedules I and II drugs are considered to have high abuse liability, but those in Schedule I are not approved for medical use in the US. Schedules III, IV, and V have medical use and abuse liability in descending degree of probability, with Schedule V available over-the-counter in some cases.

4. Benzodiazepines: A Question of Abuse vs Therapeutic Use

As will be discussed in subsequent sections, the idea that the abuse potential of benzodiazepine-type drugs may be less than other abused drugs has empirical, in addition to regulatory support. This notion has generated lively debate beyond the "Valium scare" era of the 1970s (for e.g., *see* refs. *4,5,7*). Based in large part on epidemiological findings at the time, Woods and colleagues *(5,20,21)* concluded that the vast majority of benzodiazepine use was appropriate and, combined with human laboratory data, the classification of benzodiazepine-type drugs as addictive drugs or drugs of abuse was perhaps erroneous, or at the very least, overstated. An important consequence of this classification might be that benzodiazepine-type drugs are, in fact, underprescribed; with many patients going untreated owing to a general reluctance to prescribe and use these drugs *(5,22)*. Another perspective is that classifying benzodiazepine-type drugs as abused drugs reflects a reactionary response by society at large: As stated by Tone (*4*, p. 375):

"...the stigmatization of a drug as dangerous...may tell us more about the politics of the society that passed judgment on it than about the drug's chemical ability to alter mood and behaviour".

Although provocative, the debate on the extent to which benzodiazepine-type drugs should be considered drugs of abuse may simply reflect a lack of consensus on what constitutes a dangerous drug of abuse vs a "mildly addictive" substance. The fact remains that benzodiazepine use is widespread, and recent reports have suggested that patterns of abuse might be quite complex *(23)*. Griffiths and colleagues have proposed two major patterns of problematic use of benzodiazepine-type drugs: Recreational abuse and chronic *quasi*-therapeutic use (for review, *see* refs. *23,24*). Recreational abuse refers to intermittent or long-term use of benzodiazepine-type drugs at high doses, often along with other substances (i.e., polydrug abuse). Recreational abuse is relatively rare compared with the rate of prescription, but similar to abuse of illicit drugs, such as the opioids *(23,24)*. Chronic quasi-therapeutic abuse, on the other hand, refers to long-term use of benzodiazepine-type drugs by patients, which is inconsistent with accepted guidelines for prescription or an individual's medical condition. This type of abuse is said to be relatively prevalent compared with the rates of prescription *(23,24)*.

Recent epidemiological findings also suggest that prevalence and patterns of use of benzodiazepine-type drugs may be changing over time. One example comes from recent reports from the Drug Abuse Warning Network (DAWN), which reports yearly estimates of drug abuse-related emergency department visits from a large network of hospitals in the United States. According to the most recent data available (2002), of all psychotherapeutic drugs, benzodiazepines were the most frequently reported in emergency department visits *(25)*.

However, most striking is that benzodiazepine-related emergency department visits are as prevalent as those of heroin and more so than the amphetamines (Fig. 1). Moreover, a steady upward trend was reported for benzodiazepines, with an overall increase of 41% from 1995 to 2002 (Fig. 1) (*see* ref. *25*).

Emergency room visits can occur for a variety of reasons, and in a follow-up report from the DAWN network it was apparent that most cases involved suicide attempts *(26)*. However, comparison of the motives underlying benzodiazepine-involved emergency department visits showed that the nature of the emergencies appear to be changing (Fig. 2). In this regard, the percent of emergency department visits involving benzodiazepine dependence as well as visits by patients who abused the drugs for "psychic" effects increased significantly from 1995 to 2002 (Fig. 2) (*see* ref. *26*). In fact, for patients age 12–25, suicide-related visits were not different from visits related to drug taking for psychic effects. A second disturbing trend in benzodiazepine-related emergency room visits is that in 1995, the age group with the highest incidence of emergency department visits was the 35–44-yr old category, which by 2002 had shifted to 26–29-yr olds. An even younger age group, 18–19-yr olds, showed the greatest percent increase (~66%) of all age groups in emergency department visits from 1995 to 2002 *(26)*.

Current reports based on substance abuse treatment admissions also lend support to a recent increase in benzodiazepine use. Based on findings from the Treatment Episode Data Set (TEDS), an annual compilation of patient characteristics in substance abuse treatment facilities in the US, admissions owing to "primary tranquilizer" use (including, but not limited to, benzodiazepine-type drugs) increased 79% from 1992 to 2002 *(27)*. For both the DAWN and TEDS data sets, a primary characteristic of the abuse of benzodiazepine-type drugs was that they were combined with other substances (usually alcohol) in the majority of cases.

It is of particular interest to look more closely at specific benzodiazepine-type drugs for clues to the factors contributing to the abuse of this class of drugs. According to the DAWN data sets, the benzodiazepine that has increased the most in reports from emergency room visits is alprazolam (Xanax®, Pfizer, Inc., New York, NY), which is an intermediate-duration triazolo-benzodiazepine, prescribed for the treatment of anxiety disorders, followed next by diazepam *(25)*. Interestingly, the newer drugs marketed for the treatment of insomnia (zolpidem and zaleplon) appear to contribute in a very minor way to overall benzodiazepine-associated emergency room visits, accounting for less than 10% of the overall visits. Of these hypnotics, zolpidem (Ambien®, Sanofi-Aventis, Bridgewater, NJ) is the most frequently reported; clearly surpassing the next two most frequently reported hypnotics, temazepam (Restoril®, Mallinckrodt Pharmaceuticals, St. Louis, MO) and flurazepam (Dalmane®, Roche Products, Inc., Humacao, Puerto Rico) (*see* Fig. 3). Of concern is the observation that visits involving zolpidem have increased dramatically from 1995 to 2002—a 118% increase that was

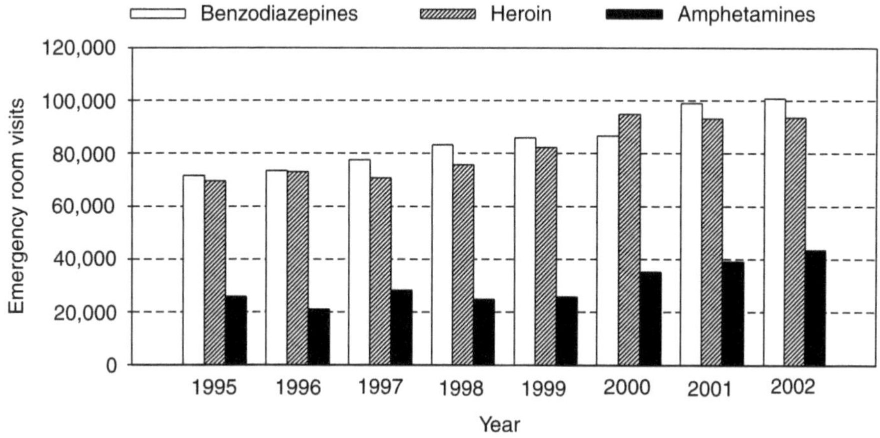

Fig. 1. Trends in benzodiazepine-involved emergency room visits in the contiguous United States. "Benzodiazepines" include all benzodiazepine-type drugs available by prescription in the US. "Amphetamines" consist of D-amphetamine, methamphetamine, and 3–4 methylenedioxymethamphetamine. Data were obtained from the DAWN (Office of Applied Studies, SAMHSA, USA) and reported in Crane and Lemanski *(25)*.

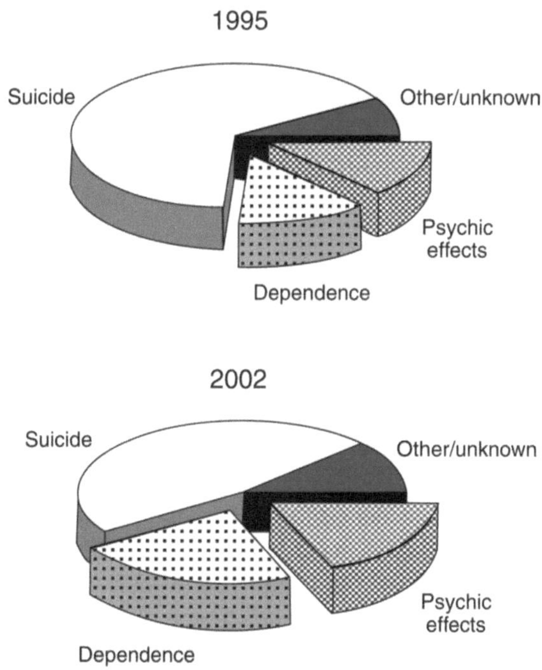

Fig. 2. Reported motives for benzodiazepine-associated emergency room visits in the contiguous US. for 1995 and 2002. Data were obtained from the DAWN (Office of Applied Studies, SAMHSA, USA, 2003 update) and reported in Crane and Lemanski *(26)*.

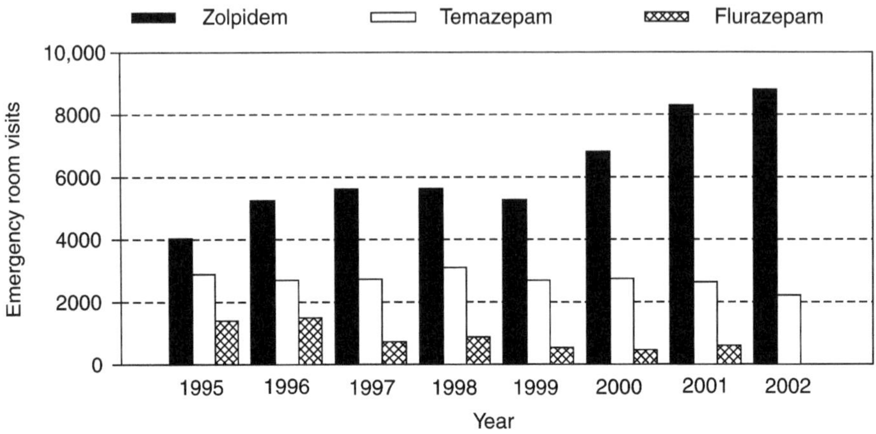

Fig. 3. Trends in hypnotic-involved emergency room visits in the contiguous United States. Shown are data for the three most frequently reported hypnotics, zolpidem (Ambien®), temazepam (Restoril®), and flurazepam (Dalmane®). Data were obtained from the 03/2002 update, DAWN, 2001, Office of Applied Studies, SAMHSA.

more than any other hypnotic or anxiolytic. A caveat of these observations is that alprazolam and zolpidem are the most frequently prescribed benzodiazepine-type anxiolytic and hypnotic, respectively, although the extent to which simple availability accounts for these findings is not known.

Regardless of controversies surrounding the extent to which abuse of benzodiazepine-type drugs warrants consideration in the same manner as other more "mainstream" drugs of abuse, there is generally a consensus that certain subpopulations are at greater risk for inappropriate benzodiazepine taking. These groups include polydrug abusers, patients with histories of alcohol abuse, and the elderly *(21,23)*. With respect to polydrug abuse, benzodiazepine-type drugs are often coabused with alcohol and opioids *(26)*, and the basis for this form of polydrug is not well understood. People with a history of moderate-to-heavy alcohol use tend to have a higher degree of long-term benzodiazepine use (often without a prescription) and appear more sensitive to the effects of these drugs *(28–30)*. And although the elderly likely do not engage in recreational abuse, prevalence of use is typically higher than in the general population *(21,23)*. Overuse in the elderly is particularly disquieting, given the potential for motor-impairing and amnestic effects of benzodiazepines and related drugs in patients with a high prevalence of motor coordination deficits and cognitive deficits. In fact, a well-documented phenomenon is a significantly higher risk of falls or hip fractures in elderly patients taking benzodiazepines (for e.g., *see* refs. *31,32*).

5. Benzodiazepine Abuse: From the Street to the Laboratory

As illustrated earlier, drug taking is a complex phenomenon influenced by a variety of historical and social factors in addition to pharmacological variables. Along with the legal and medical aspects, which contribute to a compound being classified as an abused drug, the most likely property of a compound that predicts inappropriate use is the degree to which the compound has reinforcing effects. A drug is said to have reinforcing effects if its presentation increases the probability of subsequent responses to produce it. The study of the reinforcing effects of drugs has been an important emphasis of drug abuse research for decades, and the demonstration of a drug's reinforcing effects in the laboratory forms a key component of abuse liability assessment required by worldwide regulatory agencies *(19,33,34)*.

Another major determinant of the extent to which a drug has abuse liability is the occurrence of physical dependence with repeated administration. Physical dependence is characterized by the emergence of a withdrawal syndrome on cessation of chronic drug treatment. Tolerance to some or all of the effects of a drug often accompanies the development of physical dependence. It is important to note that abuse can occur in the absence of physical dependence—the latter phenomenon is a predictor of abuse potential, but not a necessary condition. As with reinforcing effects, regulatory agencies also consider the extent to which a compound induces physical dependence following chronic treatment as part of scheduling decisions *(33)*.

A final property of a drug often considered to be a key component of abuse liability is its subjective, or interoceptive effects. These effects are often assessed with drug discrimination procedures in which subjects typically are trained to distinguish the presence and absence of a drug, i.e., a response is correct or incorrect based on whether drug or placebo is administered. In its most basic form, these procedures determine the extent to which one drug shares discriminative stimulus effects with another drug—if the latter is an abused drug of a particular class, then the likelihood that the compound of interest has subjective effects in common with the drug of abuse is high *(19,35)*.

The remaining sections of this chapter will review the current state of knowledge of the role of GABA$_A$ receptors in mediating behavior associated with the abuse liability of benzodiazepine-type drugs. Of the three properties of drugs that are considered for determination of abuse liability, the focus will be on the reinforcing effects and propensity to induce physical dependence of benzodiazepine-type drugs. The discriminative stimulus effects of benzodiazepine-type drugs have been reviewed extensively elsewhere (for e.g., *see* refs. *19,35*) and will not be discussed in detail further.

5.1. Self-Administration of Benzodiazepine-Type Drugs: Human Studies

A consistent finding in human laboratory studies is that benzodiazepine-type drugs have reinforcing effects in subjects with histories of drug or alcohol abuse, in anxious subjects, and patients with sleep disorders *(21,23)*. However, in subjects lacking these characteristics benzodiazepine-type drugs consistently do not function as reinforcers. This finding often is used to argue that benzodiazepine-type drugs are not "conventional" drugs of abuse *(5)*. The finding that the reinforcing effects of benzodiazepine-type drugs depends so strongly on subject characteristics and/or histories is unusual among abused drugs, and the implications for understanding the mechanism(s) underlying benzodiazepine reinforcement are not clear. However, one potential mechanism might be therapeutic effects, i.e., people may self-administer benzodiazepine-type drugs in order to alleviate anxiety or insomnia. In fact, Griffiths and Weerts *(23)* have proposed that the reinforcing effects of benzodiazepine-type drugs may occur solely owing to therapeutic efficacy.

Recent support for this idea comes from a study by Helmus et al. *(36)*. Subjects with social anxiety disorder and healthy controls participated in public speaking sessions (a commonly-used task for inducing anxiety), and were allowed to choose between diazepam and placebo. During public speaking, preference for diazepam was greater for subjects with social anxiety disorder (81.8%) compared with healthy controls (36.4%). Quantitative assessment of subject-rated effects indicated that diazepam attenuated anxiety when the levels of anxiety were elevated owing to public speaking in the social phobics *(36)*, providing evidence that diazepam self-administration was correlated with therapeutic effect.

The observation that benzodiazepine-type drugs do not have reinforcing effects in normal healthy subjects also suggests that polydrug abusers and alcoholics likely self-administer these compounds owing to unique characteristics of these patient populations. There are at least two possible characteristics: Polydrug abusers and alcoholics self-administer benzodiazepine-type drugs in order to reduce anxiety and/or insomnia (i.e., "self-medicate" like the social phobics described earlier) or a biobehavioral adaptation to one or more effects of benzodiazepine-type drugs occurs owing to abuse of other substances. With respect to the self-medication idea, results to date have been somewhat mixed, with some authors observing that benzodiazepines are coadministered with other substances primarily to boost a drug "high" (for e.g., *see* refs. *37,38*), whereas others have found evidence for self-medication of "emotional disturbances" or insomnia *(39,40)*. A study in an Israeli population of patients maintained on methadone for treatment of opioid dependence may shed light on these mixed findings. Gelkopf et al. *(40)* found that although large proportions of this population self-administered benzodiazepines for either recreational purposes (41.4%) or treating "emotional problems" (87.1%), about one-third of this population reported taking benzodiazepines for both reasons. These clinical findings suggest that an *interaction* between therapeutic effect and reinforcing effects following exposure to earlier abused substances should be considered as a viable mechanism underlying the abuse potential of benzodiazepine-type drugs.

Although it is likely that the reinforcing effects of benzodiazepines may be altered by previous drug history, the nature of these alterations is unclear. One known consequence of chronic drug exposure is physical dependence, and evidence exists from animal studies that benzodiazepines may suppress some of the withdrawal signs associated with chronic opioid agonist or ethanol treatment (for review, *see* ref. *21*). However, physical dependence has not been identified as a key factor in studies involving drug-experienced humans *(21)*. In addition, benzodiazepines themselves induce physical dependence, and self-administration of benzodiazepines does not require that the subjects be dependent (for review, *see* ref. *23*; physical dependence will be discussed in following sections).

5.2. Self-Administration of Benzodiazepine-Type Drugs: Nonhuman Studies

One prediction based on the idea that benzodiazepine-type drugs have reinforcing effects is that these compounds should be effective in models of self-administration that use nonhuman subjects in controlled laboratory settings. Indeed, this prediction does hold, with benzodiazepine-type compounds showing reinforcing effects under a variety of experimental conditions (for e.g., *see* refs. *12,19,41–43*). These studies employed iv self-administration procedures in which subjects are trained to press a lever in order to receive an iv drug injection through a chronic venous catheter. Reinforcing effects of the drug are affirmed if it maintains a higher degree of self-administration compared with that observed under conditions of vehicle availability.

Several features characteristic of behavior reinforced by drug self-administration are illustrated in Fig. 4. For example, increasing the unit dose of drug (mg/kg/injection) resulted in self-administration (measured in this case as the number of injections taken per daily session) that increased above levels observed when vehicle was available (i.e., diazepam functioned as a reinforcer). However, as unit dose was increased further, self-administration began to decrease to near vehicle levels, i.e., the dose-response function was biphasic (resembling a "bell-shaped" or "inverted U-shaped" function). This dose-response relationship is typical of most drugs of abuse. In addition, the peak levels of self-administration maintained by diazepam were below the peak levels maintained by the training drug methohexital, a short-acting barbiturate *(44)*. Similarly, peak diazepam self-administration as assessed by the break point measure was nearly an order of magnitude lower than maximum break

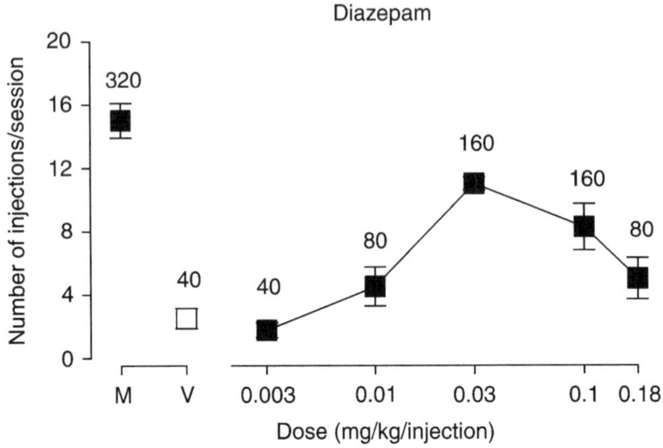

Fig. 4. Self-administration of diazepam by rhesus monkeys trained under a progressive-ratio schedule of iv methohexital injection. *Abbreviations:* M, methohexital, 0.03 mg/kg/injection; V, vehicle. Numbers above points represent median break points (last response requirement completed in the sessions). Data for number of injections/session are mean ± SEM for $N = 4$ monkeys. Experimental details can be found in Rowlett et al. *(12).*

points maintained by cocaine *(12,45)*.* These findings are consistent with the idea that benzodiazepines function as reinforcers, but their strength as reinforcers are intermediate compared with other drugs.

Another characteristic of a reinforcing drug is that it can be used to train self-administration behavior. For many years, benzodiazepines were believed to be ineffective as training drugs until Weerts et al. *(46)* demonstrated intravenous self-administration of the short-acting benzodiazepine midazolam by inexperienced baboons. As in other procedures, midazolam self-administration was described as maintaining relatively low rates of responding, consistent with benzodiazepine-type compounds having intermediate reinforcing effectiveness compared with other self-administered drugs *(46)*.

5.3. $GABA_A$-Receptor Subtypes and the Reinforcing Effects of Benzodiazepines

One of the most intriguing exceptions to the general finding that benzodiazepine-type drugs have relatively modest reinforcing effects has been

*Break point is a measure commonly used in progressive-ratio procedures similar to that illustrated in Fig. 4. A break point is the last response requirement completed in a session. Maximum break points maintained by common drugs of abuse range from 400 to 1000 under procedures similar to that of Fig. 4 (for e.g., a maximum average break point of 1000 was reported for cocaine, *see* ref. *45*).

the result with the hypnotic drug, zolpidem. In fact, zolpidem self-administration often is greater than conventional benzodiazepines and closer to barbiturate or stimulant levels *(12,47)*. Zolpidem displays selectivity for the α_1GABA_A receptor, raising the possibility that this receptor subtype plays a key role in the reinforcing effects of benzodiazepine-like compounds *(12,19,47)*. Another compound, zaleplon, that also displays selectivity for the α_1GABA_A receptor, was self-administered by baboons to a degree similar to zolpidem *(48)*, lending further support to the idea that α_1GABA_A receptors may be an important substrate for self-administration of benzodiazepine-type drugs.

The strongest support to date for a critical role for α_1GABA_A receptors in the reinforcing effects of benzodiazepines comes from a review by Ator *(19)*. In an abuse model in which baboons were trained to self-administer iv injections of the stimulant cocaine, Ator *(19)* reported that a benzodiazepine-type partial agonist, TPA123 functioned as a reinforcer, whereas another partial agonist, TPA023, was ineffective.* The primary difference between the two compounds is that TPA123 has low intrinsic efficacy in vitro at α_1GABA_A receptors, but TPA023 lacks efficacy at these receptors (i.e., was an antagonist in vitro at α_1GABA_A receptors; *see also* ref. *49*). Together with the findings obtained with zolpidem and zaleplon, these results raise the possibility that a benzodiazepine-type compound's potential for abuse may be directly related to its efficacy in vitro at α_1GABA_A receptors—certainly a promising hypothesis with respect to drug development.

Recently, findings were reported with a compound similar to TPA023 using the rhesus monkey self-administration model illustrated in Fig. 4 *(12)*. The compound L-838,417 has the in vitro efficacy profile of partial agonist activity at α_2-, α_3-, and α_5-subunit containing $GABA_A$ receptors and is an antagonist at α_1-subunit containing receptors *(10,50)*. However, in contrast to TPA023, L-838,417 was reliably self-administered, albeit to a lower degree than zolpidem or conventional benzodiazepines *(12)*. These findings are consistent with action at α_1GABA_A receptors not being necessary for benzodiazepine self-administration.

Although the findings of Ator *(19)* and our laboratory appear contradictory, as is illustrated by Table 1, reconciliation of these apparent differences may simply await further experimentation. Table 1 summarizes published receptor action and self-administration results for TPA023 and L-838,417, zolpidem, and diazepam. For comparison purposes, a scale of low, intermediate, and high degrees of self-administration based on previous work by Ator, Griffiths, and colleagues was developed (*see* Table 1 for details). As can be seen in the table, for these particular compounds the available results are concordant across the

---

*Partial agonism in this case is defined by the ability of the compound to potentiate GABA-induced chloride currents in cloned $GABA_A$-receptor subtypes, relative to chlordiazepoxide as a standard.

Table 1
Relationship of Receptor Binding, Intrinsic Efficacy, and Relative Reinforcing Effectiveness for Benzodiazepine-Type Compounds

	Binding selective?[a]	Efficacy[a]				References *19,42,47*[b]	Reference *12*[c]
		α_1	α_2	α_3	α_5		
Diazepam	No	71	81	88	57	++	++
Zolpidem	$\alpha_1 > \alpha_{2/3} >> \alpha_5$	116	103	105	–	+++	+++
TPA023	No	1	12	33	6	0	NA
L-828,417	No	1.5	42.5	42.5	38.5	NA	+

[a]Binding and efficacy data are from cloned human receptors *(10,49,75)*. Efficacy represents percentage potentiation of Cl^- currents at an EC_{20} concentration of GABA. EC_{20}-concentration resulting in 20% of maximum $C1^-$ current.

[b]Relative reinforcing effectiveness, iv self-administration in baboons, Griffiths, Ator, and colleagues: 0, not different from vehicle; +, below a mean of 4 injections/session ("low"); ++ , 4–6 injections/session (intermediate); +++, 6–8 injections/session (high).

[c]Relative reinforcing effectiveness using a scale adapted from Griffiths, Ator, and colleagues; iv self-administration in rhesus monkeys, Rowlett et al.: 0, mean 0–4 injections/session (not different from vehicle); +, 5–8 injections/session (low); ++, 9–12 injections/session (intermediate); +++, 13–20 injections/session (high).

NA, not available.

two laboratories; however, key information is missing (e.g., tests of TPA023 self-administration in rhesus monkeys, tests of L-838,417 in baboons). Irrespective of the gaps in knowledge, the data in Table 1 provides enough information for preliminary conclusions and hypotheses.

5.3.1. Intrinsic Efficacy at α_5GABA$_A$ Receptors Does Not Predict Relative Reinforcing Effectiveness

Both zolpidem and TPA023 lack binding and appreciable efficacy, respectively, at the α_5GABA$_A$ receptor. However, zolpidem was robustly self-administered, whereas TPA023 lacked reinforcing effects.

5.3.2. Action at α_1GABA$_A$ Receptors is Not Necessary for Reinforcing Effects

Although TPA023 lacked both efficacy at α_1GABA$_A$ receptors and reinforcing effects, L-838,417 lacked efficacy at α_1GABA$_A$ receptors and did function as a reinforcer.

5.3.3. Intrinsic Efficacy is a Major Determinant of the Degree of Reinforcing Effectiveness

Selective affinity and selective efficacy cannot account for differences in the reinforcing effects of the compounds shown in Table 1. At least 3 additional sub-hypotheses can be proposed with respect to intrinsic efficacy.

5.3.3.1. Efficacy at α_2GABA_A and/or α_3GABA_A Receptors Is Crucial

Zolpidem's high degree of relative reinforcing effectiveness may be attributable to its high degree of efficacy at α_2GABA_A and/or α_3GABA_A receptors, rather than selectivity at α_1GABA_A receptors.

5.3.3.2. A Threshold of Efficacy May be Required

A compound may require more than 12% efficacy at α_2GABA_A receptors and/or at least about 40% efficacy α_3GABA_A receptors in order to have reinforcing effects.

Although differences in activity at $GABA_A$ receptors provides the most intriguing hypotheses for the observed differences in self-administration shown in Table 1, some methodological factors must also be considered. For example, the subjects in the studies by Ator *(19)* were trained under a cocaine baseline, whereas the subjects in the Rowlett et al. *(12)* report were trained under a barbiturate (methohexital) baseline. As described in previous sections, the history of drug use by human subjects is a major determinant of the reinforcing effectiveness of benzodiazepines, and a similar finding has been reported in the animal literature. In this regard, Bergman and Johanson *(41)* demonstrated that the number of rhesus monkeys self-administering diazepam was significantly lower when self-administration was trained with cocaine compared with pentobarbital. The extent to which differences in baseline training conditions influenced the findings shown in Table 1 is unknown at this time.

Another important factor that must be considered in assessing the relative reinforcing strength of subtype-selective compounds is pharmacokinetics. Little is published regarding the pharmacokinetic parameters of TPA023 and L-838,417 following iv administration in monkeys; however, L-838,417 is purported to have a relatively short half-life similar to that of midazolam (ref. *12*, J.R. Atack, personal communication), whereas TPA023's duration of receptor occupancy in rodents suggests that this compound may be relatively long-acting *(49)*. These findings raise the possibility that the lack of self-administration of TPA023 may be a function of this compound's long duration of action, although long duration compounds clearly are self-administered under the procedures used by both Ator *(19)* and Rowlett et al. *(12)*, for example, diazepam engenders reliable self-administration in both (Table 1). In fact, onset of action may be the most important pharmacokinetic factor that determines the degree of reinforcing effects of abused drugs *(23)* but empirical information on the onset of action of TPA023 and L-838,417 is not yet available.

6. Physical Dependence Following Chronic Treatment With Benzodiazepine-Type Drugs

Prolonged use of benzodiazepine-type drugs can lead to physical dependence, which in turn may contribute to the abuse liability of these drugs *(51,52)*.

Table 2
Comparison of Benzodiazepine Withdrawal Signs Between Human Patients and Monkeys[a]

	Human[b]	Monkey[b]
Anxiety	+	?
Increased aggression	?	+
Self-directed behaviors	?	+
Enhanced excitability	+	+
Nausea/vomiting	+	+
Loss of coordination	+	+
Insomnia	+	?
Muscle tension/rigidity	+	+
Headaches	+	?
Tremors	+	+
Panic attacks	+	?
Perceptual disorders	+	?
Seizures	+	+

[a]Signs for human patients were obtained from refs. *21* and *54*. Signs for monkeys were obtained from the work of Griffiths, Ator, Weerts, and colleagues (for e.g., *see* ref. *61*).
[b]Symbols: +, signs are present; ?, not known.

For example, abrupt cessation of benzodiazepine use after prolonged treatment at a therapeutic dose can result in a withdrawal syndrome (for review, *see* refs. *5,23*). Benzodiazepine withdrawal is characterized by many signs that are opposite to the therapeutic effects of benzodiazepines (e.g., anxiety, insomnia) and, in more severe cases, patients may experience seizures *(23,53,54)* (*see* Table 2 for common withdrawal signs). Physical dependence to a benzodiazepine-type drug is often measured in the laboratory as the emergence of characteristic withdrawal signs on cessation of the drug that is reversed with subsequent drug administration (spontaneous withdrawal), or precipitated by administration of an antagonist, such as flumazenil (precipitated withdrawal) *(21)*.

6.1. Physical Dependence Following Chronic Benzodiazepines: Human Studies

A variety of factors may influence the development of physical dependence to chronic drug treatment in patients. These factors range from the drug's pharmacokinetic profile to the patient's personality characteristics, and are reviewed for benzodiazepine-type drugs extensively by Woods et al. *(21)*, with an update published by Woods and Winger *(5)*. As noted by Woods and Winger *(5)*, empirical evidence of physical dependence to benzodiazepines has accrued since the 1980s;

however, a complete characterization of the risk factors and consequences of benzodiazepine-induced physical dependence was lacking at the time of their review—a situation that is still essentially true today. Woods and Winger *(5)* also observed that physical dependence, or more to the point, *alleviating* physical dependence, is not a necessary condition for a benzodiazepine-type drug to be abused or self-administered.

A common laboratory approach to studying physical dependence is to use precipitated withdrawal as a means to evaluate dependence in a controlled fashion. In a recent study, Mintzer and Griffiths *(53)* administered diazepam daily at a single, relatively high therapeutic dose to healthy volunteers and assessed withdrawal symptoms by administering flumazenil after 1, 7, 14, and 28 d of chronic diazepam exposure. They found that withdrawal could be precipitated after as little as 7 d of chronic exposure. However, withdrawal severity did not increase with increased exposure (i.e., withdrawal symptoms were similar on days 7, 14, and 28), suggesting that duration of exposure to a benzodiazepine does not predict the severity of physical dependence in humans, at least over this time frame (*see also* ref. *55*). These findings also imply that the long-held assumption that the severity of withdrawal increases with therapeutic exposure should be re-examined closely *(5)*. The results of Mintzer and Griffiths *(53)* also illustrate that physical dependence to a benzodiazepine can develop at a therapeutic dose *(21)*.

Accumulating evidence from clinical studies suggests that there are several conclusions that can be drawn regarding the characteristics and risk factors of physical dependence to benzodiazepines (for review, *see* ref. *54*). A consistent finding in studies of long-term benzodiazepine use is that not all patients develop physical dependence *(21)*, although the factors contributing to this variance are unknown at present. Another emerging finding is that benzodiazepines with relatively short durations of action may engender more severe withdrawal (*5,54*, but *see also* ref. *24*). Maintenance dose of benzodiazepine also is often cited as a determining factor of the severity of withdrawal, with higher doses resulting in a more severe withdrawal syndrome; an observation with empirical support *(56)*. Other factors that may be important include patient characteristics such as age and previous/current history of drug and alcohol use—these factors await systematic study.

A recent surge in sales and interest in hypnotic drugs, in particular eszopiclone (Lunesta®, Sepracor, Inc., Marlborough, MA), potentially brings to the forefront concerns about dependence following long-term hypnotic treatment, in addition to new opportunities to study the consequences of long-term exposure to benzodiazepine-type drugs. Most hypnotic benzodiazepine-type drugs are relatively short-acting (e.g., zolpidem, zaleplon) raising concerns over the possibility of severe withdrawal after chronic treatment. However, in a recent review of hypnotic abuse liability, Griffiths and Johnson *(24)* rated the withdrawal observed after therapeutic doses of zolpidem (no information was available for zaleplon) as intermediate, i.e., similar to conventional benzodiazepines. Little information exists for the longer-acting hypnotic eszopiclone, although Griffiths and Johnson *(24)* also estimated this drug to have an intermediate degree of withdrawal, based on it being the active isomer of zopiclone (which has been shown empirically to induce an intermediate degree of withdrawal). However, the degree to which eszopiclone engenders physical dependence following long-term treatment is unknown, although the drug is reported to be well tolerated in patients during a 12-mo period of chronic treatment *(57)*.

6.2. Physical Dependence Following Chronic Benzodiazepines: Nonhuman Studies

Physical dependence can be studied in animals by chronically treating the subject and either ending treatment to assess spontaneous withdrawal or administering an antagonist or an inverse agonist to assess precipitated withdrawal (for e.g., *see* refs. *58,59*). Several approaches have been used to assess withdrawal signs, including measurement of drug-induced seizures (for e.g., *see* ref. *60*), measurement of observable withdrawal signs (for e.g., *see* ref. *61*), and drug discrimination methods (for e.g., *see* ref. *62*). Observational methods are particularly useful for evaluating signs of withdrawal in nonhuman primates *(61,63–65)*, and can provide a profile of effects remarkably similar to the signs of withdrawal seen in human subjects (Table 2). Using observational methods Lukas and Griffiths *(66)* demonstrated that severity of withdrawal increased as the dose of chronic diazepam treatment was increased (*see also* ref. *67*). In primate studies, it is not entirely clear to what extent the severity of withdrawal is altered by varying the duration of chronic treatment *(21)*. Moreover, little evidence exists for a more severe withdrawal syndrome engendered by short-acting drugs. In this regard, short-acting benzodiazepines, such as midazolam, can produce physical dependence similar in magnitude to longer-acting drugs such as chlordiazepoxide *(21)*. Latency to experience withdrawal symptoms may be shorter when compared with longer-acting benzodiazepines; however, conclusions regarding short-acting compared with long-acting benzodiazepine-type drugs and physical dependence are limited owing to a lack of direct comparisons between drugs with different durations of action.

6.3. $GABA_A$-Receptor Subtypes and Benzodiazepine Physical Dependence

Although withdrawal from benzodiazepine-type drugs has been characterized extensively in both humans and animals, the underlying mechanisms of benzodiazepine physical dependence have not been determined *(68,69)*. Using a drug discrimination model of withdrawal in rhesus monkeys,* McMahon et al. *(62)* demonstrated that the potencies of a series of benzodiazepines and related compounds to attenuate the withdrawal-inducing effects of flumazenil did not correlate with the potencies of these drugs to engender benzodiazepine-like discriminative stimulus effects in nondependent monkeys. These findings raise the possibility that distinct receptor mechanisms underlie physical dependence compared with benzodiazepine-related interoceptive effects in nondependent subjects.

A key question regarding compounds with $GABA_A$-receptor subtype selectivity is the extent to which chronic treatment with these compounds induces physical dependence. More specifically, can physical dependence be associated with a particular subtype of the $GABA_A$ receptor—a proposition that perhaps is unlikely given that physical dependence is associated with a plethora of behavioral effects. Early studies with zolpidem suggested that this $\alpha_1 GABA_A$-selective agonist did not induce physical dependence after chronic treatment in mice (for e.g., *see* refs. *69,70*), suggesting that this receptor subtype does not play a key role in the development of physical dependence. However, more recent reports have indicated that zolpidem can, in fact, engender a withdrawal syndrome in nonhuman primates that is quite similar to that observed after chronic treatment with conventional benzodiazepines *(47,61,71)*, a finding consistent with reports in humans *(24)*. Although these results suggest a lack of concordance between rodent and primate models, the primary difference may be attributable to the dependent variables used to define withdrawal. In this regard, the earlier rodent studies relied primarily on measuring seizure activity, whereas the primate studies measured a broad profile of behavioral effects in addition to seizure activity. Finally, additional evidence that the $\alpha_1 GABA_A$ receptor may play a more prominent role in physical dependence induced by benzodiazepine-type drugs comes from a report that zaleplon, similar to zolpidem, engendered a withdrawal syndrome in baboons *(72)*.

The recent development of compounds with selective efficacy at $\alpha_2 GABA_A$-, $\alpha_3 GABA_A$-, and/or $\alpha_5 GABA_A$-receptor subtypes now provides the opportunity to evaluate the role of these receptors in physical dependence induced by benzodiazepine-type drugs. Mirza and Neilsen *(60)* evaluated the degree to which chronic treatment with compounds varying in both selectivity and efficacy at $GABA_A$-receptor subtypes engendered seizures in mice following administration of the inverse agonist FG-7142. Chronic treatment with zolpidem, as well as the selective compounds L-838,417 and SL651498 (full agonist

*In drug discrimination models of physical dependence, animals are administered an agonist chronically and trained to discriminate injections of an antagonist from vehicle. Because the antagonist precipitates withdrawal, the "cue" that is the basis for the discrimination presumably consists of withdrawal signs. In this model, drugs that precipitate withdrawal "substitute" for the antagonist (i.e., share discriminative stimulus effects) and agonists can shift the dose–response function for the antagonist to the right.

at α_2GABA_A- and α_3GABA_A receptors, partial agonist at α_1GABA_A- and α_5GABA_A receptors), did not result in seizures following FG-7142 administration. Another recent study has shown that chronic treatment with TPA023 also did not result in FG-7142-induced seizures in mice *(49)*. These findings suggest that physical dependence only occurs with nonselective benzodiazepine-type compounds, i.e., action at all receptor subtypes is required for physical dependence to develop. However, Mirza and Neilsen *(60)* also observed that chronic treatment with nonselective partial agonists did not result in FG-7142-induced seizures, suggesting that relatively high efficacy also might be a requirement for the development of physical dependence.

7. Tolerance Following Chronic Treatment

In addition to the development of physical dependence, chronic benzodiazepine treatment can result in tolerance to some behavioral effects. It is important to note that the development of physical dependence does not require the development of tolerance, and that tolerance can occur in the absence of physical dependence *(21)*. Moreover, the time-course for the development of tolerance varies for different behavioral effects of benzodiazepine-type drugs. In humans, for example, tolerance develops rapidly to sedative effects and motor coordination deficits, whereas tolerance does not always develop to the anxiolytic or memory impairing effects of benzodiazepine-type drugs after long periods of use *(23,73,74)*. A clear gap in our knowledge about tolerance development is the extent to which the reinforcing effects of benzodiazepine-type drugs change over time, i.e., whether or not tolerance to the reinforcing effects of benzodiazepines develops after chronic exposure. Based on available information, tolerance to reinforcing effects of benzodiazepine-type drugs appears unlikely, as self-administration of midazolam or zolpidem was shown to be stable over relatively long durations of exposure *(61,71)*. Moreover, indirect evidence that tolerance to the reinforcing effects of benzodiazepines does not occur comes from the observation that long-term use by humans is not associated with escalation in the ingested dose of drug across time *(21,23)*.

8. Conclusions

Of the diverse types of ligands that act at the $GABA_A$ receptor, the benzodiazepines and related drugs are unique in having widespread clinical use and the liability for abuse and dependence. Recent epidemiological data suggests that abuse of benzodiazepine-type drugs may be on the upswing, with a shift in use to younger people engaging in recreational abuse of these drugs, rather than groups of people primarily misusing benzodiazepines in a therapeutic setting. Nevertheless, controversies remain regarding the extent to which benzodiazepine-type drugs represent a serious risk as "traditional" drugs of abuse.

Laboratory findings suggest that benzodiazepine-type drugs have reinforcing effects both in human and nonhuman subjects. However, benzodiazepine-type drugs appear to have lower reinforcing effectiveness compared to other drugs of abuse, such as psychomotor stimulants. This conclusion is drawn primarily from the observations that benzodiazepines typically do not function as reinforcers in normal healthy subjects and that benzodiazepine-type drugs tend to maintain self-administration at intermediate levels in nonhuman subjects. Recent research has begun to explore the role of $GABA_A$-receptor subtypes in the reinforcing effects of benzodiazepine-type drugs, and unlike other behavioral effects (e.g., motor coordination deficits), reinforcing effects are not easily attributed to a single receptor subtype. Perhaps the most firm conclusion at this point is that α_1GABA_A receptors are not necessary for self-administration of benzodiazepine-type compounds, although they might be sufficient. Research with more selective compounds that are full agonists for different subtypes clearly is needed to resolve some of the issues with our understanding of benzodiazepine reinforcement.

In addition to reinforcing effects, it is well documented that chronic exposure to benzodiazepines results in physical dependence, characterized by a withdrawal syndrome. In both human and nonhuman subjects, this withdrawal syndrome is considered to be intermediate in severity. Consensus on the most important factors determining the presence and severity of physical dependence has not yet been reached. However, dose clearly plays a role and, importantly, the length of exposure needed for dependence to develop might be relatively short. Regarding receptor mechanisms, initial studies suggested that α_1GABA_A selective agonists are devoid of physical dependence liability, whereas the most recent findings in humans and nonhuman primates indicate that long-term use of these compounds can be associated with physical dependence. Preliminary results suggest that compounds with selectivity for α_2GABA_A-, α_3GABA_A-, and/or α_5GABA_A receptors do not induce physical dependence, but these findings are complicated by the relatively low intrinsic efficacy of these ligands. As with reinforcing effects, systematic studies with selective compounds having relatively high intrinsic efficacy at particular subtypes should shed light on these important mechanistic issues.

Finally, abuse of benzodiazepines—any drug, for that matter—must always be considered in the historical context in which the compound was both developed and subsequently evolved into a drug of abuse. The history of benzodiazepine abuse in the US and Europe clearly illustrates that society at large can determine, with remarkable efficiency and often irrespective of scientific data, the relative reinforcing effectiveness of a class of drugs. Benzodiazepine-type drugs clearly have the potential to treat, and do treat a significant section of the population with anxiety and sleep disorders, but it will remain an onerous task to escape the stigma associated with these drugs brought on by past events. Given recent increases in inappropriate use of these compounds, there may be, in fact, clear reasons for concern. Regardless, the potential for the development of a new compound lacking abuse-related effects may be a reality, and if such a compound is discovered then the simple fact of having action at $GABA_A$ receptors should not be the prime determinant of scheduling, despite events of the past.

Acknowledgments

Preparation of this chapter was supported by U.S.P.H.S. Grants DA11792, DA20304, AA13850, and RR00168.

References

1. Höld, K. M., Sirisoma, N. S., Ikeda, T., Narahashi, T., and Casida, J. E. (2000) α-Thujone (the active component of absinthe): γ-Aminobutyric acid type A receptor modulation and metabolic detoxification. *Proc. Natl. Acad. Sci. USA* **97,** 3826–3831.
2. Kumar, S., Fleming, R. L., and Morrow, A. L. (2004) Ethanol regulation of γ-aminobutyric $acid_A$ receptors: genomic and nongenomic mechanisms. *Pharmacol. Ther*. **101,** 211–226.
3. Ator, N. A. and Griffiths, R. R. (1987) Self-administration of barbiturates and benzodiazepines: A review. *Pharmacol. Biochem. Behav*. **27,** 391–398.
4. Tone, A. (2005) Listening to the past: history, psychiatry, and anxiety. *Can. J. Psychiatry* **50,** 373–380.
5. Woods, J. H. and Winger, G. (1995) Current benzodiazepine issues. *Psychopharmacology* **118,** 107–115.
6. Nutt, D. J. (2005) Overview of diagnosis and drug treatments of anxiety disorders. *CNS Spectr.* **10,** 49–56.
7. Griffiths, R. R. (1995) Commentary on review by Woods and Winger. Benzodiazepines: long-term use among patients is a concern and abuse among polydrug abusers is not trivial. *Psychopharmacology* **118,** 116–117.
8. Pritchett, D. B., Lüddens, H., and Seeburg, P. H. (1989) Type I and type II $GABA_A$ benzodiazepine receptor produced in transfected cells. *Science* **245,** 1389–1392.
9. Löw, K., Crestani, F., Keist, R., et al. (2000) Molecular and neuronal substrate for the selective attenuation of anxiety. *Science* **290,** 131–134.
10. McKernan, R. M., Rosahl, T. W., Reynolds, D. S., et al. (2000) Sedative but not anxiolytic properties of benzodiazepines are mediated by the $GABA_A$ receptor alpha1 subtype. *Nature Neurosci.* **3,** 587–592.
11. Rudolph, U., Crestani, F., and Möhler, H. (2000) $GABA_A$ receptor subtypes: dissecting their pharmacological functions. *Trends Pharmacol. Sci.* **22,** 188–194.
12. Rowlett, J. K., Platt, D. M., Lelas, S., Atack, J. R., and Dawson, G. R. (2005) Different $GABA_A$ receptor subtypes mediate the anxiolytic, abuse-related, and motor effects of benzodiazepine-like drugs in primates. *Proc. Natl. Acad. Sci. USA* **102,** 915–920.

13. McKernan, R. M. and Whiting, P. J. (1996) Which $GABA_A$-receptor subtypes really occur in the brain? *Trends Pharmacol. Sci.* **19,** 139–143.
14. Collinson, N., Kuenzi, F. M., Jarolimek, W., et al. (2002) Enhanced learning and memory and altered GABAergic synaptic transmission in mice lacking the α5 subunit of the $GABA_A$ receptor *J. Neurosci.* **22,** 5572–5580.
15. Crestani, F., Keist, R., Fritschy, J. M., et al. (2002) Trace fear conditioning involves hippocampal α_5 $GABA_A$ receptors. *Proc. Natl. Acad. Sci. USA* **99,** 8980–8985.
16. Speaker, S. (1997) From "happiness pills" to "national nightmare": changing cultural assessment of minor tranquilizers in America, 1955–1980. *J. Hist. Med. Allied Sci.* **52,** 338–376.
17. Gordon, B. (1979) *I'm Dancing as Fast as I Can.* Harper & Row, New York, NY.
18. US Committee on Labor and Human Resources (1979) Use and misuse of benzodiazepines. Washington DC, US. Government Printing Office.
19. Ator, N. A. (2005) Contributions of $GABA_A$ receptor subtype selectivity to abuse liability and dependence potential of pharmacological treatments for anxiety and sleep disorders. *CNS Spectr.* **10,** 31–39.
20. Woods, J. H., Katz, J. L., and Winger, G. (1987) Benzodiazepines: Use, abuse, and consequences. *Pharmacol. Rev.* **39,** 251–419.
21. Woods, J. H., Katz, J. L., and Winger, G. (1992) Benzodiazepines: Use, abuse, and consequences. *Pharmacol. Rev.* **44,** 151–347.
22. Rosenbaum, J. F. (2005) Attitudes toward benzodiazepines over the years. *J. Clin. Psychiatry* **66,** 4–8.
23. Griffiths, R. R. and Weerts, E. M. (1997) Benzodiazepine self-administration in humans and laboratory animals – implications for problems of long-term use and abuse. *Psychopharmacology* **134,** 1–37.
24. Griffiths, R. R. and Johnson, M. W. (2005) Relative abuse liability of hypnotic drugs: A conceptual framework and algorithm for differentiating among compounds. *J. Clin. Psychiatry* **66,** 31–41.
25. Crane, E. H. and Nemanski, N. (2004) Benzodiazepines in drug abuse-related emergency department visits: 1995–2002. The DAWN Report, April 2004, Office of Applied Studies, US Substance Abuse and Mental Health Services Administration (available online at http://dawninfo.samhsa.gov/). Last accessed: January 3, 2007.
26. Crane, E. H. and Nemanski, N. (2004) Demographic characteristics of benzodiazepine-involved ED visits. The DAWN Report, July 2004, Office of Applied Studies, US Substance Abuse and Mental Health Services Administration (available online at http://dawninfo.samhsa.gov/). Last accessed: January 3, 2007.
27. The DASIS Report (2005) Characteristics of primary tranquilizer admissions: 2002. Drug and Alcohol Services Information System, Office of Applied Studies, Substance Abuse and Mental Health Services Administration (available online at http://www.oas.samhsa.gov). Last accessed: January 3, 2007.
28. Ciraulo, D. A., Sands, B. F., and Shader, R. I. (1988) Critical review of liability for benzodiazepine abuse among alcoholics. *Am. J. Psychiatry* **145,** 1501–1506.
29. de Wit, H. and Doty, P. (1993) Preference for ethanol and diazepam in light and moderate social drinkers: a within-subjects study. *Psychopharmacology* **115,** 529–538.
30. Evans, S. M., Griffiths, R. R., and de Wit, H. (1996) Preference for diazepam, but not buspirone, in moderate drinkers. *Psychopharmacology* **123,** 154–163.

31. Sobel, K. G. and McCart, G. M. (1983) Drug use and accidental falls in an intermediate care facility. *Drug Intell. Clin. Pharm.* **17,** 539–542.
32. Herings, R. M., Stricker, B. H., de Boer, A., Bakker, A., and Sturmans, F. (1995) Benzodiazepines and the risk of falling leading to femur fractures. Dosage more important than elimination half-life. *Arch. Intern. Med.* **155,** 1801–1807.
33. Ator, N. A. and Griffiths, R. R. (2003) Principles of drug abuse liability assessment in laboratory animals. *Drug Alcohol Depend.* **70,** S55–S72.
34. Griffiths, R. R., Bigelow, G. E., and Ator, N. A. (2003) Principles of initial experimental drug abuse liability assessment in humans. *Drug Alcohol Depend.* **70,** S41–S54.
35. Lelas, S., Spealman, R. D., and Rowlett, J. K. (2000) Using behavior to elucidate receptor mechanisms: A review of the discriminative stimulus effects of benzodiazepines. *Exp. Clin. Psychopharmacol.* **8,** 294–311.
36. Helmus, T. C., Tancer, M., and Johanson, C. E. (2005) Reinforcing effects of diazepam under anxiogenic conditions in individuals with social anxiety. *Exp. Clin. Psychopharmacol.* **13,** 348–356.
37. Iguchi, M. Y., Handelsman, L., Bickel, W. K., and Griffiths, R. R. (1993) Benzodiazepine and sedative use/abuse by methadone maintenance clients. *Drug Alcohol Depend.* **32,** 257–266.
38. Darke, S. G., Ross, J. E., and Hall, W. D. (1995) Benzodiazepine use among injecting heroin users. *Med. J. Aust.* **162,** 645–647.
39. Perera, K. M., Tulley, M., and Jenner, F. A. (1987) The use of benzodiazepines among drug addicts. *Br. J. Addict.* **82,** 511–515.
40. Gelkopf, M., Bleich, A., Hayward, R., Bodner, G., and Adelson, M. (1999) Characteristics of benzodiazepine abuse in methadone maintenance treatment patients: a 1 year prospective study in an Israeli clinic. *Drug Alcohol Depend.* **55,** 63–68.
41. Bergman, J. and Johanson, C. E. (1985) The reinforcing properties of diazepam under several conditions in the rhesus monkey. *Psychopharmacology* **86,** 108–113.
42. Griffiths, R. R., Lamb, R. J., Sannerud, C. A., Ator, N. A., and Brady, J. V. (1991) Self-injection of barbiturates, benzodiazepines and other sedative-anxiolytics in baboons. *Psychopharmacology* **103,** 154–161.
43. Broadbear, J. H., Winger, G., and Woods, J. H. (2005) Self-administration of methohexital, midazolam and ethanol: effects on the pituitary-adrenal axis in rhesus monkeys. *Psychopharmacology* **178,** 83–91.
44. Winger, G., Stitzer, M. L., and Woods, J. H. (1975) Barbiturate-reinforced responding in rhesus monkeys: comparisons of drugs with different durations of action. *J. Pharmacol. Exp. Ther.* **195,** 505–514.
45. Rowlett, J. K., Rodefer, J. S., and Spealman, R. D. (2002) Self-administration of cocaine, alfentanil, and nalbuphine under progressive-ratio schedules: Consumer demand and labor supply analysis of relative reinforcing effectiveness. *Exp. Clin. Psychopharmacol.* **10,** 367–375.
46. Weerts, E. M., Kaminski, B. J., and Griffiths, R. R. (1998) Stable low-rate midazolam self-injection with concurrent physical dependence under conditions of long-term continuous availability in baboons. *Psychopharmacology* **135,** 70–81.
47. Griffiths, R. R., Sannerud, C. A., Ator, N. A., and Brady, J. V. (1992) Zolpidem behavioral pharmacology in baboons: self-injection, discrimination, tolerance and withdrawal. *J. Pharmacol. Exp. Ther.* **260,** 1199–1208.

48. Ator, N. A. (2000) Zaleplon and triazolam: drug discrimination, plasma levels, and self-administration in baboons. *Drug Alcohol Depend.* **24,** 55–68.
49. Atack, J. R., Wafford, K. A., Tye, S. J., et al. (2006) TPA023 [7-(1,1-Dimethylethyl)-6-(2-ethyl-2H-1,2,4-triazol-30ylmethoxy)-3-(2-fluorophenyl)-1,2,4-triazolo[4,3-b] pyridazine], an agonist selective for α2 and α3-containing $GABA_A$ receptors, is a nonsedating anxiolytic in rodents and primates. *J. Pharmacol. Exp. Ther.* **316,** 410–422.
50. Dawson, G. R., Collinson, N., and Atack, J. R. (2005) Development of subtype selective $GABA_A$ modulators. *CNS Spectr.* **10,** 21–27.
51. Ashton, H. (1991) Protracted withdrawal syndromes from benzodiazepines. *J. Subst. Abuse Treat.* **8,** 19–28.
52. Petursson, H. (1994) The benzodiazepine withdrawal syndrome. *Addiction.* **89,** 1455–1459.
53. Mintzer, M. Z. and Griffiths, R. R. (2005) Flumazenil-precipitated withdrawal in healthy volunteers following repeated diazepam exposure. *Psychopharmacology* **178,** 259–267.
54. O'Brien, C. P. (2005) Benzodiazepine use, abuse, and dependence. *J. Clin. Psychiatry* **66,** 28–33.
55. Orsini, A., Brancato, V., and Smeraldi, E. (1990) Length of anxiolytic treatment and benzodiazepine withdrawal symptoms: a double blind study with clotiazepam. *New Trends Exp. Clin. Psychiatry* **6,** 45–52.
56. Mintzer, M. Z., Stoller, K. B., and Griffiths, R. R. (1999) A controlled study of flumazenil-precipitated withdrawal in chronic low-dose benzodiazepine users. *Psychopharmacology* **147,** 200–209.
57. Roth, T., Walsh, J. K., Krystal, A., Wessel, T., and Roehrs, T. A. (2005) An evaluation of the efficacy and safety of eszopiclone over 12 months in patients with chronic primary insomnia. *Sleep Med.* **6,** 487–495.
58. Martin, J. R., Moreau, J. L., Jenck, F., and Pieri, L. (1998) Sarmazenil-precipitated withdrawal: A reliable method for assessing dependence liability of benzodiazepine receptor ligands. *Pharmacol. Biochem. Behav.* **59,** 939–944.
59. Kaminski, B. J., Sannerud, C. A., Weerts, E. M., Lamb, R. J., and Griffiths, R. R. (2003) Physical dependence in baboons chronically treated with low and high doses of diazepam. *Behav. Pharmacol.* **14,** 331–342.
60. Mirza, N. R. and Nielsen, E. Ø. (2006) Do subtype-selective $GABA_A$ receptor modulators have a reduced propensity to induce physical dependence in mice? *J. Pharmacol. Exp. Ther.* **316,** 1378–1385.
61. Weerts, E. M., Ator, N. A., Grech, D. M., and Griffiths, R. R. (1998) Zolpidem physical dependence assessed across increasing doses under a once-daily dosing regimen in baboons. *J. Pharmacol. Exp. Ther.* **285,** 41–53.
62. McMahon, L. R., Gerak, L. R., and France, C. P. (2001) Potency of positive γ-aminobutyric $acid_A$ modulators to substitute for a midazolam discriminative stimulus in untreated monkeys does not predict potency to attenuate a flumazenil discriminative stimulus in diazepam-treated monkeys. *J. Pharmacol. Exp. Ther.* **298,** 1227–1235.
63. Lukas, S. E. and Griffiths R. R. (1982) Precipitated withdrawal by a benzodiazepine receptor antagonist (Ro 15-1788) after 7 days of diazepam. *Science* **217,** 1161–1163.

64. Grant, S. J., Galloway, M. P., Mayor, R., et al. (1985) Precipitated diazepam withdrawal elevates noradrenergic metabolism in primate brain. *Eur. J. Pharmacol.* **107,** 127–132.
65. Gallager, D. W., Heninger, K., and Heninger, G. (1986) Periodic benzodiazepine antagonist administration prevents benzodiazepine withdrawal symptoms in primates. *Eur. J. Pharmacol.* **132,** 31–38.
66. Lukas, S. E. and Griffiths, R. R. (1984) Precipitated diazepam withdrawal in baboons: effects of dose and duration of diazepam exposure. *Eur. J. Pharmacol.* **100,** 163–171.
67. Sloan, J. W., Martin, W. R., and Wala, E. (1993) Effect of the chronic dose of diazepam on the intensity and characteristics of the precipitated abstinence syndrome in the dog. *J. Pharmacol. Exp. Ther.* **265,** 1152–1162.
68. Allison, C. and Pratt, J. A. (2003) Neuroadaptive processes in GABAergic and glutamatergic systems in benzodiazepine dependence. *Pharmacol. Ther.* **98,** 171–195.
69. Perrault, G., Morel, E., Sanger, D. J., and Zivkovic, B. (1992) Lack of tolerance and physical dependence upon repeated treatment with the novel hypnotic zolpidem. *J. Pharmacol. Exp. Ther.* **263,** 298–303.
70. Elliot, E. E. and White, J. M. (2000) Precipitated and spontaneous withdrawal following administration of lorazepam but not zolpidem. *Pharmacol. Biochem. Behav.* **66,** 361–369.
71. Weerts, E. M. and Griffiths, R. R. (1998) Zolpidem self-injection with concurrent physical dependence under conditions of long-term continuous availability in baboons. *Behav. Pharmacol.* **9,** 285–297.
72. Ator, N. A., Weerts, E. M., Kaminski, B. J., Kautz, M. A., and Griffiths, R. R. (2000) Zaleplon and triazolam physical dependence assessed across increasing doses under a once-daily dosing regimen in baboons. *Drug Alcohol Depend.* **61,** 69–84.
73. Cowley, D. S., Roy-Byrne, P. P., Radant, A., et al. (1995) Benzodiazepine sensitivity in panic disorder: Effects of chronic alprazolam treatment. *Neuropsychopharmacology* **12,** 147–157.
74. Stoops, W. W. and Rush, C. R. (2003) Differential effects in humans after repeated asministratios of zolpidem and triazlolam. *Am. J. Drug Alcohol Abuse.* **29,** 281–299.
75. Smith, A. J., Alder, L., Silk, J., et al. (2001) Effect of a subunit on allosteric modulation of ion channel function in stably expressed human recombinant γ-aminobutyric acid$_A$ receptors determined using ^{36}Cl ion flux. *Mol. Pharmacol.* **59,**1108–1118.

8

Mechanisms of GABA$_A$ and GABA$_B$ Receptor Gene Regulation and Cell Surface Expression

David H. Farb, Janine L. Steiger, Stella C. Martin, Maria C. Gravielle, Terrell T. Gibbs, and Shelley J. Russek

Summary

The γ-aminobutyric acid (GABA) neurotransmitter acting through ionotropic and metabotropic receptor classes exerts the major inhibitory control in the central nervous system. Therapeutic agents targeting GABA receptors (GABA-R), such as benzodiazepines and baclofen, are used to treat many nervous system conditions, including anxiety and spasticity. The subunit composition of GABA-Rs at the cell surface plays a critical role in determining their physiological and pharmacological properties, and alteration of GABA-R subunit expression has been associated with a number of diseases including schizophrenia, temporal lobe epilepsy, and alcoholism. The ionotropic type A GABA receptor (GABA$_A$R) and type C GABA receptor (GABA$_C$R) are pentameric complexes that comprise a ligand gated chloride channel. The metabotropic type B GABA receptor (GABA$_B$R) is a heterodimer that couples G protein-signaling to GABA binding. There are eight classes of ionotropic receptor subunits and only two metabotropic receptor subunit classes. Most of the GABA$_A$R subunit genes are localized in syntenic β-α-γ gene clusters on four chromosomes but the two GABA$_B$R genes are localized on distinct chromosomes. Control over subunit expression in different brain regions and during development is orchestrated at the genomic level by the use of multiple promoter regions and through the alternative splicing of GABA-R subunit RNAs. This chapter examines current GABA-R research relevant to the many levels of control over receptor gene regulation and cell surface receptor expression that may be relevant to both health and disease.

From: *The Receptors: The GABA Receptors, Third Edition*
Edited by: S. J. Enna and H. Möhler © Humana Press Inc., Totowa, NJ

Key Words: GABA receptor; $GABA_A$ receptor; $GABA_B$ receptor; genes; transcription; gene expression; alternative promoters; alternative splicing; uncoupling; protein turnover.

1. Introduction

γ-Aminobutyric acid (GABA) is the primary inhibitory transmitter in the adult nervous system and can activate three classes of receptors: ionotropic type A GABA ($GABA_A$), metabotropic type B GABA ($GABA_B$), and ionotropic type C GABA ($GABA_C$) (for review *see* refs. *1–3*). The ligand-gated $GABA_A$ receptor ($GABA_AR$) is a pentameric complex assembled from eight subunit families defined by amino acid (aa) sequence homology: α_{1-6}, β_{1-3}, γ_{1-3}, δ, ε, π, θ, and ρ_{1-3}. With the availability of powerful molecular tools, the cloning of $GABA_ARs$ led to the discovery that each subunit is encoded by a distinct gene and most genes are organized in *β-α-α-γ* and *β-α-γ* gene clusters on different chromosomes *(4,5)*. Alternative splicing and alternative promoters generate additional heterogeneity in $GABA_AR$ subunit variants that may allow for the assembly of thousands of possible pentameric combinations. In contrast, the metabotropic $GABA_BR$ is made up of two distinct subunits $GABA_BR1$ and $GABA_BR2$. It is now clear that the most abundant $GABA_BR1$ isoforms ($GABA_BR1a$ and $GABA_BR1b$) are produced by alternative promoters in the *$GABA_BR1$* gene *(6)*, and there is no evidence for $GABA_BR2$ variants *(7)*.

Whereas the ligand-gated $GABA_CR$ is predominantly found in the retina *(8–10)*, $GABA_ARs$ and $GABA_BRs$ are broadly expressed throughout the embryonic and adult nervous systems *(11–15)*. Moreover, selective expression of $GABA_AR$ and $GABA_BR$ subunits during normal physiology and in response to a variety of neurological and psychiatric diseases may play a major role in the choice of subunits available for assembly. Consistent with this hypothesis, multiple receptor subtypes have been described based on pharmacological analyses of recombinant receptors and native tissue *(16,17)*. Differential gene regulation of *GABA-R* expression may contribute to the particular function of GABA-Rs in a cell.

The precise mechanisms that account for alterations in *$GABA_AR$* and *$GABA_BR$* subunit gene expression are only now beginning to be revealed. Exciting new discoveries indicate that GABA-R subunit expression is controlled at multiple levels ranging from subunit gene expression to control of protein turnover by cell-signaling pathways. Several mechanisms that underlie regulation of subunit expression include: chromatin remodeling, transcription initiation, alternative splicing, messenger RNA (mRNA) stability, translation, post-translational modification, intracellular trafficking, and protein degradation. The purpose of this review is to discuss evidence for genomic (and nongenomic) mechanisms regulating *$GABA_AR$* and *$GABA_BR$* gene expression. It is intriguing to remember that GABA-induced uncoupling of the allosteric interactions between the GABA and benzodiazepine recognition sites on the $GABA_AR$ was first described in 1990 *(18)*. Subsequently, experiments using primary neuronal cultures demonstrated that the occupancy of the $GABA_AR$ can induce downregulation of receptor number *(19–21)* through a transcriptional mechanism *(22,23)*. Rapid progress in the molecular determinants of the $GABA_BR$ system was prompted by the cloning of *$GABA_BR1$* in 1997 *(24)*. Only 1 yr later several groups cloned the *$GABA_BR2$* subunit gene and determined that functional $GABA_BRs$ assemble as a heterodimer of $GABA_BR1$ and $GABA_BR2$ subunits *(25–30)*. Moreover, recent advances indicate that regulation of GABA-R function in the brain is a highly compensating system. The authors hope that this review will illustrate the molecular mechanisms for underlying adaptive changes in *$GABA_AR$* and *$GABA_BR$* gene expression that may ultimately impact receptor function.

2. $GABA_A$ Receptors

Fast inhibitory neurotransmission in the central nervous system (CNS) is primarily mediated by the $GABA_AR$. This receptor consists of a ligand-gated chloride channel that is opened by the binding of GABA. $GABA_AR$ function can be regulated by binding of different allosteric modulators to the receptor such as benzodiazepines, barbiturates, neurosteroids, anesthetics, and convulsants *(31)*. $GABA_ARs$ have a pentameric structure and can be made up of a variety of subunit combinations with different pharmacological properties *(32,33)*. Nineteen $GABA_AR$ subunits are present in the mammalian CNS and the majority of receptors are made up of two α, two β, and one γ subunits *(2,34–37)*. The δ, ε, and π subunits may substitute for the γ subunits *(34)*, whereas the θ subunit promotes the action of the anesthetic etomidate on β_1 containing receptors *(38)*. The ρ subunits form homo- and hetero-oligomeric receptors with different ρ subunit subtypes and these receptors are often classified as $GABA_CRs$ *(2)*.

2.1. The $GABA_AR$ Family of Genes and Gene Products

2.1.1. Chromosomal Localization and Gene Clusters

Most of the *$GABA_AR$* subunit genes are organized in syntenic clusters localized to four different human chromosomes: 4, 5, 15, and X (Fig. 1) (for review *see* refs. *39,40*). Genes within the clusters have a conserved transcriptional orientation that is consistent with duplication events stemming from an ancestral-gene cluster containing α-, β-, and γ-like subunits. It has been proposed that a tandem duplication of an α-subunit within the ancestral cluster *β-α-γ* generated *β-α-α-γ* after a whole genome duplication and further before a whole genome

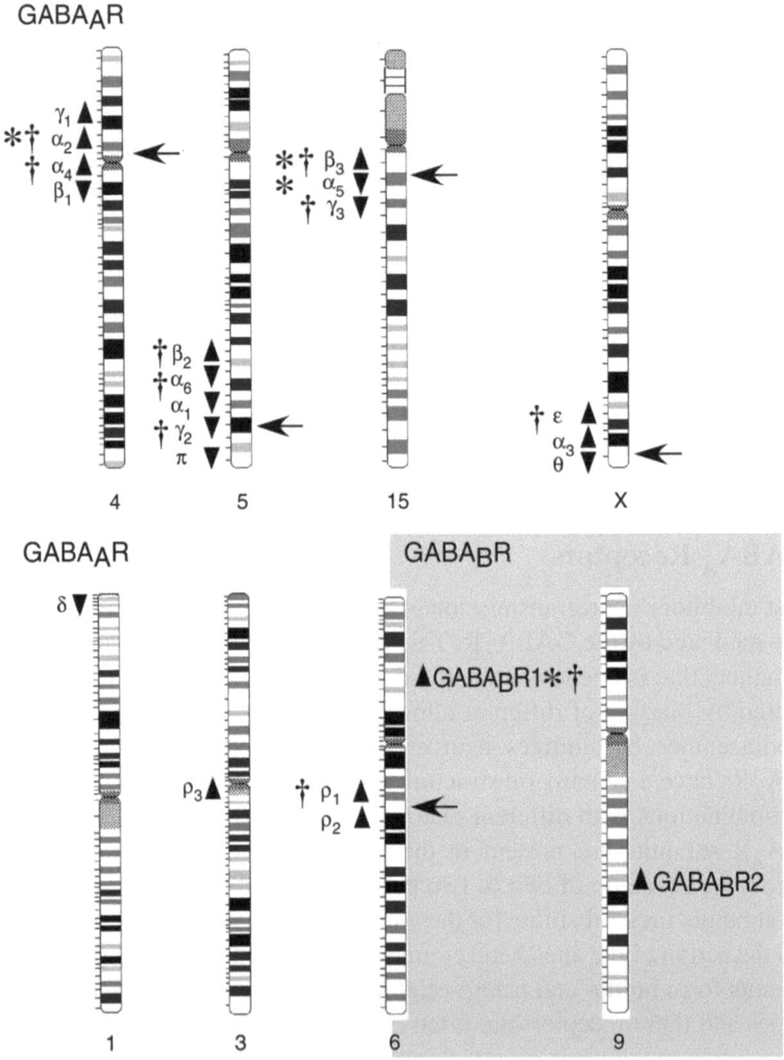

Fig. 1. Multiplicity of human $GABA_AR$ and $GABA_BR$ subunits: multiple genes with additional diversity generated by alternative R promoters and RNA splicing. The $GABA_AR$ and $GABA_BR$ subunit genes are localized to human chromosomes 1, 3, 4, 5, 6, 9, 15, and X. Genes with alternative promoters (*) and alternative splicing (†) are indicated. $GABA_AR$ gene clusters are indicated by an arrow on chromosomes 4, 5, 6, 15, and X. The orientation of each gene is shown where the up arrowhead (▲) indicates the gene is transcribed on the positive strand and the down arrowhead (▼) indicates the gene is transcribed on the negative strand. This information is from the NCBI Map Viewer, Human Genome Build 36.1. The receptor subunit gene names are α_1 *(GABRA1)*; α_2 *(GABRA2)*; α_3 *(GABRA3)*; α_4 *(GABRA4)*; α_5 *(GABRA5)*; α_6 *(GABRA6)*; β_1 *(GABRB1)*; β_2 *(GABRB2)*;

duplication. The same transcriptional organization is present in both four-gene and three-gene clusters *(4,5,41–44)*. The two β-α-α-γ $GABA_A R$ clusters contain the γ_1-α_2-α_4-β_1 genes that are localized to chromosome 4p14-q12 (Fig. 1) *(4,45–48)* and (although in the reverse order) the β_2-α_6-α_1-γ_2 genes are found at chromosome 5q31.2-q35 *(4,45,48–51)*. The three-gene clusters containing a single α-subunit are the β_3-α_5-γ_3 genes that are localized to chromosome 15q11-q13 *(4,42,43,52–55)*, and ε-α_3-θ at chromosome Xq28 (Fig. 1) *(4,5,41,44)*.

A tandem gene duplication of an ancestral ρ gene may have given rise to ρ_1 ρ_2 genes that have been localized to chromosome 6q15 *(56)* after the duplication and translocation of another ρ gene, ρ_3 that has been localized to 3q12.2 (Fig. 1) *(57)*. In contrast, there is a single δ gene that is localized to chromosome 1p36.3 (Fig. 1) *(58,59)*. The π gene is localized to chromosome 5q35.1 and is separated from the β_2-α_6-α_1-γ_2 gene cluster (Fig. 1) *(60)*. Gene duplication has generated a diverse set of $GABA_A R$ subunits with different functions as a consequence of different aa sequences. Another consequence of gene duplication is the flexibility that different promoter regions bring to the level and pattern of gene expression for each subunit. Gene clusters can provide a mechanism for coordinate expression. For example, the embryonically expressed α_4, β_1, and γ_1 all reside in the chromosome 4 cluster *(61)*. In addition, the most abundantly expressed subunits are α_1, β_2, and γ_2, and these subunits are all found in the chromosome 5 cluster. Locus control regions have been described for erythroid specific and developmental switching of the human-globin gene clusters *(62)*. Certain $GABA_A R$ clusters may be subject to a locus control region that is distant from but acts as a regulator of the cluster. Many aspects of $GABA_A R$ subunit gene expression do not correlate with the location of the subunit gene within the clusters, reflecting the complexity of expression patterns and the presence of individual as well as coordinate control of many subunits.

2.1.2. Alternative Promoters

Alternative promoters for a single gene provide additional diversity of receptor subunit sequence and expression involving the 5′ noncoding sequence and amino (N) terminal aa sequence of the subunit. The activity of alternative promoters has been described for human α_2, α_5, and β_3 $GABA_A R$ subunit genes (Table 1). Two alternate 5′ noncoding exons in the α_2 gene, termed 1A and 1B, result from the activity of two different promoters, and higher expression of 1B compared to 1A is detected in all brain regions

---

Fig. 1. *(Continued)* β_3 *(GABRB3)*; γ_1 *(GABRG1)*; γ_2 *(GABRG2)*; γ_3 *(GABRG3)*; δ *(GABRD)*; ε *(GABRE)*; θ *(GABRQ)*; ρ_1 *(GABRR1)*; ρ_2 *(GABRR2)*; ρ_3 *(GABRR3)*; $GABA_B R1$ *(GABBR1)*, and $GABA_B R2$ *(GABBR2)*.

Table 1
$GABA_AR$ and $GABA_B$ Receptor Subunits are Formed by Alternative Promoters and Alternative Splicing

Subunit	Isoform	Alters	References
Alternative promoter			
α_2	1A, 1B	5′ Noncoding	*63*
α_5	1A, 1B, 1C	5′ Noncoding	*64*
β_3	1, 1a	5′ Noncoding signal peptide	*65*
$GABA_BR1$	$GABA_BRIa$	5′ Noncoding N-terminus	*6*
	$GABA_BRIb$	5′ Noncoding N-terminus	*6*
Alternative splicing			
α_2	i–vii	3′ Noncoding C-terminal truncation transmembrane region	*63*
α_4	α_4L, $\alpha_4\Delta$	C-terminal truncation	*68*
α_5	α_5S, α_5L	C-terminal truncation	*64*
α_6	α_6S, α_6L	C-terminal truncation	*71*
β_2	β_2S, β_2L	Truncated	*73*
γ_2	γ_2S,γ_2L	Phosphorylation site	*74*
	γ_2XL	N-terminal domain	*77*
γ_3	γ_3S, γ_3L	Phosphorylation site	*75*
ε	εS, εL	Truncated	*70*
ρ_1	ρ_1, $\rho_1\Delta51$, $\rho_1\Delta450$	N-terminal domain	*72*
$GABA_BRI$	$GABA_BRIc$	One SD	*7*
	$GABA_BRle$	Soluble EC domain	*262*

examined *(63)*. Three different promoter regions in the α_5 gene produce transcripts that differ in the 5' noncoding region and correspond to the first exon, termed 1A–C, and may serve to control α_5-subunit expression by transcriptional control and differential stability (Table 1) *(64)*. The β_3-gene contains two distinct promoters which produce transcripts, from exon 1 or exon 1a, that differ in noncoding and coding regions resulting in two alternative signal peptide sequences (Table 1) *(65)*. Transcripts containing exon 1a are the minor product of the β_3 gene in the brain. Different N-terminal signal peptide sequences could affect receptor assembly, processing or distribution of the β_3 subunit in neurons.

2.1.3. Alternative Splicing

Unlike alternative promoters that can alter the expression level and nature of N-terminal sequences only, alternative splicing can alter the expression level and composition of any exon in the gene transcript. Eight $GABA_AR$ subunit genes α_2, α_4, α_5, α_6, β_2, γ_2, γ_3, ε, and ρ_1 are subject to alternative splicing (Table 1). Six $GABA_AR$ α_2 subunit isoforms that differ in their 3′ noncoding regions can be produced by alternative splicing mechanisms *(63,66)*. Linkage and association analyses suggest that these α_2 subunit variants are associated with brain oscillations in the β frequency range and the risk of alcoholism *(67)*.

A truncated α_4 subunit mRNA product results from alternative splicing (Table 1) and is more abundant in embryonic whole brain and adult cerebellum suggesting a developmental and regional regulation *(68)*. Alternative splicing of human α_5, α_6, ρ_1, and ε subunit mRNAs also leads to truncated transcripts that apparently lack function (Table 1) *(5,69–72)*. Two splice variants of the β_2 subunit differ by a 38aa insertion containing a phosphorylation site for calmodulin-dependent protein kinase II (Table 1) *(73)*. Similarly, two splice variants have been described for γ_2 and γ_3 subunits that differ by the presence or absence of an insert containing a protein kinase C (PKC) phosphorylation site (Table 1) *(74,75)*. The γ_2L variant contains a 24-bp insertion from a separate exon encoding eight additional aa not present in γ_2S in the cytoplasmic domain between the third- and fourth putative membrane spanning regions. The two variants of the γ_2 subunit, γ_2S and γ_2L, are present in different regional and developmental patterns *(76)*. An additional variant of γ_2, γ_2XL, has been described that contains an alternatively spliced exon containing an additional 40 aa in the N-terminal extracellular region and which is not expressed at the cell surface *(77)*.

2.2. Differential Expression of $GABA_AR$ in the CNS

2.2.1. Developmental Expression

Analyses of mRNA and protein levels indicate that most $GABA_AR$ subunits exhibit distinct patterns of expression in the embryonic and adult nervous systems, suggesting a change in receptor subunit composition during development *(14,78,79)*. The expression of α_2, α_3, α_5, and β_3 subunits is prominent at embryonic and early postnatal stages and decreases during development *(14,78,79)*. A similar developmental pattern is observed for the β_1 subunit but with much lower expression levels. In contrast, α_1, α_4, and β_2 subunits exhibit low levels at early developmental stages and their expression increases with development *(14,78,79)*. The expression levels of γ_1 and γ_3 subunits decrease during development, whereas the level of the γ_2 subunit remains relatively constant during ontogeny *(78)*. RNase protection assays indicate that γ_2S is the predominant splice variant at early postnatal stages and that during development there is an increase in γ_2L *(80)*.

A developmental switch in the expression of α_1 and α_2 subunits may contribute to changes in receptor function *(78,80–83)*. The α_1-subunit is present at low concentrations and in very few areas of the brain at birth and increases dramatically during the first postnatal week, exhibiting a ubiquitous distribution in adult brain *(78)*. Parallel to the increase in α_1 levels, the α_2 subunit is more abundant and widespread at birth and its expression and distribution within the CNS is reduced during development *(78)*. These developmental changes in α subunit expression are consistent with a modification in the pharmacological properties of $GABA_A$Rs *(80–83)*. At early developmental stages the benzodiazepine type II receptors containing α_2 subunits are predominant, whereas there is a delayed development of benzodiazepine type I receptors that contain $\alpha_1$subunits *(81–83)*. Moreover, the postnatal increase in α_1 subunit expression correlates with an increase in the binding affinity of zolpidem, a compound that exhibits selectivity for α_1 containing receptors *(80)*.

2.2.2. *Regional Expression*

Immunohistochemical and *in situ*-hybridization studies demonstrate that the α_1, β_{1-3}, and γ_2 subunits display a widespread pattern of expression, whereas the localization of most subunits is limited to certain areas of the brain *(13,15,84)*. Moreover, α_1, β_2, and γ_2 are the most widely co-localized subunits supporting the notion that this subunit combination is the most abundant receptor subtype in the adult brain *(13,15,84)*. The α_1 subunit is present in almost all brain areas and is considered the most ubiquitous α subunit *(13,15,84)*. Low levels of α_1 expression are found in the striatum, reticular thalamic nucleus, and the internal granular layer of olfactory bulb. The α_2 subunit is most abundant in the accessory olfactory bulb, dentate molecular layer, Cornue Ammonis (CA)3 region of hippocampus, amygdaloid nuclei, septum, striatum, accumbens, and hypothalamus. In contrast, the α_3 subunit is predominant in the main olfactory bulb (external plexiform and glomerular layers). Expression of the α_4 subunit is found in thalamus, striatum, nucleus accumbens, tuberculum olfactorium, and dentate gyrus. The expression of the α_5 subunit is highest in the olfactory bulb, inner layers of cerebral cortex, endopiriform nucleus, subculiculum, Ammon's horn, and ventromedial hypothalamic nucleus. The α_6 subunit exhibits the most restricted localization and is only found in the granule cells of cerebellum and cochlear nucleus *(13,15,84)*.

Immunoprecipitation studies indicate that β_2 is the most abundant β subunit in the brain *(85)*. The three β subunits are widely expressed, especially in cerebral cortex *(13,15,84,85)*. Their distribution overlaps in most of the brain *(13,15,84)* suggesting the existence of receptors with more than one β subunit *(86)*. However, the expression of β subunits is restricted in some brain regions. For example, the striatum contains low β_1 and β_2, but high β_3 expression *(13,15,84)*.

The γ_2 subunit is the most widely distributed γ subunit and is highly expressed in olfactory bulb, cerebral cortex, hippocampus, amygdala, septum and basal brain, pallidum, and hypothalamus *(13,15,84)*. A difference in the relative distribution of γ_2S and γ_2L subunits was demonstrated using antibodies specific to each splice variant *(87)*. The γ_2S subunit is more abundant than γ_2L in hippocampus, cerebral cortex, and olfactory bulb. In contrast, γ_2L is more abundant than γ_2S in the inferior colliculus, medulla, and cerebellar Purkinje cells *(87)*. The highest expression levels of γ_1 subunit are found in the pallidum, substantia nigra, septum, medial, and central amygdaloid nuclei and in the bed nucleus of the stria terminalis *(13,15,84)*. The γ_3 subunit is rarely expressed in most brain regions, with higher concentrations in cerebral cortex and medial geniculate nucleus *(13,15,84)*.

The δ subunit is codistributed with α_4 and is most abundant in cerebellar granule cells, thalamus, dentate molecular layer, subculiculum, cortex, and striatum *(13,15,84)*. The ε subunit is localized in septal and preoptic areas and various hypothalamic nuclei *(88)*. The π subunit is present in hippocampus and nonneuronal tissues such as uterus and ovaries *(89,90)*. Whereas ρ subunits are present in the superior colliculus and cerebellar Purkinje cells *(91)*, the highest levels of ρ_{1-3} subunits are found in retina *(92)*. The θ subunit also displays a restricted pattern of expression in monoaminergic neurons and is enriched in the locus ceruleus *(93,94)*.

2.2.3. *Subcellular Expression*

Using light-microscopy immunofluorescence and electron microscopy immunogold methodologies, several groups have localized $GABA_A$ Rs at synaptic, perisynaptic, and extrasynaptic sites *(95,96)*. The α_{1-3}, α_6, β_2, β_3, and γ_2 subunits are found at both postsynaptic and extrasynaptic plasma membranes *(95–99)*. In contrast, δ subunits are localized almost exclusively at extrasynaptic sites *(100)*. The α_5 subunit presents in hippocampus a diffuse staining that is not co-localized with gephyrin, a postsynaptic clustering protein, suggesting an extrasynaptic localization *(98,99)*. It is believed that $GABA_A$Rs made up of α_1, α_2, or α_3 subunits in combination with β_2/β_3 are the principal receptor subtypes localized at synaptic sites that mediate phasic inhibition *(101)*. Receptors that contain α_4, α_5, or α_6 in combination with δ subunits are mainly present at extrasynaptic sites and are responsible for tonic inhibition *(101)*.

The presynaptic localization of $GABA_A$R subunits is controversial *(102)*. Although expression of α_2, α_3, β_2, β_3, and γ_2 subunits in primary afferent terminals in the spinal cord could be important for mediating presynaptic inhibition *(102–104)*, contradictory data suggests that β_2 and β_3 subunits are primarily found at sites apposed by glutamic acid decarboxylase (GAD)-immunoreactive terminals *(105)*. In addition, it is difficult to directly record presynaptic membrane potentials *(102)*. However, it is clear that activation of presynaptic $GABA_A$ Rs induces inhibition of neurotransmitter release *(102)*.

2.3. Use-Dependent Regulation of $GABA_A$Rs

Long-term potentiation and long-term depression of GABAergic synapses have been described both in neonatal and adult brain *(106–108)*. Although different mechanisms are involved depending on the brain region and developmental stage, an increase in the intracellular Ca^{2+} concentration is crucial to induce all the changes in the strength of inhibitory synapses. These forms of activity-dependent synaptic plasticity may have a role in the development and function of neuronal networks. Exposure of early postnatal cortical neurons in culture to the positive endogenous modulator allopregnanolone produces changes in the expression of specific $GABA_A$R subunit genes *(109)*. These results suggest that an activation-dependent plasticity mechanism regulates the development of $GABA_A$Rs.

Several lines of evidence indicate that $GABA_A$Rs can be persistently activated by high GABA concentrations under physiological conditions. As a result of a single-action potential, GABA released reaches concentrations of 1–5 m*M* in the synaptic cleft and is then cleared biphasically with time constants of 100 μs and 2 ms *(110)*. Moreover, GABAergic neurons in neocortex and cerebellum are fast-spiking cells with basal firing rates of 5–50 Hz (equivalent to one action potential every 200 ms and 20 ms, respectively) that can last many minutes *(111–113)*. Furthermore activation of postsynaptic $GABA_A$Rs long outlasts the presence of free GABA because the neurotransmitter is "trapped" in the receptor. GABA responses decay biphasically with time constants (τ) of approx 50 and 171 ms *(114,115)*. Taken together, the high frequency firing of GABAergic neurons and the slow decay of GABA responses indicate that activation of $GABA_A$Rs can last several minutes.

Exposure of $GABA_A$Rs to high GABA concentrations can be even more evident under certain pathological situations. For instance, the firing frequency of GABAergic neurons in cat neocortex increases up to 800 Hz during electrographic seizures *(116)*. In human brain the extracellular concentrations of GABA increase 600-fold during an ischemic episode *(117)*. This prolonged $GABA_A$R activation suggests the existence of regulatory mechanisms. Tolerance to the anticonvulsant and sedative/hypnotic effects of benzodiazepines has been observed after chronic treatments *(118–120)*. Although the molecular bases of this tolerance have not been elucidated, chronic in vivo administration of diazepam to rats induces a subsensitivity of $GABA_A$Rs to GABA and a decrease in the potentiation of GABA responses by benzodiazepines *(121)*. Chronic administration of flurazepam produces tolerance and changes in the

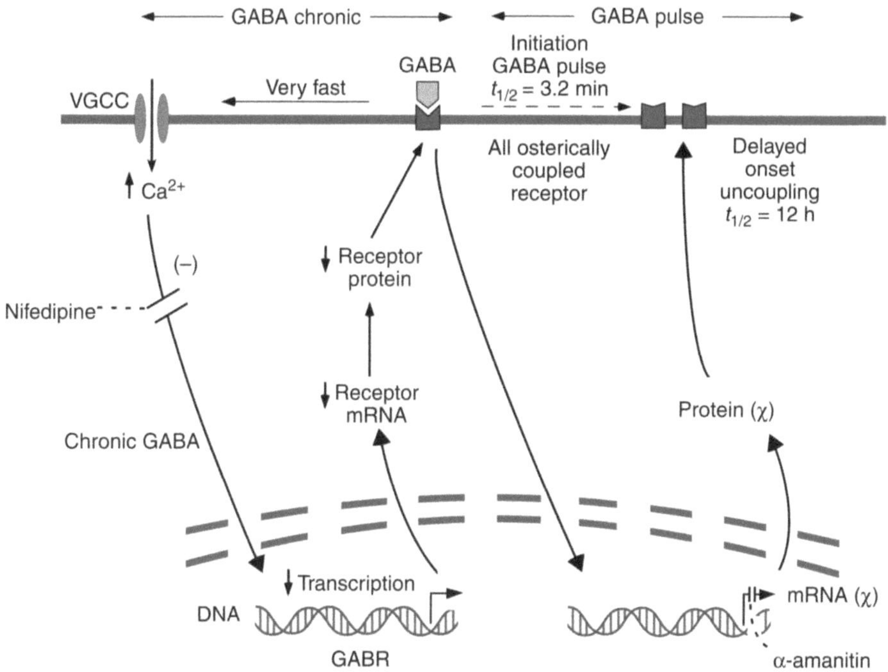

Fig. 2. Cellular mechanisms involved in the use-dependent regulation of $GABA_A$Rs. Chronic GABA exposure induces downregulation of receptor number and uncoupling of the allosteric interactions between GABA and benzodiazepine binding sites *(18)*. The mechanism of downregulation involves transcriptional repression of $GABA_A$R subunit genes and is mediated by activation of VGCC *(22,23,123)*. Brief GABA exposure (GABA pulse, $t_{1/2} = 3$ min) induces uncoupling many hours later ($t_{1/2}$ onset = 12 h) without downregulation. This delayed-onset uncoupling does not depend on VGCC activation and involves transcriptional activation of an unknown gene(s) (χ). GABA pulse also produces transcriptional repression of specific $GABA_A$R subunit genes *(129)*.

protein and mRNA levels of specific $GABA_A$R subunits in certain brain areas. The relationship between these changes in receptor subunits and tolerance has not been determined. Interestingly, Holt et al. *(122)* reported a decrease in the allosteric interactions between GABA and benzodiazepines 4–12 h after a single injection of diazepam, which precedes the changes in the mRNA levels of γ_2 subunit.

The molecular mechanisms of $GABA_A$R regulation have been investigated in primary neuronal cultures and cell lines expressing recombinant receptors. Chronic exposure of cultured neurons to GABA induces a decrease in the receptor number and a reduction in the allosteric interactions between GABA and benzodiazepine-binding sites, a phenomenon termed homologous uncoupling *(18)*, with a half-life of 24–25 h (Figs. 2–4). Moreover, a decrease in the

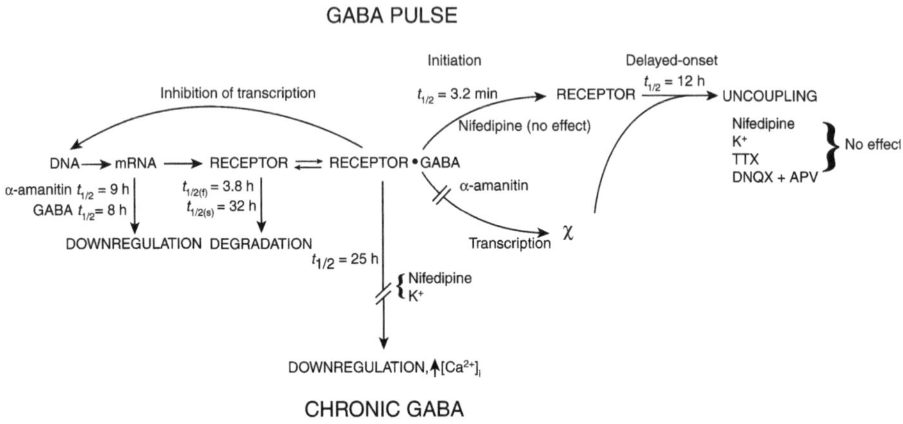

Fig. 3. Life cycle for downregulation and uncoupling of the $GABA_AR$ in neurons. GABA-induced downregulation of $GABA_AR$ subunit mRNAs has a half-life of 8 h. This time constant is similar to the apparent half-life of $GABA_AR$ subunit mRNAs in control neurons (9 h) as determined by the addition of α-amanitin, supporting the idea that the mechanism of downregulation involves transcriptional repression *(123)*. The $GABA_AR$ is degraded biphasically with time constants of 3.8 and 32 h *(151,152)*. Downregulation is blocked by nifedipine, a VGCC blocker, and elevated K^+. In contrast, uncoupling is resistant to nifedipine, elevated K^+, tetrodotoxin (TTX), and 6,7-dinitroquinoxaline-2,3-dione (DNQX) plus 2-amino-5-phosphonopentanoic acid (APV) but is blocked by α-amanitin *(22,129)*. (Modified from ref. *129*.)

interactions between benzodiazepines and barbiturates (heterologous uncoupling) is induced by chronic GABA administration *(18,21)*. The mechanism of downregulation of receptor number involves transcriptional repression of $GABA_AR$ subunit genes *(23,123)* and is mediated by activation of voltage-gated Ca^{2+} channels (VGCC) (Figs. 2–4) *(22)*.

Exposure of neurons to benzodiazepines, neurosteroids, or barbiturates produces uncoupling between the recognition sites for allosteric modulators of the $GABA_AR$ without a change in receptor number *(19,21,124,125)*. These results suggest that downregulation and uncoupling are mediated by independent mechanisms (Figs. 2 and 3). Surprisingly, in contrast Follesa et al. studies by *(125)* show that a chronic treatment of cultured neurons with progesterone induces a reduction in the modulation of GABA currents by diazepam and a β-carboline that is accompanied by a decrease in the mRNA levels of α_1, α_3, α_5, and γ_2 subunits. These observations may be due to the upregulation of diazepam insensitive receptors that contain α_4 subunits.

Benzodiazepine-induced uncoupling has also been reported in cell lines expressing recombinant $GABA_ARs$. The reported time-courses of uncoupling

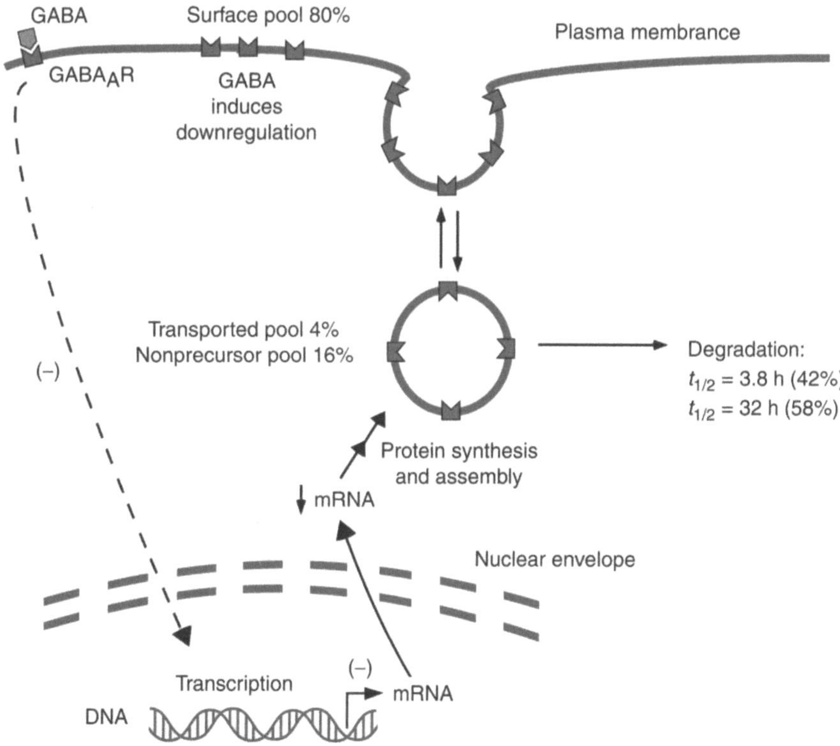

Fig. 4. $GABA_AR$ turnover. Eighty percent of $GABA_ARs$ are present at the membrane surface (surface pool) *(147)*. The intracellular receptor population is made up of a non-precursor pool (16%) and a transported pool (4%) *(150)*. Degradation of internalized receptors and nonprecursor receptor pool exhibits a biphasic kinetic *(151,152)*. Persistent activation of $GABA_ARs$ by GABA induces repression of receptor subunit transcription *(123)*.

vary from minutes in PA3 cells *(126)* to hours in WSS-1 cells *(127)*. These studies suggest that uncoupling is the consequence of a post-transcriptional process. For instance, studies performed in an insect cell line suggest that diazepam-induced uncoupling is produced by the exposure of receptors to an acidic environment inside of intracellular compartments as a result of a receptor internalization process *(128)*.

A recent report demonstrates in primary cortical cultures that uncoupling is a two-step process: a brief initiation step (half-life of 3.2 min) that requires the activation of $GABA_ARs$ by GABA followed by a delayed-onset (half-life of 12 h) that can occur in the absence of neurotransmitter (Figs. 2 and 3) *(129)*. This delayed-onset uncoupling is not accompanied by receptor downregulation and is independent of VGCC activation. This brief activation of $GABA_ARs$ also induces a decrease in the mRNA levels of α_1, α_3, and β_{1-3} subunits but no change is produced in α_2, α_4, γ_1, and γ_2 subunits *(129)*. As this treatment failed to alter the receptor number, uncoupling might have been the result of a change in receptor subunit composition leading to the expression of $GABA_A$Rs with a decreased strength in the coupling between GABA and benzodiazepine sites *(129)*.

Although the mechanism of uncoupling has not been elucidated it has been shown that this regulatory process can be blocked by α-amanitin, suggesting that uncoupling is contingent on the transcription of a regulatory factor *(129)*. The function of this factor has not been demonstrated yet, but it might be a protein kinase or phosphatase that alters the phosphorylation state of the $GABA_A$R. Run-down of the $GABA_A$R responses and its modulation by different positive allosteric modulators such as benzodiazepines, barbiturates, and steroids is the result, at least in part, of the loss of phosphorylation factors *(130,131)*. This regulatory protein might, alternatively, be a $GABA_A$R-associated protein. For instance, p130, a protein that binds to the $GABA_A$R γ_2-subunit, seems to regulate the modulation of GABA currents by diazepam and zinc (*see* "$GABA_A$ R-Associated Protein") *(132)*.

2.4. Transcriptional Gene Regulation

Neurons must continually adapt to their changing environment by altering their gene expression patterns to direct normal brain function as well as the etiology of neurological and psychiatric diseases. Gene expression can be regulated at multiple levels from DNA to RNA to protein. These levels include chromatin remodeling, transcription initiation, alternative splicing, mRNA stability, translation, post-translational modification, intracellular trafficking, and protein degradation. The fact that certain $GABA_AR$ and $GABA_BR$ subunit genes are transcribed in distinct neuronal subtypes during development and in diseased states suggests that transcriptional control plays a major role in the abundance and choice of subunit variants for receptor assembly. The complex transcriptional regulatory networks that control *GABA-R* gene expression have been specified by interactions of the promoter with a shifting array of transcription factors. Several recent studies illustrate how complementary approaches of expression studies in cell cultures, transgenic models, and computational methods elucidate the mechanisms of transcriptional regulation for the $GABA_AR$ subunit genes.

Promoters mark the start of every transcript and are recognized by sequence-specific and general transcriptional regulators during transcription initiation and serve to integrate multiple cellular signals to generate diversity in gene expression. Sequence motifs such as the TATA box and/or initiator (Inr) element are found in the promoter region of many $GABA_A$R subunit genes (for review *see* ref. *40*). Results of transient transfection in primary neuronal

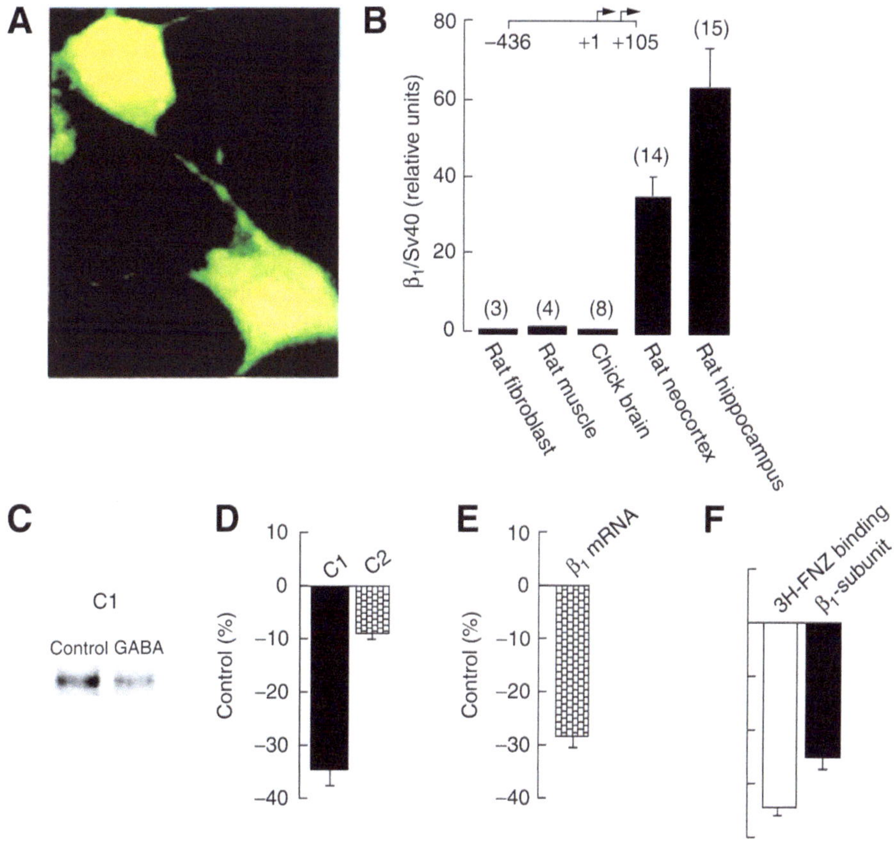

Fig. 5. An Inr element in the $GABA_AR$ β_1 promoter mediates neural specificity and GABA-induced downregulation. **(A)** Primary cultures of rat hippocampal neurons can be successfully transfected with a reporter construct containing the immediate 5′ flanking region of the human β_1 gene upstream of the gene encoding β-galactosidase. **(B)** The β_1 promoter construct β_1-P (–436/+105) displays neural-specific activity. In addition, a 72bp construct that contains three copies of the Inr and its flanking sequence $[(-5/+19)_3]$ produces activity that is equal to or more than β_1-P in primary rat neocortical neurons and exhibits negligible activity in primary rat fibroblast cells *(23)*. **(C,D)** GABA (500 μ*M*; 48 h) reduces sequence-specific binding of nuclear proteins from neocortex to the Inr in the β_1 promoter. Moreover, the Inr produces promoter activity that is autologously downregulated by GABA *(23)*. **(E,F)** In GABA treated neocortical cultures, decreased nuclear protein binding and promoter activity is consistent with decreased levels of β_1 mRNA, [^{3}H] flunitrazepam binding, and β_1 subunit protein. (Modified from ref. *23*, Copyright 2000 National Academy of Science, USA.)

cultures indicate that the TATA-less β_1 promoter contains a functional Inr that is critical for neural specific activity (Fig. 5) *(23)*. In addition, three concatenated copies of the Inr $(-5/+19)_3$ produces promoter activity that is autologously regulated, consistent with GABA-induced downregulation of mRNA, subunit protein and functional receptor (Fig. 5) *(23)*. The activity-dependent regulation of $GABA_AR$ β_1 subunit gene expression is mediated by a decrease in sequence-specific binding of general transcription factors to the β_1 *Inr* (Fig. 5) and raises the possibility that regulation of the preinitiation complex on a receptor subunit gene may in turn control the availability of subunits for receptor assembly *(23)*.

Moreover, deletional analysis in primary neuronal cultures and/or cell lines has defined the proximal promoter regions for the α_3 *(133)*, α_5 *(64)*, α_6 *(134)*, β_3 *(65)*, γ_2 *(135)*, and δ subunit genes *(136)* that are important for tissue and cell-specific transcriptional activity. Additional characterization of the α_6 *(136–138)* promoter using transgenic mice indicates the DNA sequences that are capable of reproducing the neuronal subtype specific and spatiotemporal expression patterns of the endogenous gene.

Given the importance of *cis*-regulatory sequences to modulate promoter activity, it is not surprising that more recent studies have focused on the content of the DNA sequences to identify transcription factorbinding sites. Two DNA sequence-based approaches have been used for identification of transcriptional regulatory networks that control $GABA_AR$ subunit gene expression. The former relies on prediction of transcriptional elements specific to individual promoters, whereas the latter develops a conceptional framework of the consensus sequences across promoters for the $GABA_AR$ subunit genes. Using a transcription factor database and searches based on promoter sequence and binding site preferences, potential transcription factor binding sites have been localized in 5′ regulatory regions of the α_1*(139)*, α_3*(133)*, α_5*(64)*, α_6*(134,138)*, and β_1 subunit genes *(23)*. Furthermore, functional promoter analysis has identified a GA repeat that contributes to transcriptional activation of the α_3 promoter *(133)* and a novel 60-bp element in the α_6 promoter that is essential for cerebellar granule cell-specific gene expression *(134)*.

Neural-specific gene expression can also be mediated through a sequence-specific interaction between the neuron-restrictive silencer element (NRSE) and the neuron-restrictive regulatory factor (NRSF) *(140–142)*. Whereas NRSE-like sequences are found in many $GABA_AR$ subunit genes *(40)*, the NRSE/NRSF regulatory complex has been demonstrated to direct expression of the γ_2 *(135)* and δ subunit genes *(142)* to neuron-like cells. Additional neural-specific activators required for neuron-specific gene regulation include the γ_2 promoter element 1 (GPE1) located downstream of the γ_2 transcriptional start site *(135)*, the brain-specific factor 1 (BSF1) which recognizes a novel 22-bp purine repeat element in the δ promoter *(143)*, and the Sp3- and Sp4-transcription factors that interact with two Sp1-binding sites in the α_4 promoter *(144)*.

In contrast, promoter prediction and cross comparison of multiple promoter regions is a powerful tool to identify potential consensus sites and transcription factors that participate in regulatory pathways that may not have been immediately accessible through analysis of a single promoter (for review *see* ref. *40*). Importantly, bioinformatic analysis *(40)* coupled with biologically relevant in vivo and in vitro assays *(145)* has been used to identify early growth response factor-3 (Egr3) as a major regulator of the $GABA_AR$ α_4 subunit gene in developing neurons and in epilepsy. The human α_4 minimal promoter in an AAV/eYFP viral construct produces region-specific transcription consistent with the expression pattern of endogenous α_4 mRNA and controls condition-specific upregulation of α_4 mRNA in a pilocarpine model of temporal lobe epilepsy *(145)*. Studies of α_4 promoter activity in primary neurons and in vivo chromatin immunoprecipitation indicate that PKC, PKA, and Egr3 are critical for α_4-transcriptional activation. Moreover, increased levels of α_4 and *Egr3*-mRNAs are accompanied by increased binding of Egr3 to the α_4 promoter after status epilpeticus *(145)*. It is anticipated that in the future complementary approaches that combine biological and computational methods will reveal the regulatory networks that participate in normal physiology as well as the impaired nervous system and define novel targets for therapeutic intervention.

2.5. Receptor Turnover

Regulation of $GABA_AR$ expression and stability at synapses is important for synaptic plasticity. $GABA_ARs$ at postsynaptic sites are subjected to a continuous dynamic turnover comprising the following processes: synthesis, assembly, targeting to plasma membrane, clustering, internalization, recycling and degradation (Figs. 4 and 6) *(146)*. Eighty percent of $GABA_ARs$ are localized at the surface and 20% are internal *(147)*. $GABA_ARs$ are synthesized and assembled in the endoplasmic reticulum (ER) and then transported to the plasma membrane through the Golgi apparatus *(148,149)*. Only 4% of the total $GABA_AR$ population (20% of internal receptors) is destined for trafficking to the membrane surface, whereas 16% represent a nonprecursor pool that is subjected to degradation *(150)*. $GABA_AR$ degradation exhibits biphasic kinetic with time constants of 3.8 and 32 h *(151,152)*. Receptors form clusters at the postsynaptic membrane, a crucial step for efficient synaptic neurotransmission *(153,154)*. The $GABA_AR$ turnover is regulated by the interaction of the receptor with different proteins including gephyrin, $GABA_A$ receptor-associated protein (GABARAP), Plic-1, adaptin protein-2 (AP2), brefeldin A-inhibited GDP/GTP exchange factor 2 (BIG2), and Golgi-specific DHHC (aspartate-histidine-histidine-cysteine) zinc finger protein (GODZ) (*see* Section 2.6.) (Fig. 6).

2.6. Proteins That Interact With the $GABA_AR$

2.6.1. Gephyrin

The inhibitory synaptic scaffold protein gephyrin is important for the composition of certain postsynaptic inhibitory specializations *(155,156)*. Studies

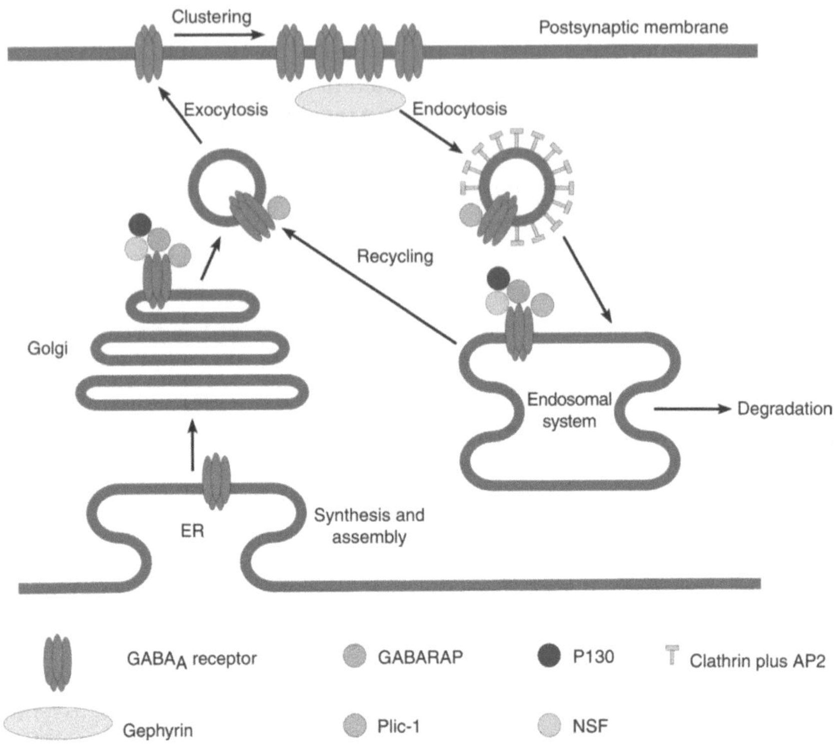

Fig. 6. Schematic representation of $GABA_AR$ trafficking. The intracellular $GABA_AR$ transport depends on the association of the receptor with different proteins. $GABA_ARs$ are assembled in the ER and transported to the cell surface through the Golgi apparatus *(148,149)*. $GABA_ARs$ are internalized by endocytosis into clathrin-coated vesicles and this process depends on the association of β and γ $GABA_AR$ subunits with the adaptin complex AP2 *(170)*. $GABA_ARs$ colocalize at postsynaptic membranes with the scaffolding protein gephyrin that seems to participate in receptor clustering *(157)*. The intracellular transport of $GABA_ARs$ is regulated by the interaction between γ_2 subunits and GABARAP *(162)*. GABARAP also interacts with *N*-ethylmaleimide sensitive factor (NSF), a protein involved in vesicle-fusion events *(165)*. The binding of GABARAP with the $GABA_AR$ is competitively inhibited by P130, a protein that can also regulate receptor function *(132)*. Plic-1, an ubiquitin-like protein, interacts with α and β $GABA_AR$ increasing the stability of receptor intracellular pools *(168)*.

using gene knockout and antisense oligonucleotide approaches indicate that the $GABA_AR$ γ_2 subunit and gephyrin facilitate the clustering of $GABA_ARs$ at inhibitory synapses (Fig. 6) *(157,158)*. Inhibition of gephyrin expression by RNA interference in hippocampal neurons demonstrates a role for gephyrin in reducing the mobility of $GABA_AR$ clusters *(159)*. The lack of a direct interaction between gephyrin and $GABA_AR$ subunits *(160)* suggests that a linker protein is needed for a physical association. Moreover, a gephyrin-independent mechanism of $GABA_AR$ clustering is evident for receptors containing α_1 and α_5, but not α_2, α_3, $\beta_{2/3}$, and γ_2 subunits in the spinal cord of gephyrin knockout mice and implicates additional clustering proteins in the synaptic localization of selected $GABA_AR$ subtypes *(161)*.

2.6.2. *$GABA_AR$-Associated Protein*

Yeast-two hybrid experiments demonstrate the interaction of the $GABA_AR$ γ_2 subunit intracellular loop with the intermediate region of GABARAP (Fig. 6) *(162)*. In addition, the N-terminus of GABARAP binds soluble tubulin and polymerized microtubules, suggesting a role linking the receptors to the cytoskeleton. An analysis of the crystal structure when coupled with structure-based mutagenesis studies reveals that GABARAP presents two conformations, a monomeric and an oligomeric form *(163)*. It is proposed that the oligomerization process can induce tubulin polymerization and thus facilitate $GABA_AR$ clustering.

Although GABARAP can interact with gephyrin *(164)*, it is not detected at GABAergic synapses and is mainly localized in intracellular compartments arguing against a role for GABARAP in synaptic anchoring *(164,165)*. GABARAP can also interact with the NSF, a protein involved in vesicle-fusion processes, suggesting a role in intracellular trafficking of $GABA_ARs$ (Fig. 6) *(165)*. Moreover, a role in sorting or targeting is also suggested from the homology between GABARAP and the protein p16, an intra-Golgi trafficking factor *(164)*. Heterologous expression of GABARAP in Cos7 cells and cultured hippocampal neurons increases the levels of $GABA_ARs$ at the plasma membrane and this effect is reduced by point mutations in the γ_2 binding domain of GABARAP *(166)*. This finding supports a role for GABARAP in trafficking $GABA_ARs$ to the plasma membrane. However, an immunocytochemical study of GABARAP deficient mice shows that GABARAP is not essential for trafficking of $GABA_ARs$ containing γ_2 subunits to the neuronal plasma membrane or a synaptic localization *(167)*.

P130, a protein structurally related to phospholipase (PL) C that lacks catalytic activity, also interacts with GABARAP (Fig. 6) and competitively inhibits the binding of the $GABA_AR$ γ_2 subunit to GABARAP *(132)*. Electrophysiological studies performed in hippocampal neurons show that the modulation of GABA currents by diazepam and zinc is reduced in p130 knockout mice *(132)*. These results suggest a role for p130 and GABARAP in regulating the function of $GABA_ARs$.

2.6.3. *Plic-1*

Plic-1 is an ubiquitin-like protein that has been shown to interact with the second intracellular loop of both α and β subunits of $GABA_ARs$ through a

C-terminal ubiquitin-associated domain *(168)*. Plic-1 is localized at GABAergic synapses and intracellular compartments. Inhibition of the interaction between Plic-1 and $GABA_AR$ subunits induces a decrease, whereas overexpression of Plic-1 produces an increase in the number of cell surface receptors without affecting receptor internalization. In addition, Plic-1 increases the half-life of $GABA_AR$ subunits probably by the inhibition of receptor polyubiquination that leads to reduction of $GABA_AR$ proteosomal degradation. Taken together, these results suggest that Plic-1 facilitates $GABA_AR$ expression at the plasma membrane by increasing the stability of receptor intracellular pools *(168)*.

2.6.4. Adaptin Protein-2 (AP2)

The presence of $GABA_ARs$ in clathrin-coated vesicles isolated from rat brain was first reported by Barnes and colleagues *(169)*. It was subsequently demonstrated that $GABA_ARs$ in hippocampal neurons and recombinant receptors in A293 cells are constitutively internalized through clathrin-dependent endocytosis *(170)*. $GABA_AR$ β and γ subunits interact with AP2, which is important for membrane protein internalization in clathrin-coated pits *(170,171)* (Fig. 6). Furthermore, a dileucine AP2 adaptin-binding motif within the $GABA_AR$ β_2 subunit is critical for receptor endocytosis *(171)*. Phosphorylation of serine residues within the β_3 subunit inhibits the interaction with the AP2 complex, suggesting a phospho-dependent mechanism to regulate $GABA_AR$ internalization *(172)*. Importantly, inhibition of clathrin-dependent endocytosis results in an increase in the amplitude of miniature inhibitory postsynaptic currents (mIPSCs) *(170,171)*. These findings indicate that the interaction of the AP2 complex with $GABA_ARs$ can modulate inhibitory synaptic transmission through recruitment of receptors into clathrin-coated pits.

2.6.5. Brefeldin A-Inhibited GDP/GTP Exchange Factor 2 (BIG2)

Several groups have used yeast two hybrid screens to identify proteins including BIG2 that interact with the $GABA_AR$. The C-terminus of BIG2 interacts with all $GABA_AR$ β subunits and may be involved in vesicular and protein trafficking *(173)*. BIG2 is concentrated in the *trans*-Golgi network in the rat brain and stimulates the translocation of newly synthesized $GABA_ARs$ to the plasma membrane in a cell line expressing recombinant receptors, suggesting a critical role for BIG2 in the exocytosis of $GABA_ARs$ *(173)*.

2.6.6. Golgi-Specific DHHC Zinc Finger Protein (GODZ)

$GABA_ARs$ are palmitoylated on multiple cysteine residues of the γ_2 subunit both in heterologous expression systems and in neurons *(174,175)*. Inhibition of palmitoylation leads to a reduction in the clustering of $GABA_ARs$ at synapses and total cell-surface receptor number *(174)*. Moreover, the interaction of

$GABA_A$Rs with GODZ, a neuronal-specific thioacyltransferase, mediates the palmitoylation of the γ_2 subunit *(175)*. Thus, palmitoylation of γ_2 subunit containing $GABA_A$Rs by GODZ might represent a post-translational modification that regulates trafficking of specific postsynaptic $GABA_A$R subtypes.

2.7. Modulation of $GABA_A$R Turnover by Cell Signaling Pathways

2.7.1. Insulin

The processes that facilitate insertion of $GABA_A$Rs into neuronal membranes are critical in the regulation of inhibitory synaptic strength. Insulin causes a rapid translocation of functional $GABA_A$Rs from intracellular compartments to the plasma membrane and thus increases the amplitude of mIPSCs *(176)*. This translocation depends on the β_2 subunit of the $GABA_A$R and is mediated by activation of the insulin receptor tyrosine kinase. A recent study also shows that the insulin-induced insertion of $GABA_A$Rs to postsynaptic sites can be mediated by the serine/theronine kinase Akt (also known as PKB) signaling pathway *(177)*. Moreover insulin can prevent D3-receptor-induced internalization of $GABA_A$Rs by a rapid recruitment of receptors to the membrane surface *(178)*. Importantly, insulin provides neuroprotection by counteracting decreases in cell-surface $GABA_A$Rs after oxygen-glucose deprivation in cultured cortical neurons *(179)*.

2.7.2. Brain-Derived Neurotrophic Factor

Activation of receptor tyrosine kinase TrkB by the neurotrophic factor brain-derived neurotrophic factor (BDNF) produces a decrease in the amplitude of mIPSCs by downregulation of $GABA_A$R surface-expression in cultured hippocampal neurons *(180)*, in cultured cerebellar granule cells *(181)*, and in brain slices *(182,183)*. This reduction in synaptic strength is accompanied by a selective decrease in α_1-, α_2-, and α_3-containing receptors *(180)* and $\beta_{2/3}$-containing receptors *(181)*. Furthermore, the decrease in surface expression of functional $GABA_A$Rs in the paraventricular nucleus of the hypothalamus occurs through a dynamin-mediated endocytosis of postsynaptic receptors *(184)*.

In contrast, Mizoguchi et al. *(185)* reported that BDNF increases the surface expression of $GABA_A$Rs in visual cortex. BDNF can also enhance the strength of GABAergic synapses on cultured hippocampal neurons with preserved neuronal activity and can prevent the reduction of $GABA_A$R clusters in activity-deprived neurons *(186)*. Moreover, when hippocampal neurons are cultured with astrocytes, only the activation of BDNF and TrkB signaling within neurons is required for astroctye-mediated increases in postsynaptic $GABA_A$R clusters *(187)*.

Studies from Jovanovic et al. *(188)* also suggest that BDNF can bidirectionally modulate $GABA_A$R function in cultured cortical and hippocampal neurons. Thus, BDNF first induces an increase followed by a decrease in the amplitude of mIPSC. This process is accompanied by an enhancement in

$GABA_AR$ phosphorylation by PKC and a subsequent receptor dephosphorylation step mediated by the protein phosphatase 2A. This regulatory mechanism seems to depend on the differential interaction of PKC, receptor for activated C-kinase, and protein phosphatase 2A with the $GABA_AR$ *(188)*.

Overall, BDNF has different effects on $GABA_AR$ trafficking and warrants further investigation. The finding that BDNF produces a stabilizing action on repetitively active "epileptic" $GABA_ARs$ through PLC and PKC-mediated signaling *(189)* highlights the importance of this system in regulating the efficacy of GABAergic inhibition and the stability of $GABA_ARs$ at synapses.

2.7.3. Protein Kinase C (PKC)

Pharmacological studies with drugs that activate or inhibit PKC indicate that PKC plays a role in $GABA_AR$ trafficking, GABA-gated Cl^- conductance and $GABA_AR$ sensitivity to allosteric modulators. In recombinant systems such as *Xenopus* oocytes *(190)* and A293 cells *(191)*, PKC produces a reduction in cell-surface $GABA_ARs$ through a mechanism that is independent of receptor phosphorylation. Several groups have presented opposing mechanisms that mediate PKC-induced changes in cell-surface stability of $GABA_ARs$. Connolly et al. *(191)* has found that PKC downregulation of cell-surface receptors requires a γ subunit and is mediated by a decrease in receptor recycling to the plasma membrane. Alternatively, PKC activation may stimulate endocytosis that is not dependent on the γ subunit and may occur through a dynamin-independent pathway *(192)*. It has also been shown that PKC-mediated decreases in $GABA_AR$ function may be owing to an increase in endocytosis that requires the γ_2 subunit, occurs through a dynamin-dependent pathway, and is dependent on a dileucine motif within the β_2 subunit *(193)*. The physiological significance of $GABA_AR$ endocytosis in neurons awaits further investigation, but this process could possibly allow neurons to selectively modify cell surface levels of $GABA_ARs$ on distinct cues from the neuronal environment.

2.8. *Diffusion of $GABA_ARs$ Within the Plasma Membrane*

Stability of the $GABA_AR$ population in the plasma membrane is vital for controlling neuronal excitability *(148)* and involves both the insertion of new receptors from intracellular pools and the movement of $GABA_ARs$ within the plasma membrane *(155)*. Evidence for the mobility of functional $GABA_ARs$ in the plasma membranes of hippocampal neurons was obtained using a functional "knockdown" approach *(194)*. $GABA_ARs$ that are irreversibly inhibited by a sulfhydryl reagent were obtained by expressing α_1 subunits with a mutation conferring activation-dependent binding to the inhibitor that blocks the ion channel *(194)*. These studies suggest that the function of synaptic receptors is rapidly recovered after inhibition most likely by importing receptors from extrasynaptic sites through a diffusion process within the membrane plane.

2.9. Interaction Between $GABA_A$Rs and $GABA_B$Rs

Surprisingly, hetero-oligomerization of the $GABA_A$R γ_2S subunit with the $GABA_B$R subunit $GABA_B$R1 stimulates the cell-surface expression of $GABA_B$R1 in the absence of its usual partner $GABA_B$R2 *(195)*. Although agonist stimulation of the $GABA_B$R1/γ_2S complex does not mediate phosphorylation of extracellular signal-regulated kinase (ERK) or G protein-gated potassium channel activity as does $GABA_B$R1/$GABA_B$R2 heterodimers, coexpression of $GABA_A$R γ_2S with $GABA_B$R1 enhances agonist-promoted internalization of $GABA_B$R1/$GABA_B$R2 receptors. This report suggests that a cross-talk between $GABA_A$Rs and $GABA_B$Rs may regulate an aspect of $GABA_A$R trafficking and endocytosis.

2.10. Regulated Gene Expression and Implications for Disease

2.10.1. Epilepsy

Alterations in $GABA_A$R-mediated inhibitory synaptic transmission have been implicated in the generation of seizures. Mutations in α_1, γ_1, and γ_2 subunits are associated with familial forms of idiopathic generalized epilepsy (IGE) *(196–199)*. The α_1 subunit epilepsy mutation α_1(A322D) reduces α_1 cell-surface expression and alters α_1 subunit trafficking after translation but before receptor assembly *(200,201)*. Moreover, hippocampal cell-surface expression of $\alpha_1\beta_2\gamma_2$ receptors containing γ_2 epilepsy mutations [γ_2(R43Q), γ_2(K289M), and γ_2(Q351X)] is reduced with elevated temperature and may be attributable to accelerated endocytosis and/or increased retention in the endoplasmic reticulum *(202)*. This finding suggests that febrile seizures resulting from mutations in the γ_2 subunit gene may be provoked in response to fever.

Animal studies suggest that changes in the levels of $GABA_A$R subunit expression and alterations in $GABA_A$R subunit composition may contribute to the induction and maintenance of epileptogenesis *(203–215)*. The results of these studies reveal increased and decreased $GABA_A$R expression (for review *see* ref. *40*). For example, in the electrical kindling model for temporal lobe epilepsy, the postsynaptic insertion of new $GABA_A$Rs containing $\beta_{2/3}$ subunits is accompanied by an increase in postsynaptic GABA responses *(216)*. Moreover, changes in the mRNA levels of specific $GABA_A$R subunits following pilocarpine-induced status epilepticus are associated with a decrease in potentiation by zolpidem and an increase in zinc inhibition of GABA currents in dentate granule cells *(212)*. The selective downregulation of α_1 and β_1 mRNAs and upregulation of α_4, β_3, δ, and ε subunit mRNAs are long-lasting and precede the onset of spontaneous seizures *(212)*. A recent study suggests that the stimulation of α_4 subunit transcription by status epilepticus is mediated by increased binding of the transcription factor Egr3 to the α_4 promoter *(145)*.

Unlike the former studies in adult rats, increased levels of α_1 subunit expression are detected after status epilepticus in postnatal day 10 (P10) rat pups *(203)*. Coupled together with the observation that the P10 rats do not become epileptic, age-dependent alterations in hippocampal *$GABA_AR$* gene expression may contribute to the selective resistance of the immature brain to epileptogenesis. Overall, these results suggest that the marked disruption in postsynaptic $GABA_AR$ function may result from reorganization of receptor subunits as a causative and/or compensatory response of the genome to maintain homeostasis.

2.10.2. Ethanol

The effects of ethanol consumption are similar to those of positive allosteric modulators of the $GABA_AR$ such as anxiolysis, sedation/hypnosis, anticonvulsant, and motor-incoordination actions. However, a direct in vivo action of physiological concentrations of ethanol on $GABA_AR$ function has not been proved *(217)*. The molecular and cellular adaptations of $GABA_ARs$ following chronic ethanol exposure and withdrawal are unclear *(217)* but may include changes in cell-surface density *(218)*, in post-translational modification by PKC *(219,220)*, or in subunit expression *(221–227)*.

Several groups have reported that chronic ethanol exposure and subsequent ethanol withdrawal produces subunit- and region-specific changes in *$GABA_AR$* gene expression (for review *see* refs. *40,217,228*). A significant decrease in α_1 subunit expression with a concomitant increase in α_4 and δ is observed in primary hippocampal neurons *(229,230)* and a switch in α subunit expression is found in the hippocampus, cerebellum, and cortex of ethanol-dependent rats *(223,227,231)*. Moreover, withdrawal from chronic intermittent ethanol treatment downregulated α_1 and δ subunit expression and upregulated α_4, γ_1, and γ_2 subunits in the hippocampus *(232)*. These changes are accompanied by a corresponding increase in anxiety, a decrease in the behavioral effects of positive allosteric modulators and a decrease in the decay time of mIPSCs *(232)*.

However, in cerebellar granule cells, ethanol withdrawal produces an upregulation of the α_2 subunit *(233)* and downregulation of δ subunit expression *(229)* that can be prevented by treatment with diazepam or γ-hyroxybutyric acid. The alterations in α_2 and δ subunit expression are consistent with changes in $GABA_AR$ function *(229,233)*. In addition, ethanol-induced α_6 gene expression in the cerebellum may be mediated by specific transcription factor binding to distinct regulatory regions of the α_6 promoter *(234)*.

Studies performed with recombinant $GABA_ARs$ show that receptor subunit combinations $\alpha_4\beta_3\delta$ and $\alpha_6\beta_3\delta$ are more sensitive to ethanol, suggesting that extrasynaptic receptors represent the main targets of ethanol *(235)*. Moreover, low ethanol concentrations appear to stimulate tonic inhibition through the activation of δ-containing receptors in dentate gyrus granule cells *(236)*. Together these findings indicate that complex mechanisms of regulation control the expression, localization, and function of distinct $GABA_AR$ populations induced by prolonged exposure to and withdrawal of ethanol.

2.10.3. Schizophrenia

An impairment of certain cognitive functions, such as working memory, is a core feature of schizophrenia. Moreover, individuals with schizophrenia exhibit reduced synthesis of GABA in the parvalbumin-containing subpopulation of inhibitory neurons of the dorsolateral prefrontal cortex, a brain region involved in working memory *(237)*. The significant decrease in GABA levels is accompanied by a marked increase in the cellular expression of the $GABA_AR$ α_1 subunit mRNA in prefrontal cortex of schizophrenics *(238)*. A compensatory increase in $GABA_AR$-binding activity has been described in prefrontal cortex of schizophrenia patients *(239)*. An increase in [^{3}H] muscimol binding to $GABA_ARs$ is also found in the schizophrenic superior temporal gyrus, a region of the brain involved in auditory hallucinations *(240)*. The abnormalities in $GABA_AR$-mediated inhibitory neurotransmission may have a modulatory effect to contribute to the dysfunction of cortical regions linked to deficits in working memory *(241)*. It also remains unclear if the subunit composition of $GABA_ARs$ is altered in schizophrenia and/or specific schizophrenic symptoms such as working memory, delusions, or hallucinations *(242)*.

3. $GABA_B$ Receptors

The $GABA_BR$ mediates slow inhibitory synaptic transmission and disruption of $GABA_BR$-mediated synaptic pathways has been implicated in many diseases including neuropathic pain, spasticity, schizophrenia, epilepsy, and drug addiction (for review *see* refs. *1,11,16,243*). Activation of presynaptic $GABA_BRs$ inhibits neurotransmitter release by suppressing neuronal Ca^{2+} conductance. Postsynaptic $GABA_BR$-mediated responses inhibit adenylate cyclase and stimulate inwardly rectifying K^+ channels. It is believed that formation of a fully functional receptor requires the coassembly of $GABA_BR1$ and $GABA_BR2$ subunits *(25–29)*. Multiple isoforms of human $GABA_BR1$ have been described ($GABA_BR1a$, $GABA_BR1b$, $GABA_BR1c$, and $GABA_BR1e$) but only one $GABA_BR2$ has been identified *(7)*.

Pharmacological analyses of native $GABA_BRs$ have revealed evidence for $GABA_BR$ subtypes, suggesting heterogeneity in $GABA_BR$ isoforms as well as subcellular targeting and/or coupling to distinct effector systems *(16)*. A striking example of differential expression is found in the neonatal brain as well as the adult striatum and olfactory bulb where $GABA_BR1$ is present in greater amount than $GABA_BR2$ *(12)*. In addition, $GABA_BR2$ knockout mice ($GABA_BR2^{-/-}$) display a selective redistribution of $GABA_BR1$ protein to the soma and proximal dendrites, possibly representing expression at the synaptic plasma membrane *(244)*. In $GABA_BR2^{-/-}$ mice, atypical postsynaptic $GABA_B$ R-mediated electrophysiological responses are detected in hippocampal neurons, whereas knockout of $GABA_BR1$ ($GABA_BR1^{-/-}$) eliminates the response. This, along with the fact that $GABA_BR1^{-/-}$ and $GABA_BR2^{-/-}$ suffer from spontaneous seizures, highlights the importance of identifying mechanisms that control $GABA_BR$ subunit expression and receptor assembly.

The $GABA_BR1$ variants $GABA_BR1a$ and $GABA_BR1b$ differ in their N-terminal aa sequence. Thus far, most reports have demonstrated that there are no differences in the pharmacological profile of $GABA_BR1a$- and $GABA_BR1b$-containing receptors *(24,245–248)*. However, differential expression of $GABA_BR1$ isoforms in embryonic and adult neurons suggests that regulation of their expression may define the particular function of $GABA_BRs$ in a cell. $GABA_BR2$ is expressed at the cell surface *(27,28)* and targets $GABA_BR1$ to the membrane *(29,30,249–252)*. The C-terminal sequences of $GABA_BR1$ encode an ER retention signal that prevents it from being expressed at the cell surface. Interaction of $GABA_BR1$ and $GABA_BR2$ through their coiled-coil regions masks the ER retention signal, allowing expression of the assembled receptors at the cell surface.

3.1. The $GABA_B$ R Family of Genes and Gene Products

3.1.1. Chromosomal Localization

A requirement of both $GABA_BR1$ and $GABA_BR2$ subunits for $GABA_BR$ function suggests that the expression of both genes may be coordinately regulated, perhaps in tandem as duplicated genes in the same chromosomal environment. Contrary to this expectation, the human *$GABA_BR1$* gene has been mapped to chromosome 6p21.3 *(253–255)* and the human *$GABA_BR2$* gene has been mapped to a distinct chromosome 9q22.1 *(28,30)* (Fig. 1). The majority of exon locations and sizes are conserved between orthologues and paraloques in human and *Drosophila,* suggesting an ancestral gene *(7)*. The *$GABA_BR1$* coding region spans 22 kb, whereas the *$GABA_BR2$* coding region spans 418 kb of genomic DNA. Despite their location on different chromosomes, the two $GABA_BR$ subunit genes may achieve complementary expression by having some control in common such as being in a CpG island or being subject to similar transcriptional control *(6)* (Fig. 7).

The *$GABA_BR1$* chromosomal location is within the susceptibility region for IGE, comprising juvenile myoclonic epilepsy, iodiopathic absence epilepsies, and IGE with generalized tonic clonic seizures on awakening *(254,255)*. The region of chromosome 9 containing *$GABA_BR2$* is found within a susceptibility locus for hereditary neuropathy type-1. This disease is a rare neurological

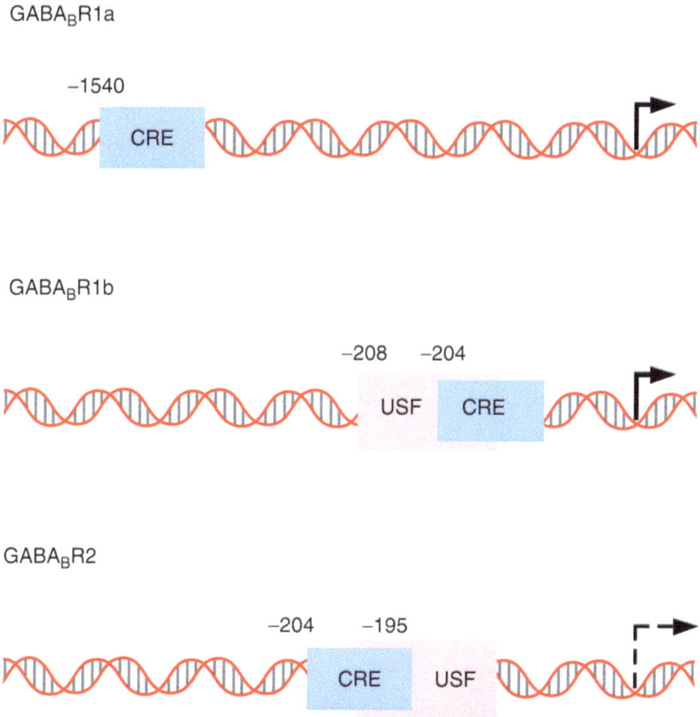

Fig. 7. Identification of CRE and USF transcriptional regulatory elements in the *GABA$_B$R1a*-, *GABA$_B$R1b*-, and *GABA$_B$R2*-promoters. Solid arrows indicate the major transcriptional starts determined by RNase protection *(6)*. The transcriptional start site for *GABA$_B$R2* has not been identified. Dashed arrow indicates the *GABA$_B$R2* translational start site. Bioinformatics was performed using MatInspector software (http://www.genomatix.de) *(350)*.

disorder manifested from early childhood by diminished or absent sensibility to pain, touch, and temperature. To date, no studies have reported a link between *GABA$_B$R2* and hereditary neuropathy type-1.

3.1.2. Alternative Promoters

Complexity and diversity of the human genome can be achieved by gene number and by the production of multiple proteins that are encoded by a single gene locus. Approximately 35–50% of human genes have evidence of alternative RNA splicing of a parent heteronuclear transcript *(256,257)*. Additional heterogeneity of protein isoforms can be achieved by alternative promoters. Recent bioinformatics analyses estimate that 18% of all human genes contain multiple promoters and transcriptional start sites *(258,259)*. The structural organization of genes containing alternative promoters can be organized by class: (a) multiple

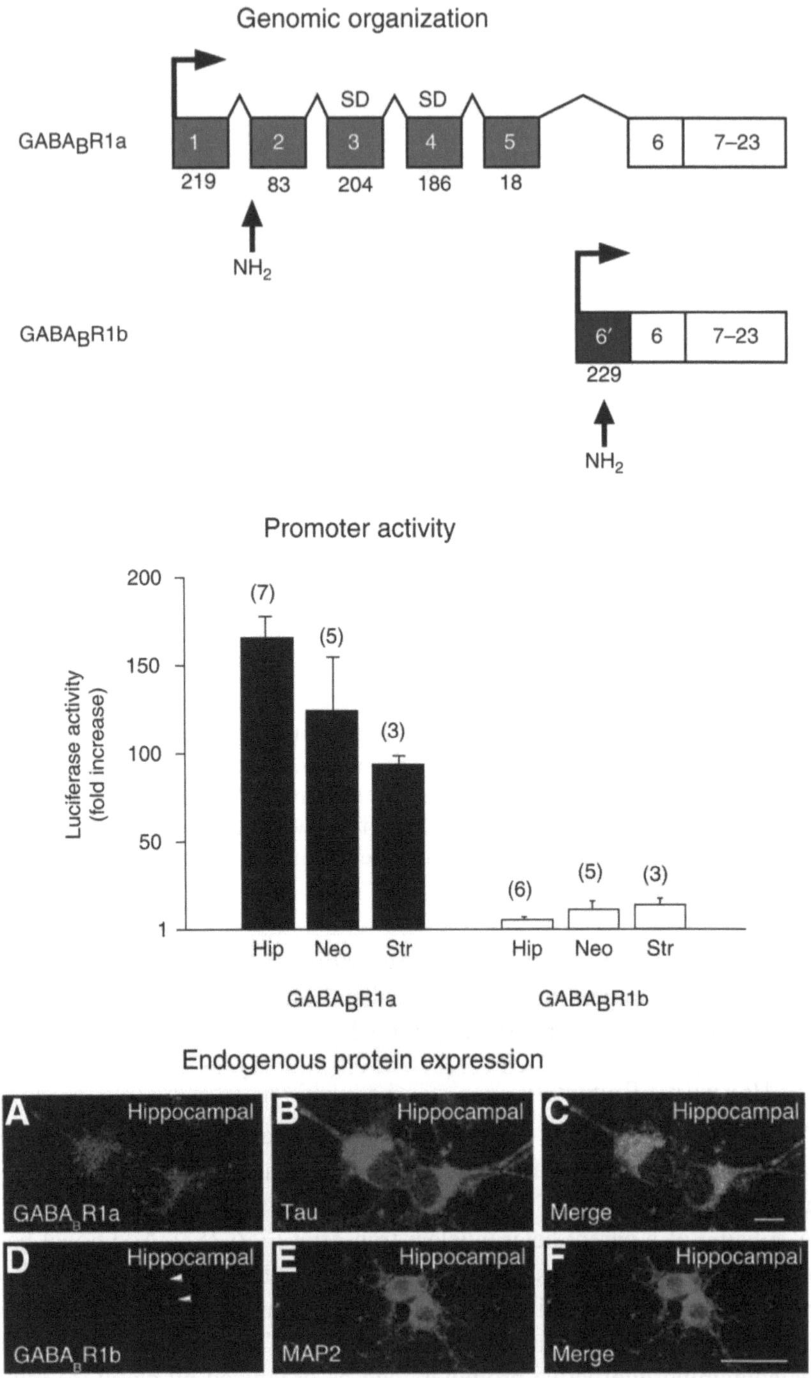
Genomic organization
SD
SD
GABABR1a
1
2
3
4
5
6
7–23
219
83
204
186
18
NH2
GABABR1b
6′
6
7–23
229
NH2
Promoter activity
200
150
100
50
1
Luciferase activity (fold increase)
(7)
(5)
(3)
(6)
(5)
(3)
Hip
Neo
Str
Hip
Neo
Str
GABABR1a
GABABR1b
Endogenous protein expression
A
Hippocampal
GABABR1a
B
Hippocampal
Tau
C
Hippocampal
Merge
D
Hippocampal
GABABR1b
E
Hippocampal
MAP2
F
Hippocampal
Merge

tandemly arranged promoters to generate mRNAs with alternative first noncoding exons and a common downstream exon that contains the translation initiation site; (b) alternative promoters to generate mRNAs with different 5′ exons that contain alternative translation initiation sites and are spliced to a common exon with the same open reading frame; and (c) alternative promoters to generate mRNAs with different first exons that are spliced to a common second exon but are read in different frames *(259,260)*. Alternative promoters create diversity in regulatory control of gene expression that is needed for developmental and tissue-specific gene expression, subcellular localization, differential levels of expression, and the specific capacity to respond to extracellular stimuli *(259,260)*. For some genes, the use of alternative promoters may also influence mRNA stability, translational efficiency, and protein function *(259,260)*.

The human $GABA_BR1$ gene encodes at least four subunit variants. The most abundant $GABA_BR1$ isoforms $GABA_BR1a$ and $GABA_BR1b$ contain different N-termini where the first 164 aa of $GABA_BR1a$ are replaced with 47 aa to generate $GABA_BR1b$. The N-terminal sequence of $GABA_BR1a$, but not $GABA_BR1b$, encodes two Sushi domains (SD) that may interact with the fibulin family of extracellular matrix proteins *(261)*. The 5′ ends of $GABA_BR1a$ and $GABA_BR1b$ have recently been identified and examination of the genomic sequence immediately upstream does not reveal the presence of consensus splice junctions that are necessary to produce the different transcripts by alternative slicing *(6)*. The 5′ ends of $GABA_BR1a$ and $GABA_BR1b$, rather, represent unique sites of transcriptional initiation and the genomic sequence immediately upstream of $GABA_BR1a$ and $GABA_BR1b$ exhibits significant promoter activity (Fig. 8). Together, these

Fig. 8. *(Opposite page)* $GABA_BR1a$ and $GABA_BR1b$ are produced by alternative promoters. (Left panel) Identification of independent transcription initiation sites for $GABA_BR1a$ and $GABA_BR1b$. Transcription initiation sites (arrows) were identified by RNase protection analysis *(6)*. The SD and N-terminus (NH_2) are indicated. The $GABA_BR1a$-specific exons are shown in red, the $GABA_BR1b$-specific exons are shown in blue, and the size of the exons (base pairs) is shown below. (Middle panel) $GABA_BR1a$- and $GABA_BR1b$-promoter activity in primary embryonic hippocampal (Hip), neocortical (Neo), and striatal (Str) neurons. Data shown are mean ± SEM. *Right panel*, levels of $GABA_BR1a$ protein exceed $GABA_BR1b$ in primary embryonic hippocampal neurons. The endogenous $GABA_BR1a$ (A, C) and $GABA_BR1b$ (D, F; red) protein was detected using an antiserum raised against the N-termini of $GABA_BR1a$ and $GABA_BR1b$, respectively. Hippocampal neurons were identified using the axonal marker Tau (B) or the dendritic marker MAP2 (E), and colocalization with the $GABA_B$-R subunit proteins is shown as merged images (C, F). Expression was monitored in fixed cells after 7 DIV. The images were obtained as 1 μm optical sections by confocal microscopy. Expression of $GABA_BR1b$ is low and barely detectable (arrowheads). Bar for A–C, 10 μm. Bar for D-F, 20 μm.

findings indicate that *$GABA_BR1a$* and *$GABA_BR1b$* transcripts are under the control of alternative promoters (class b) and not alternative splicing (Table 1) *(6)*.

Differential regulation of *$GABA_BR1a$* and *$GABA_BR1b$* during development is achieved by the differential use of the alternative transcription initiation sites. Both *$GABA_BR1$* transcripts are present in the human adult brain, whereas *$GABA_BR1a$* but not *$GABA_BR1b$* is detected in the human fetal brain *(6)*. This result is consistent with levels of promoter activity and protein expression in the embryonic brain. The *$GABA_BR1a$*-promoter is 33-times more active than *$GABA_BR1b$* in primary embryonic hippocampal neurons at 7 d in vitro (DIV) *(6)* (Fig. 8). This difference in promoter activity is mirrored in the endogenous $GABA_BR1a$ and $GABA_BR1b$ protein levels in primary hippocampal neurons at this stage in development (Fig. 8). Promoter activity also parallels endogenous gene expression in striatum. The differences are smaller, but *$GABA_BR1a$* promoter activity is seven times greater than *$GABA_BR1b$* in striatal neurons (Fig. 8). Differential use of the alternative *$GABA_BR1$* promoters most likely reflects modulation of transcription factor binding to unique regulatory elements in the *$GABA_BR1a$* and *$GABA_BR1b$* 5′ flanking regions *(6)*. Additionally, these independent promoters may have different regulatory specificity to control subcellular localization and genomic responses to synaptic input.

3.1.3. Alternative Splicing of $GABA_BR1$ But Not $GABA_BR2$

3.1.3.1. Alternative Splicing of Human $GABA_BR1$

Two alternatively spliced transcripts arise from the human *$GABA_BR1$* gene, *$GABA_BR1c$* and *$GABA_BR1e$* (Fig. 9). Both of these transcripts contain 5′ sequences that are consistent with both isoforms being under the control of the *$GABA_BR1a$*-promoter.

3.1.3.1.1. Human $GABA_BR1c$

The human $GABA_BR1c$ isoform results from exon skipping of the $GABA_BR1$ gene exon 4 that encodes the second SD (Table 1) *(7)*. The human $GABA_BR1c$ isoform is predicted to encode a protein with only the first of the two SD present in $GABA_BR1a$ (Fig. 9). Similar to patterns of $GABA_BR1a$ expression, $GABA_BR1c$ mRNA is more highly expressed in early nervous system development as compared with adult levels *(7)*, suggesting that the two isoforms are coordinately regulated (Fig. 10). The functional significance of two SD in $GABA_BR1a$ or one sushi domain in $GABA_BR1c$ remains to be determined in future studies.

3.1.3.1.2. Human *$GABA_BR1e$*

$GABA_BR1e$ was isolated from human prostate cDNA and is a truncated isoform encoding only the signal peptide, SD, and extracellular ligand-binding

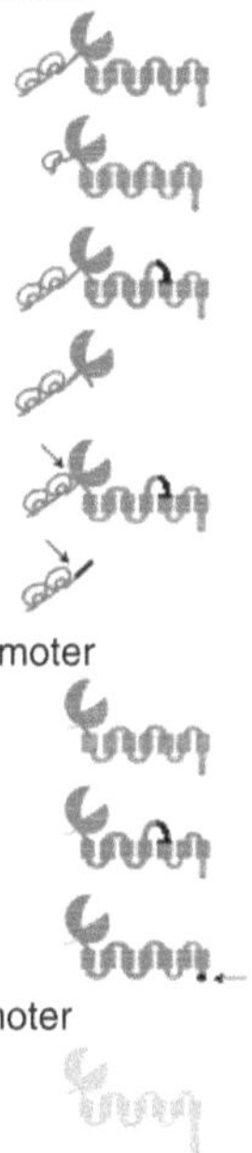

Isoform	Species	Splicing	References
GABA$_B$R1a promoter			
GABA$_B$R1a	Human, rat	None	*6,24*
GABA$_B$R1c*	Human	Exon skipping of sushi domain	*7*
GABA$_B$R1h*	Rat		*12*
GABA$_B$R1c-a	Rat	Inserion between ECL2-TM5	*247*
GABA$_B$R1e	Human, rat	Exon skipping, new stop codon	*262*
GABA$_B$R1f	Rat	Exon skipping of 7aa in ECD insertion between ECL2-TM5	*264*
GABA$_B$R1g	Rat	Insertion, new stop codon	*265*
GABA$_B$R1b promoter			
GABA$_B$R1b	Human, rat	None	*6,24*
GABA$_B$R1c-b	Rat	Insertion between ECL2-TM5	*247,263*
GABA$_B$R1d	Rat	Insertion, new stop codon	*263*
GABA$_B$R2 promoter			
GABA$_B$R2	Human, rat	None	*7,25–30,266*

Fig. 9. Human and rat GABA$_B$R1 and GABA$_B$R2 subunit variants. Schematic representations of GABA$_B$R1a-h and GABA$_B$R2 are shown. The GABA$_B$R1 isoforms are produced by alternative promoters and alternative RNA splicing, whereas only one GABA$_B$R2 has been identified. The N-terminal aa-binding domain, transmembrane, C-terminal, intracellular, and extracellular loops are indicated. The alternatively spliced regions are shown in black and indicated by arrows, the N-terminal SD produced by the *GABA$_B$R1a*-promoter are shown in red, and the alternative N-terminal region produced by the *GABA$_B$R1b*-promoter is shown in blue. The asterisks indicate that the human GABA$_B$R1c isoform has been identified as GABA$_B$R1h in rat. ECL2, second extracellular loop; TM5, fifth transmembrane domain.

domain of *GABA$_B$R1* (Fig. 9) (Table 1) *(262)*. Exon 11 of the *GABA$_B$R1* gene, which encodes the first transmembrane domain, is skipped in the *GABA$_B$R1e* transcript. The reading frame of exon 12 is altered by the omission of exon 11 and encodes nine new aa and an inframe stop codon consequently producing the truncation. *GABA$_B$R1e* transcripts are more highly expressed in peripheral tissues than the brain, when measured by reverse transcriptase polymerase chain reaction (RT-PCR) *(262)*. In a heterologous system, the truncated receptor is both secreted and membrane associated, but the functional role of the GABA$_B$R1e isoform is not known.

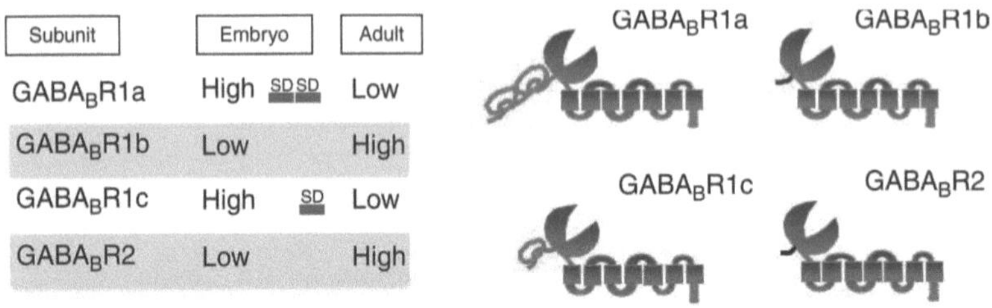

Fig. 10. A summary of *GABA$_B$R* subunit gene expression during development. (Left panel) A comparison of *GABA$_B$R* subunit gene expression in the embryonic and adult nervous systems. The GABA$_B$R1a and GABA$_B$R1c isoforms, which contain 1 or 2 SD respectively, are abundantly expressed during embryonic development *(12)*. The presence of sushi domain(s) may relate to a novel functional requirement for the GABA$_B$R in the embryo. (Right panel) schematic representations of human GABA$_B$R1a, GABA$_B$R1b, GABA$_B$R1c, and GABA$_B$R2. The SD are shown in red, the GABA$_B$R1b-specific N-terminal region is shown in blue, and the N-terminus of GABA$_B$R2 is shown in black. The two lobes of the aa-binding domain, transmembrane domains, C-terminus, intracellular, and extracellular loops are also indicated.

3.1.3.2. Alternative Splicing of Rat *GABA$_B$R1*

In addition to the two alternatively spliced *GABA$_B$R1* transcripts described in human, *GABA$_B$R1c* and *GABA$_B$R1e*, rat has five further alternatively spliced transcripts bringing the total number of alternatively spliced transcripts in rat to seven (Fig. 9). Five isoforms are generated by alternative splicing of the transcript arising from the *GABA$_B$R1a*-promoter: GABA$_B$R1h, GABA$_B$R1c-a, GABA$_B$R1e, GABA$_B$R1f, and GABA$_B$R1g. Two isoforms are generated by alternative splicing of the transcript arising from the *GABA$_B$R1b*-promoter: GABA$_B$R1c-b and GABA$_B$R1d.

3.1.3.2.1. Rat *GABA$_B$R1c-a* and *GABA$_B$R1c-b*

The rat GABA$_B$R1c isoform was isolated from a hippocampal library and differs from GABA$_B$R1a and GABA$_B$R1b by an inframe 31 aa insertion between the second extracellular loop and the fifth transmembrane region (Fig. 9) *(247,263)*. The rat GABA$_B$R1c-a and GABA$_B$R1c-b variants contain divergent N-terminal regions that are respectively produced by the alternative *GABA$_B$R1a* and *GABA$_B$R1b* promoters.

3.1.3.2.2. Rat GABA$_B$R1*d*

The GABA$_B$R1d isoform contains an aa sequence identical to GABA$_B$R1b except for the C-terminus (Fig. 9) *(263)*. This suggests that the *GABA$_B$R1d* transcript is generated under the control of the *GABA$_B$R1b*-promoter and may be more highly expressed in the adult than the embryo. GABA$_B$R1d differs from

$GABA_BR1b$ by a 566-bp insertion of intron 22 sequences, generating a new stop codon that results in a C-terminal truncation *(263)*. The 57 C-terminal aa of $GABA_BR1b$ are replaced by 25 different aa in $GABA_BR1d$. The $GABA_BR1d$ cDNA was isolated from a cerebellum cDNA library and RT-PCR was used to detect transcripts for $GABA_BR1d$ in forebrain, cerebellum, eye, kidney, and urinary bladder *(263)*.

3.1.3.2.3. Rat *$GABA_BR1e$*

A transcript corresponding to rat *$GABA_BR1e$* lacking exon 11 is detected by RT-PCR in brain and peripheral tissues *(262)*. This provides evidence that the same N-terminal extracellular domain truncation described for human *$GABA_BR1e$* is conserved in rat (Fig. 9).

3.1.3.2.4. Rat *$GABA_BR1f$*

The rat $GABA_BR1f$ isoform isolated from a hippocampal cDNA library has a 93-bp inframe insertion encoding 31 new aa between the second extracellular loop and the fifth transmembrane domain (Fig. 9) *(264)*. The 93-bp insertion is sequence from rat intron 17 and, although the intron/exon boundary is conserved between rat and human, no sequence similar to the 93-bp insertion is found in human intron 17. In addition, the rat $GABA_BR1f$ isoform has an inframe deletion of exon 5 *(264)*. Importantly, exon 5 encodes a seven aa hinge sequence that links the second sushi domain to the aa-binding domain *(7)*. Skipping exon 5 results in a 21-bp deletion and the encoded seven aa hinge sequence is removed from the N-terminal extracellular region in the $GABA_BR1f$ isoform (Fig. 9) *(264)*.

Results of RT-PCR demonstrate that transcripts without exon 5 sequences are present at low levels in rat cortex, midbrain, cerebellum, brainstem, thalamus, and hippocampus *(264)*. This contrasts with RT-PCR results that demonstrate the presence of the $GABA_BR1f$ transcript in rat skeletal muscle, testis, kidney, colon, small intestine, spleen, stomach, liver, lung, and heart, suggesting a role for $GABA_BR1f$ in peripheral tissues.

3.1.3.2.5. Rat *$GABA_BR1g$*

The $GABA_BR1g$ cDNA was isolated from a rat hippocampal library and the predicted protein sequence is 239 aa *(265)*. The first 157 aa of $GABA_BR1g$ are identical to $GABA_BR1a$ followed by an 82 aa insertion that generates a new stop codon. The truncated protein contains the signal peptide, two SD and the new sequence but does not contain the hinge sequence encoded by exon 5, the extracellular ligand-binding domain, transmembrane regions, or the C-terminus (Fig. 9). The frame-shift is achieved by the insertion of 124 bp after exon 5. $GABA_BR1g$ is expressed in both the rat brain and peripheral tissues *(265)*.

3.1.3.2.6. RAT *GABA$_B$R1h*

A transcript for the rat isoform lacking the second sushi domain in the N-terminal extracellular region and corresponding to the GABA$_B$R1c isoform described in human *(7)* is detected in mRNA isolated from rat embryonic and adult nervous system tissues by RT-PCR *(12)*. As the term rat GABA$_B$R1c has already been used for a different isoform, the rat isoform lacking the second SD is referred to as GABA$_B$R1h, the next unused letter for rat GABA$_B$R1 isoforms.

3.1.3.3. *GABA$_B$R2*

GABA$_B$R2 encodes 941 aa in human and 940 aa in rat and the same transcript was independently cloned by a number of groups *(25–30)*. The transcript encoding human GABA$_B$R2 is split into 19 exons *(7)*. Two further low abundance transcripts of GABA$_B$R2 differing in the last exon encoding the C-terminal tail were isolated by PCR cloning *(266)*. Analysis of genomic sequence reveals no consensus splice sites at the sequence junctions required to generate these spliced forms within the GABA$_B$R2 C-terminus *(7)*. Furthermore, no transcripts in human fetal or adult brain mRNA were detected that correspond to the size expected for C-terminal alternative forms of GABA$_B$R2 *(7)*. Therefore in contrast to the many alternative isoforms of the GABA$_B$R1 protein, to date there are no alternative isoforms of the GABA$_B$R2 protein.

3.2. *Differential Expression of GABA$_B$R1 and GABA$_B$R2 in the CNS*

3.2.1. *Developmental Expression*

GABA has been proposed to act as a developmental signal to influence the formation of neuronal networks *(267)*. At early developmental stages when the function of GABA$_A$Rs is excitatory *(268)*, GABA$_B$Rs are likely the major inhibitory receptor in the CNS *(269,270)*. Evidence for multiple GABA$_B$R subtypes comes from pharmacological analyses of native receptors *(16)*. In principle, the molecular heterogeneity in GABA$_B$R isoforms, subcellular localization, and differences in effector systems could lead to the functional heterogeneity of receptor subtypes.

Despite the functional requirement for a heterodimeric receptor complex of GABA$_B$R1 and GABA$_B$R2 subunits, these proteins are not coordinately regulated in the embryonic brain (Fig. 10). At embryonic day (E)10 levels of GABA$_B$R1 mRNA greatly exceed GABA$_B$R2 mRNA *(12)*. By E14, GABA$_B$R1 and GABA$_B$R2 mRNAs are coexpressed in the brain, cranial, and dorsal root ganglia *(12)*. Moreover, GABA$_B$R1 and GABA$_B$R2 immunoreactivity is found in the hippocampal primordium at E14 *(271)*. GABA$_B$R1 and GABA$_B$R2 transcripts and protein are also present in the E18 brain and E18 cultured cortical and hippocampal neurons *(12,272)*. At this developmental stage, GABA$_B$Rs are negatively coupled to adenylyl cyclase *(12)* (Fig. 11).

During early postnatal development, the alternative $GABA_BR1$ isoforms $GABA_BR1a$ and $GABA_BR1b$ exhibit regional differences in developmental regulation. In cerebral cortex, thalamus, and cerebellum, expression of the $GABA_BR1a$ isoform is most abundant between P0 and P14 *(273–275)*. Parallel to the downregulation of $GABA_BR1a$, levels of $GABA_BR1b$ expression peak at P10–20 and continue to be more abundant than $GABA_BR1a$ in the adult brain *(273–275)*. Given the independent maturation of $GABA_BR1a$, $GABA_BR1b$, and $GABA_BR2$ isoforms, $GABA_BR1a$–$GABA_BR2$ heterodimers may predominate during early embryonic development whereas $GABA_BR1b$–$GABA_BR2$ are most abundant in the adult (Figs. 10 and 12). No differences in the levels of $GABA_BR1a$ and $GABA_BR1b$ expression are found in brainstem, striatum, and hippocampus at P21, P28, and adulthood *(273,275)*.

Consistent with expression of $GABA_BR1a$, the alternatively spliced variant $GABA_BR1c$, which is produced by the *$GABA_BR1a$* promoter, is most abundant in the embryonic brain *(7)*. Similarly, expression of the equivalent rat $GABA_BR1$ isoform $GABA_BR1h$ is three times more highly expressed in the embryo than in the adult *(12)*. Differential expression of human $GABA_BR1a$ and $GABA_BR1c$ most likely reflects differences the chromatin environment and/or transcription factor binding to specific response elements in the *$GABA_BR1a$* promoter. Moreover, selective expression of $GABA_BR1a$ and $GABA_BR1c$ subunits containing one or two SD, respectively, may reflect a functional requirement of the SD during development (Fig. 10). Although SD is believed to be involved in protein–protein interactions, the specific function of the SD in $GABA_BR1a$ and $GABA_BR1c$ await further characterization.

3.2.2. Regional Expression

3.2.2.1. Hippocampal Formation

$GABA_BR1$ and $GABA_BR2$ are widely expressed in the hippocampus throughout development. By E14, both $GABA_BR1$ and $GABA_BR2$ are detected in the hippocampal primordium *(271)*. In addition, levels of $GABA_BR1$ and $GABA_BR2$ protein expression increase at E18 where $GABA_BR1$ expression is concentrated in the cell body and $GABA_BR2$ is present in neuronal processes and neuropil *(271)*. The somatic distribution of $GABA_BR1$ can be ascribed to the accumulation of receptors in the ER *(271,276)*. Similar expression patterns have been observed in cultured E18 hippocampal neurons. At 3 DIV $GABA_BR1$ and $GABA_BR2$ immunoreactivity is restricted to the cell body and nucleus with no labeling in the developing neurites *(272)*. Between 7 DIV and 21 DIV, when baclofen-evoked inwardly rectifying K^+ channels are first detected, there is a dramatic redistribution of $GABA_BR1$ and $GABA_BR2$ subunits to a punctate pattern throughout the neuronal processes *(272)*.

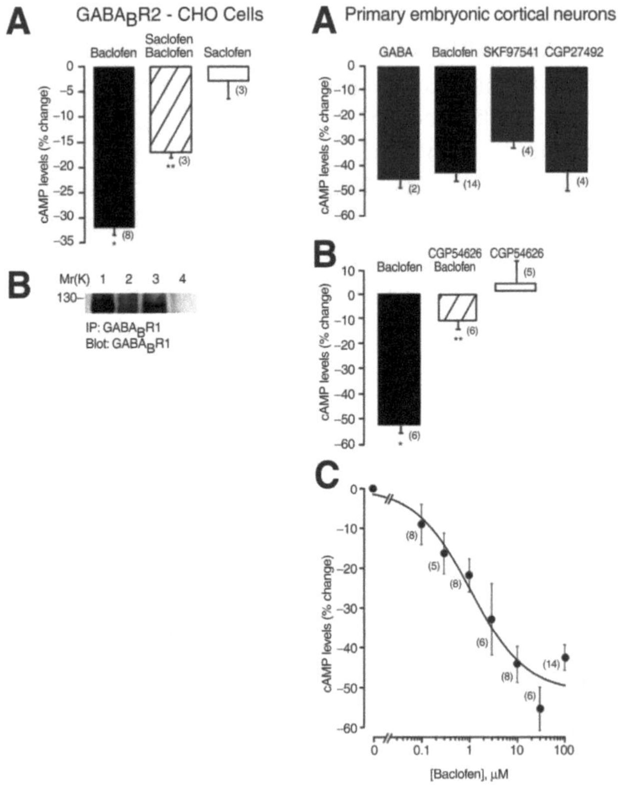
A
GABABR2 - CHO Cells
Baclofen
Saclofen Baclofen
Saclofen
cAMP levels (% change)
B
Mr(K) 1 2 3 4
130–
IP: GABABR1
Blot: GABABR1
A Primary embryonic cortical neurons
GABA
Baclofen
SKF97541
CGP27492
B
Baclofen
CGP54626 Baclofen
CGP54626
C
[Baclofen], μM

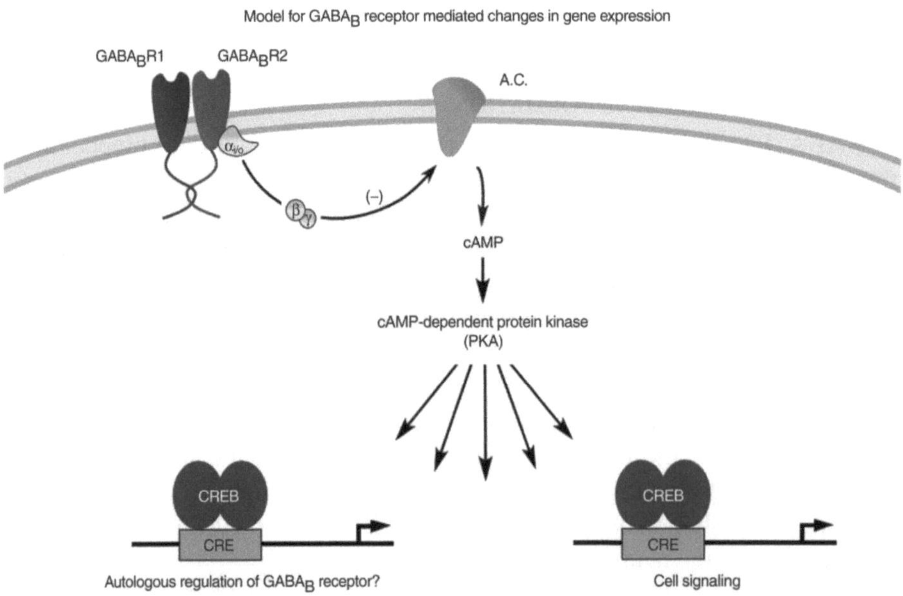
Model for GABAB receptor mediated changes in gene expression
GABABR1
GABABR2
A.C.
(–)
cAMP
cAMP-dependent protein kinase (PKA)
CREB
CRE
Autologous regulation of GABAB receptor?
CREB
CRE
Cell signaling

During postnatal development, $GABA_BR1$ is distributed in pyramidal cells of the CA1 and CA3 subfields as well as interneurons throughout the hippocampus *(271,274,276)*. Moreover, $GABA_BR1a$, but not $GABA_BR1b$, is present at birth in CA1 and CA3 and a prominent upregulation of $GABA_BR1b$ can be observed in at P20 *(274)*. In addition, $GABA_BR1a$ is most abundant in CA3 pyramidal cells, whereas $GABA_BR1b$ is highly expressed in the CA1 subfield *(277)*. In contrast to $GABA_BR1$, $GABA_BR2$ immunoreactivity is stronger in CA3 than CA1, and is found mostly in the neuropil and weaker in the pyramidal cell layer *(274)*. Also, $GABA_BR1$, but not $GABA_BR2$, is expressed in glial cells throughout the hippocampus *(271)*.

$GABA_BR$ subunits are localized to pre- and postsynaptic elements throughout the hippocampus *(271,276)*. Presynaptically, $GABA_BR1$ and $GABA_BR2$ are localized along the extrasynaptic membrane of axon terminals as well as the presynaptic active zones of asymmetrical and symmetrical synapses *(271,276)*. Postsynaptically, $GABA_BR1$ and $GABA_BR2$ are detected in the extrasynaptic and perisynaptic membranes of dendritic spines and shafts in pyramidal and nonpyramidal cells *(271,276)*. In both locations, $GABA_BRs$ could be activated by spillover of GABA after its synaptic release. Moreover, the presence of $GABA_BR1$ and $GABA_BR2$ on excitatory (glutamatergic) and inhibitory (GABAergic) terminals suggests a potential mechanism for selective GABAergic control of synaptic transmission at glutamatergic inputs *(276)*.

3.2.2.2. Olfactory Bulb and Striatum

$GABA_BR1$ and $GABA_BR2$ are expressed together in the majority of the adult nervous system consistent with the functional requirement of a presumed heterodimer of the $GABA_BR$ subunits *(25,26,28,29)*. Multiple lines of evidence indicate that $GABA_BR1$ and $GABA_BR2$ are not coordinately regulated in all

---

Fig. 11. *(Opposite page)* $GABA_BR$ activation reduces cAMP levels in CHO cells stably expressing $GABA_BR2$ ($GABA_BR2$-CHO) and primary embryonic cortical neurons expressing $GABA_BR1$ and $GABA_BR2$. (Left panel) $GABA_BR2$-CHO cells are negatively coupled to adenylyl cyclase in the absence of $GABA_BR1$. (Reprinted with permission from ref. *28*.) (Middle panel) E18 cortical neurons cultured for 7 DIV contain functional $GABA_BR$ negatively coupled to adenylyl cyclase. (Reprinted with permission from ref. *12*.) (Right panel) A schematic model for $GABA_BR$-mediated cell signaling and autologous gene regulation. $GABA_BR$ activation is negatively coupled to the cAMP/protein kinase A (PKA) signaling pathway that can modulate CREB-mediated transcription. Given that *$GABA_BR1a$* and *$GABA_BR1b$* gene expression is controlled by CREB-binding to distinct CRE sites in the alternative *$GABA_BR1$* promoters, $GABA_BR$ activation, and the subsequent inhibition of cAMP/PKA signaling, may provide a mechanism to control expression of its own gene product by autologous gene regulation.

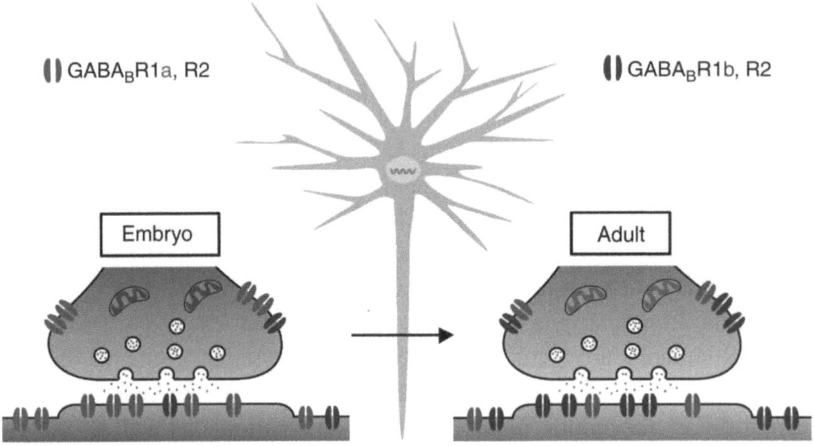

Fig. 12. A schematic diagram for the developmental switch of GABA$_B$R1a and GABA$_B$R1b receptor subtypes. Developmental changes in *GABA$_B$R1a* and *GABA$_B$R1b* mRNAs and proteins have been demonstrated by RNA and protein analyses, which show high expression of GABA$_B$R1a in the embryonic brain and a postnatal increase in GABA$_B$R1b *(6,274,275)*. Furthermore, results of functional promoter analyses indicate that the *GABA$_B$R1a* promoter is much stronger than *GABA$_B$R1b* in primary embryonic hippocampal neurons *(6)*. This developmental switch is represented here as high levels of GABA$_B$R1a, GABA$_B$R2 heterodimers (red) in the embryo and high levels of GABA$_B$R1b, GABA$_B$R2 heterodimers (blue) in the adult. A model in which selective control of the alternative *GABA$_B$R1* promoters during development may control the abundance of specific GABA$_B$R subtypes at the cell surface is proposed. The functional significance of this developmental switch in GABA$_B$R populations awaits further characterization.

regions of the nervous system including regions within the olfactory bulb and striatum *(12,25,26,29,278–280)*. The accessory olfactory bulb coexpresses GABA$_B$R1 and GABA$_B$R2 but in the main olfactory bulb *GABA$_B$R1* mRNA is highly expressed in the granular and glomerular cell layers and *GABA$_B$R2* mRNA is barely expressed in these layers *(12)*. Subcellular GABA$_B$R1 and GABA$_B$R2 protein distribution patterns in the main olfactory bulb are not completely overlapping *(281)*. In the striatum the most abundant cell type, medium spiny neurons, express either low GABA$_B$R2 protein levels *(280)* or no detectable GABA$_B$R2 protein levels *(279)* and high GABA$_B$R1 protein levels *(279,280,282,283)*. The differential expression of GABA$_B$R1 and GABA$_B$R2 in the olfactory bulb and striatum is set up early in development *(12,281)*.

The functional consequences of differential GABA$_B$R subunit expression in certain nervous system regions have not been determined. If no GABA$_B$R2 protein exists in striatal medium spiny neurons, for example, then it is possible these neurons do not contain functional GABA$_B$Rs. Alternatively GABA$_B$R1 may function in a different way in striatal medium spiny neurons, such as with a different partner for membrane localization and signaling. A third possibility is that low levels of $GABA_BR2$ protein are sufficient for function. Regardless of the functional consequences it is likely that the control of the expression of $GABA_BR2$ is different in a subset of regions of the nervous system. The expression studies imply that $GABA_BR1$ and $GABA_BR2$ are not necessarily expressed coordinately. This is consistent with both subunits being expressed on different chromosomes and indicates that their promoter regions are likely to be under the control of distinct transcription factors.

3.3. $GABA_BR$ Agonists and Antagonists: Effects on Gene Expression

3.3.1. Agonists

3.3.1.1. Baclofen-Mediated Changes in Gene Expression

Several groups have shown that $GABA_BR$ activation by the specific agonist baclofen inhibits forskolin-stimulated increases in cAMP levels in cortical brain slices *(284)*, transfected CHO cells *(28,285)*, and primary embryonic cortical neurons (Fig. 11) *(12)*. This effect is reversed by co-application with a $GABA_BR$ antagonist. $GABA_BRs$ are also described to have stimulatory or inhibitory effects on Ca^{2+} signaling pathways *(286,287)*. Responses to cAMP and Ca^{2+} signaling are ultimately mediated by transcription factor binding to discrete regulatory regions in many genes, and alterations in $GABA_BR$ activity may initiate long-term effects on protein synthesis, including autologous regulation of $GABA_BR$ gene expression (Fig. 11).

Several groups have reported the effects of baclofen on $GABA_BR$ subunit gene expression. In the gerbil hippocampus, baclofen produces a dose-dependent decrease in $GABA_BR1$ and $GABA_BR2$ protein, whereas the $GABA_BR$ antagonist phaclofen enhances receptor subunit immunoreactivity *(288)*. In contrast, repeated administration of baclofen to rats does not alter the levels of $GABA_BR$ subunit mRNA and protein in the cerebral cortex *(289)* and lumbar spinal cord *(290)*. Whereas microarray analysis of baclofen-treated primary hippocampal neurons does not reveal changes in $GABA_BR$ subunit gene expression, $GABA_BR$ activation produces selective changes in the expression of 20 genes that encode proteins for signal transduction, endocytosis/trafficking, and cytoskeletal organization *(291)*. Expression of these genes remains unchanged when baclofen is applied in the presence of the $GABA_BR$ antagonist CGP-55845A. Moreover, $GABA_BR$ activation blocks amphetamine-induced neuropeptide gene expression and behavioral activity *(292,293)*.

A well-characterized target of the cAMP and Ca^{2+} signaling pathways is the cyclic ADP (cAMP) response element (CRE) that binds members of the cAMP responsive element-binding protein (CREB) family of transcription factors

(294–296). Acute exposure of cerebellar granule neurons to GABA (1–100 μ*M*) and baclofen (1–100 μ*M*) inhibits cAMP and Ca^{2+}-mediated expression of a CRE-driven reporter gene *(297)*. Moreover, the $GABA_BR$ antagonist CGP-35348 (50 μ*M*) attenuates the inhibitory effects of GABA and baclofen. $GABA_BR$ activity inhibits CRE-dependent transcription by negative modulation of CREB *(297)*. In contrast, exposure of cerebellar granule neurons to 10 μ*M* baclofen has also been demonstrated to increase the DNA-binding activity of CREB to a consensus CRE *(298)*. Furthermore, acute in vivo administration of baclofen (10 mg/kg) increases CRE-binding in the hippocampus *(299)*. The discrepancies between the effects of baclofen on $GABA_BR$ subunit gene expression and CREB-mediated transcription may be attributed to different treatment paradigms and/or differential regulation in different species and different brain regions.

3.3.1.2. Baclofen-Mediated Changes in the Translocation of a CREB-Related Protein

The mechanism-of-action of $GABA_BRs$ continue to challenge the thinking by interacting in their C-terminal regions with transcription factors such as CREB2, also termed activating transcription factor -4 (ATF-4), and C/EBP homologous protein (CHOP), a downstream target of ATF4 *(300–303)*. CHOP interacts selectively with $GABA_BR1a$–$GABA_BR2$ receptors to regulate cell-surface expression *(303)*. In contrast, $GABA_BR1a/b$ can bind to $GABA_BR2$ or ATF4 but not to both simultaneously *(301,302)*. ATF4, therefore, may prevent $GABA_BR$ dimerization and inhibit the formation of functional heterodimeric $GABA_BRs$. Conversely, the C-terminus of $GABA_BR1$ may mask the ATF4 nuclear localization signal to control its concentration in the nucleus *(300)*. $GABA_BR$ activation can also stimulate translocation of ATF4 from one cellular compartment to another, regulating ATF4-directed transcription *(300–302)*. $GABA_BRs$, therefore, may influence gene expression both by modulating the levels of cAMP *(297)* and by direct interaction with ATF4 *(300–302)*.

3.3.2. *Antagonists*

In models of learning memory, increased binding of CREB to CRE sequences enhances the expression of proteins important for memory consolidation. ATF4, on the other hand, acts as a suppressor of gene transcription mediated by CREB and inhibition of ATF4 activity has been found to enhance memory processes *(304–306)*. Recent evidence suggests that the $GABA_BR$ antagonist SGS742 (CGP-36742), which is currently in a clinical trial of patients with mild cognitive impairment, improves spatial learning and reduces basal and baclofen-induced ATF4-binding to a consensus CRE *(299)*. It is, therefore, believed that

SGS742 may modulate postsynaptic $GABA_B$R-mediated repression of ATF4 activity to enhance long-term memory.

Generalized absence seizures induced by γ-butyrolactone in mice is associated with increased nuclear protein binding to CRE and AP-1 regulatory sequences and inhibition of $GABA_B$Rs by CGP 35348 attenuates the genomic response *(307)*. Consistent with this finding, increased CRE and AP-1 binding activity is inhibited by the $GABA_B$R antagonist CGP 46831 at a dose that suppresses seizure behavior in a genetic model of absence epilepsy *(308,309)*.

3.4. Transcriptional Gene Regulation

The alternative *$GABA_B$R1a* and *$GABA_B$R1b* promoters contain consensus CRE sequences located 1540 and 202 bp upstream of their respective transcriptional start sites (Fig. 7). The 5′ flanking regions of *$GABA_B$R1a* and *$GABA_B$R1b* have been analyzed in vitro by reporter-gene expression and nuclear protein-binding assays targeting the *$GABA_B$R1a* and *$GABA_B$R1b* CREs *(6)*. These studies demonstrate that recruitment of CREB to the CRE sites is essential for activation of *$GABA_B$R1a* and *$GABA_B$R1b* promoter activity (Fig. 13). Moreover, CRE decoy oligonucleotides that compete for binding of endogenous CREB family members inhibit *$GABA_B$R1a* and *$GABA_B$R1b* gene expression, indicating a critical role for CREB in the activation of *$GABA_B$R1* transcription from the alternative promoters in vivo *(6)*.

Differential regulation of the *$GABA_B$R1* promoters is mediated by the CREB-related protein ATF4 and the depolarization-sensitive upstream stimulatory factor USF *(6)*. Overexpression of ATF4 relative to CREB stimulates *$GABA_B$R1a* and inhibits *$GABA_B$R1b* promoter activity (Fig. 13). Consistent with this finding, $CREB^{\alpha/\Delta}$ knockout mice, exhibit increased *$GABA_B$R1a* and decreased *$GABA_B$R1b* gene expression *(6)*. The novel observation that $GABA_B$R1 associates with the transcription factor ATF4 in the cytoplasm of neurons, and the genomic regulation of $GABA_B$R1 by such factors, suggest a surprising additional potential role for $GABA_B$R subunits in the trafficking of ATF4 to mediate autologous gene regulation and in the role that ATF4 may play in the trafficking of $GABA_B$R subunits to control heterodimerization.

Recent evidence also suggests that overlapping USF- and CREB-binding sites define a novel *$GABA_B$R1b* regulatory region *(6)*. USF decoy oligonucleotides selectively increase $GABA_B$R1b protein in embryonic hippocampal cells that normally contain very low levels of $GABA_B$R1b, indicating that USF may inhibit endogenous levels of $GABA_B$R1b subunits (Fig. 13). Only $GABA_B$R1b transcription is under control of USF-mediated repression, as $GABA_B$R1a is not regulated by USF decoy oligonucleotides. The USF-CRE regulatory element in the *$GABA_B$R1b*-promoter is preferentially occupied by CREB; however, in the absence of CREB (as measured by a mutant CRE site) it is occupied by USF,

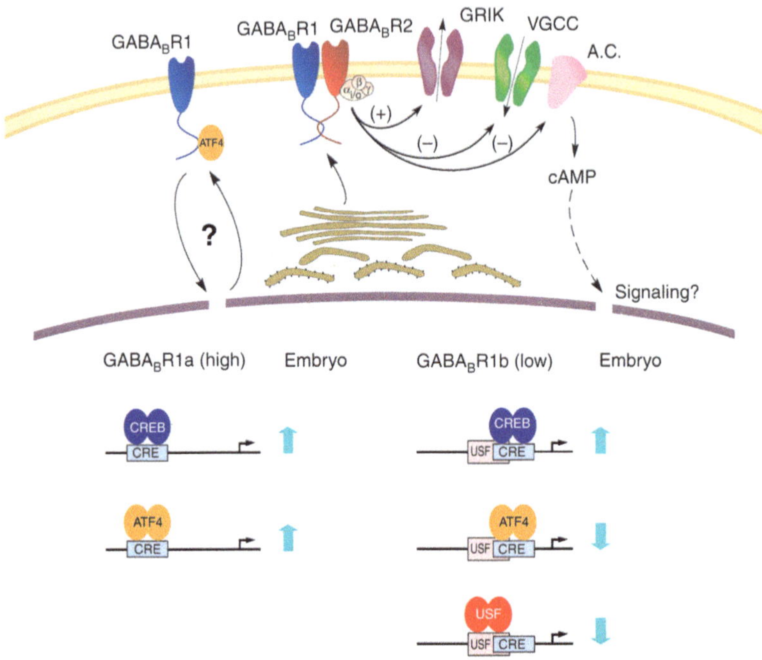

Fig. 13. A model for combinatorial regulation of the alternative *GABA$_B$R1* promoters. Whereas it is known that GABA$_B$R1 interacts directly with the transcription factor ATF4 in the cytoplasm of neurons, GABA$_B$R activation may stimulate translocation of ATF4 from one cellular compartment to another *(300–302)*. In primary embryonic hippocampal neurons, high levels of *GABA$_B$R1a* and low levels of *GABA$_B$R1b* promoter activity and endogenous gene expression (Fig. 8) indicate a transcriptional mechanism to control differential expression of these subunits during development. In contrast to the embryo, levels of GABA$_B$R1b increase postnatally and continue to be more abundant than GABA$_B$R1a in the adult *(273–275)*. A model in which activator and repressor proteins selectively recognize DNA regulatory elements specific to the *GABA$_B$R1a* and *GABA$_B$R1b* promoters, independently controlling expression of these receptor subunit subtypes during development is proposed. Regulation of *GABA$_B$R1a* gene expression in the embryonic and adult nervous systems may be meditated by the competitive binding of ATF4 and CREB to the *GABA$_B$R1a* CRE site. In contrast, binding of ATF4 and USF repressor proteins may inhibit *GABA$_B$R1b* expression in the embryo. Derepression of *GABA$_B$R1b* gene expression during postnatal development may be regulated by a loss of ATF4- and USF-binding to the *GABA$_B$R1b* promoter. CREB-activator proteins may then be recruited to the GABA$_B$R1b CRE site to stimulate transcriptional activation. Although the possibility that chromatin remodeling may also play a role cannot be excluded, differential regulation by CREB may be a control point for influencing the subunit composition of receptor subtypes in the developing and mature brain. (Modified from ref. *6*. Reproduced with permission, © 2004 by the Society for Neuroscience.)

suggesting that the USF/CRE site may be dynamically regulated by the availability of these two different transcription factors *(6)*.

During early development levels of $GABA_BR1b$ and $GABA_BR2$ are low, whereas $GABA_BR1a$ is abundantly expressed (Figs. 10 and 12). Given that CREB, ATF4, and USF differentially regulate *$GABA_BR1a$* and *$GABA_BR1b$* transcription in embryonic hippocampal neurons, these transcription factors may participate in differential gene regulation during early development. Similar to the *$GABA_BR1b$*-promoter, an overlapping CRE-USF site is also found in *$GABA_BR2$* 5′ flanking region (Fig. 7), which is consistent with the hypothesis that these response elements mediate repression of $GABA_BR1b$ and $GABA_BR2$ in embryonic tissue followed by induction of $GABA_BR1b$ and $GABA_BR2$ in adult brain. Functional promoter analyses will provide insight into the possible role that USF transcription factors may play in de-repression of $GABA_BR1b$ and $GABA_BR2$ during CNS development.

3.5. Receptor Turnover

Desensitization and receptor turnover are important aspects of receptor expression and termination of signaling. Presynaptic and postsynaptic $GABA_BRs$ desensitize at different rates in rat hippocampal neurons indicating a distinction between these two receptor populations *(310)*. The plasma membrane stability of $GABA_BRs$ has been measured in rat cortical and hippocampal neurons *(311)*. Minutes after stimulation with agonist, $GABA_BRs$ are not internalized and do not associate with arrestin family proteins indicating that $GABA_BRs$ have an atypical G protein-coupled receptor desensitization mechanism that may allow a functionally stable pool of receptors in the presence of agonist. Chronic exposure to baclofen (60 h, 100 µ*M*) results in a 30% downregulation of cell surface $GABA_BR$ expression indicating a greater stability compared with other G protein coupled receptors. $GABA_BR$ turnover is believed to be initiated indirectly by an agonist-mediated decrease of cAMP levels and cAMP-dependent PKA activity that results in the degradation and removal of $GABA_BRs$ from the cell surface. Increased cAMP levels and PKA activity results in $GABA_BR2$ phosphorylation on serine 892 and greater $GABA_BR$ stability at the cell surface *(311)*. Although tolerance has been reported in patients receiving baclofen, the fact that increasing doses with time is not a limiting factor for long-term treatment may reflect $GABA_BR$ cell-surface stability in response to agonist *(312–315)*.

3.6. Regulated Gene Expression and Implications for Disease

3.6.1. Neuropathic Pain and the Antinociceptive Response to Antidepressants

$GABA_BR$ expression is altered as a consequence of nociception *(290,316)*. Whereas the levels of *$GABA_BR1$* mRNA are increased bilaterally in the dorsal lumbar spinal cord, an enhancement is also observed in the ipsilateral lumbar dorsal root ganglion at 24 h following formalin injection into the right hindpaw *(316)*. Furthermore, antidepressants used for the treatment of neuropathic pain selectively alter the expression of *$GABA_BR1$* and *$GABA_BR2$* mRNAs in the hippocampus *(317)*. Fluoxetine, desipramine, tranylcypromine, and phenelzine increase the levels of *$GABA_BR1a$* but not *$GABA_BR1b$* mRNAs, whereas tranylcypromine also produces an increase in *$GABA_BR2$*. In the dorsal lumbar spinal cord, the fluoxetine produces a selective upregulation of $GABA_BR1a$ and downregulation of $GABA_BR2$ *(318)*. Amitriptyline has nonselective effects on *$GABA_BR$* subunit mRNA levels, increasing both *$GABA_BR1a$* and *$GABA_BR1b$*, whereas decreasing *$GABA_BR2$*. In contrast, desipramine upregulates *$GABA_BR1a$* and *$GABA_BR2$* with no change in *$GABA_BR1b$* mRNA levels *(319)*. These changes in gene expression are accompanied by increased pain threshold and increased $GABA_BR$ function *(318,319)*.

Together these findings suggest that, other than upregulation of *$GABA_BR1a$* mRNAs, there is no consistent pattern of $GABA_BR$ gene regulation in response to antidepressants. However, when considered with the fact that $GABA_BRs$ are made up of a heterodimer of $GABA_BR1$ and $GABA_BR2$ proteins, the selective increase in $GABA_BR1a$ could modify the population of receptor subtypes in the spinal cord to increase the ratio of $GABA_BR1a$–$GABA_BR2$ heterodimers relative to $GABA_BR1b$–$GABA_BR2$ *(318)*.

3.6.2. Epilepsy

$GABA_BRs$ have also been implicated in the pathophysiology of epilepsy. In animal models of epilepsy, complex alterations in *$GABA_BR$* subunit gene expression are observed in surviving neurons. Perforant pathway kindling in rats transiently increases *$GABA_BR1a$*, but not *$GABA_BR1b$*, mRNA in the dentate gyrus *(320)*. Kainic acid (KA)-induced seizures produce different changes in expression. A rapid decrease in $GABA_BR1$ and $GABA_BR2$ mRNA and protein is observed in the CA1 and CA3 subfields of the hippocampal formation *(321,322)*. Whereas expression levels in CA3 partially recover post-KA *(321)*, the decreases persist in CA1 presumably as a result of neurodegeneration *(321,322)*. Similarly, levels of *$GABA_BR1$* and *$GABA_BR2$* mRNAs are downregulated in dentate gyrus within 4–6 h post-KA and gradually recover to baseline after 24 h *(321)*. In contrast, $GABA_BR1$ and $GABA_BR2$ protein gradually increase in the dentate gyrus indicating a compensatory mechanism in surviving neurons *(322)*.

Human temporal epilepsy is associated with significant increases in $GABA_BR$ subunit mRNA and receptor binding per surviving neuron in hippocampal subfields *(321,323,324)*. When correcting for cell loss, *$GABA_BR1b$* and *$GABA_BR2$* mRNAs are upregulated in CA1, *$GABA_BR2$* mRNA is upregulated in CA3, and *$GABA_BR1a$*, *$GABA_BR1b$*, and *$GABA_BR2$* mRNAs are unregulated in the dentate gyrus *(324)*. These findings suggest that changes in $GABA_BR$ mechanisms may play an important role as a compensatory change to maintain homeostasis.

Sustained stimulation of neurons associated with epilepsy may be associated with the overactivation of nuclear Ca^{2+}-regulated processes *(325)*. Increased expression of total CREB protein as well as phosphorylated CREB is found in the human epileptic hippocampus *(326)*. This, when considered with the observation that CREB mediates transcription from the $GABA_BR1a$ and $GABA_BR1b$ promoters that epileptic rats exhibit elevated levels of subunit mRNAs, and that $GABA_BR$ antagonists inhibit CREB-DNA binding activity at doses that suppress seizure behavior, points to a novel feedback mechanism for $GABA_BRs$ in the generation and/or propagation of absence seizures and suggests a role for $GABA_BR$ antagonists as a pharmacotherapy to treat absence seizures.

3.6.3. Drug Addiction

Recent preclinical and clinical studies suggest that $GABA_BR$ agonists may be useful to treat psychostimulant addiction *(327,328)*. The selective $GABA_BR$ agonist baclofen attenuates methamphetamine *(329)* and cocaine *(330)* self-administration in rats without affecting food-reinforced responding. Moreover, baclofen is considered a putative pharmacotherapy for opiate *(331,332)* and cocaine dependence *(333,334)* in humans. Whereas the acute behavioral effects of drugs of abuse are owing to enhanced neurotransmitter release, the long-term adaptive changes in the brain that lead to drug addiction are believed to involve regulation of gene expression. The specific alterations in $GABA_BR$ gene expression in response to repeated cocaine administration are controversial. $GABA_BR1a$, $GABA_BR1b$, and $GABA_BR2$ mRNAs are upregulated in the hippocampal formation, but not in the caudate putamen and nucleus accumbens, of cocaine-kindled rats 1 and 10 d after withdrawal *(335)*. Consistent with the change in $GABA_BR1a$ mRNA, the levels of $GABA_BR1a$ protein are also increased. $GABA_BR$ gene expression returns to baseline 30 d after the last injection of cocaine. In contrast, $GABA_BR$ gene expression is not altered in the hippocampus, caudate putamen, cortex, and thalamus of cocaine-sensitized rats that are challenged with the same dose of cocaine, 1 or 20 d after withdrawal *(336)*. Yet, in a third animal model of repeated cocaine administration, levels of $GABA_BR1$ mRNA are unregulated in the nucleus accumbens, hippocampal CA1 subfield, and thalamus 1 d after the final injection of cocaine *(337)*.

The induction of immediate early gene expression in striatum consequent to cocaine and amphetamine treatment *(338–342)* is dependent on the transcription factor CREB *(341,343)*. Repeated cocaine and amphetamine treatment increases CREB phosphorylation and CRE-mediated transcription in the ventral tegmental area (VTA), nucleus accumbens, and striatum *(344–346)*. Furthermore, viral-mediated overexpression of CREB within the nucleus accumbens decreases the reinforcing properties of cocaine *(347)*, whereas overexpression of dominant-negative CREB increases cocaine reward *(348)*. Similarly, CREB-deficient mice show an increased response to the rewarding properties of cocaine *(349)*. The observation that $GABA_B$Rs play a major role in cocaine-induced neuroadaptations, that cocaine-dependant rats exhibit elevated levels of phosphorylated CREB, and that repeated cocaine treatment increases *$GABA_BR1$* mRNA *(335,337)* when taken together with an association of CREB with the *$GABA_BR1$* alternative promoters points to a novel role of $GABA_B$Rs in the initial genomic response to drugs of abuse.

4. Conclusion

The reconstruction of the molecular mechanisms that underline regulated expression of $GABA_A$Rs and $GABA_B$Rs is a formidable challenge. This challenge arises, in part, from the diversity of genes but also from the generation of additional gene products by alternative splicing and alternative promoters. Whereas post-translational modifications can be rapid but short-lived effects to alter receptor function, transcriptional changes reflect a delayed response of the genome to mediate long-lasting changes in subunit expression that controls the availability of subunits for receptor assembly. The molecular determinants that underlie such transcriptional events have not been resolved. However, it is clear that changes in the rates of transcription are dependent on intracellular signaling pathways that modulate the activity of specific transcription factors, a change in the expression of transcription factors themselves, and/or an alteration in chromatin structure modulating the accessibility of DNA regulatory elements for the transcriptional apparatus. Together with alterations in receptor trafficking, subcellular localization, and receptor turnover, regulated expression of *$GABA_AR$* and *$GABA_BR$* subunit genes may contribute to alterations in GABAergic neurotransmission.

Endogenous GABA and clinically significant compounds including benzodiazepines ($GABA_A$) and baclofen ($GABA_B$) may also play a role to modulate gene expression. Recent studies have demonstrated that transcriptional processes mediate activity dependent downregulation of specific $GABA_A$R subunits *(22,23,123,129)*. It is also well established that $GABA_B$Rs are negatively coupled to G proteins to inhibit adenylyl cyclase activity and decrease the levels of cAMP, an important regulator for the activation of transcription factors such as CREB. The consequence of $GABA_B$R signaling to gene transcription is not yet known. Given the regulation of $GABA_B$R subunits by the CREB family of transcription factors *(6)*, future studies will be aimed at determining whether $GABA_B$R activity may initiate long-term effects on protein synthesis including autologous regulation of $GABA_B$R gene expression.

In summary, considerable progress has been made in identification of the mechanisms controlling $GABA_AR$ and $GABA_BR$ gene expression. However, this progress is only the first step toward understanding how complex neuronal networks regulate gene expression in response to developmental and environmental cues. The recent application of bioinformatic methods to study DNA sequence content and regulatory networks should generate a dynamic picture of complex processes including gene expression, promoter sequences, protein–DNA binding, and protein–protein interactions. An integrated approach of targeted animal models and studies of primary neuronal cultures coupled together with new imaging technologies will allow an analysis of signaling pathways and transcription in individual neurons. Lastly, it will be essential to translate the molecular signature of regulatory events into a physiological context to determine whether changes in gene expression underlie changes in GABA-R subtype composition and receptor function. The authors hope that this molecular knowledge will reveal new information concerning the basic mechanisms underlying regulated gene expression in the brain and define novel targets for therapeutic intervention in diseases that are associated with alterations in $GABA_AR$ and $GABA_BR$ expression, including epilepsy, neuropathic pain, Alzheimer's disease, anxiety, depression, schizophrenia, and drug addiction.

Acknowledgments

We would like to express our gratitude to all members of the Laboratory of Molecular Neurobiology. We acknowledge the contributions of researchers in the $GABA_AR$ and $GABA_BR$ fields that we were unable to cite owing to space restrictions.

References

1. Bowery, N. G., Bettler, B., Froestl, W., et al. (2002) International Union of Pharmacology. XXXIII. Mammalian γ-aminobutyric $acid_B$ receptors: structure and function. *Pharmacol. Rev.* **54,** 247–264.
2. Barnard, E. A., Skolnick, P., Olsen, R. W., et al. (1998) International Union of Pharmacology. XV. Subtypes of γ-aminobutyric $acid_A$ receptors: classification on the basis of subunit structure and receptor function. *Pharmacol. Rev.* **50,** 291–313.
3. Bormann, J. (2000) The 'ABC' of GABA receptors. *Trends Pharmacol. Sci.* **21,** 16–19.
4. Russek, S. J. (1999) Evolution of $GABA_A$ receptor diversity in the human genome. *Gene* **227,** 213–222.
5. Wilke, K., Gaul, R., Klauck, S. M., and Poustka, A. (1997) A gene in human chromosome band Xq28 (*GABRE*) defines a putative new subunit class of the $GABA_A$ neurotransmitter receptor. *Genomics* **45,** 1–10.
6. Steiger, J. L., Bandyopadhyay, S., Farb, D. H., and Russek, S. J. (2004) cAMP response element-binding protein, activating transcription factor-4, and upstream

stimulatory factor differentially control hippocampal $GABA_BR1a$ and $GABA_BR1b$ subunit gene expression through alternative promoters. *J. Neurosci.* **24,** 6115–6126.
7. Martin, S. C., Russek, S. J., and Farb, D. H. (2001) Human $GABA_BR$ genomic structure: evidence for splice variants in $GABA_BR1$ but not $GABA_BR2$. *Gene* **278,** 63–79.
8. Enz, R., Brandstatter, J. H., Wassle, H., and Bormann, J. (1996) Immunocytochemical localization of the $GABA_C$ receptor ρ subunits in the mammalian retina. *J. Neurosci.* **16,** 4479–4490.
9. Koulen, P., Brandstatter, J. H., Kroger, S., Enz, R., Bormann, J., and Wassle, H. (1997) Immunocytochemical localization of the $GABA_C$ receptor ρ subunits in the cat, goldfish, and chicken retina. *J. Comp. Neurol.* **380,** 520–532.
10. Bormann, J. and Feigenspan, A. (1995) $GABA_C$ receptors. *Trends Neurosci.* **18,** 515–519.
11. Bettler, B., Kaupmann, K., Mosbacher, J., and Gassmann, M. (2004) Molecular structure and physiological functions of $GABA_B$ receptors. *Physiol. Rev.* **84,** 835–867.
12. Martin, S. C., Steiger, J. L., Gravielle, M. C., Lyons, H. R., Russek, S. J., and Farb, D. H. (2004) Differential expression of γ-aminobutyric acid type B receptor subunit mRNAs in the developing nervous system and receptor coupling to adenylyl cyclase in embryonic neurons. *J. Comp. Neurol.* **473,** 16–29.
13. Wisden, W., Laurie, D., Monyer, H., and Seeburg, P. (1992) The distribution of 13 $GABA_A$ receptor subunit mRNAs in the rat brain. I. Telencephalon, diencephalon, mesencephalon. *J. Neurosci.* **12,** 1040–1062.
14. Laurie, D. J., Wisden, W., and Seeburg, P. H. (1992) The distribution of thirteen $GABA_A$ receptor subunit mRNAs in the rat brain. III. Embryonic and postnatal development. *J. Neurosci.* **12,** 4151–4172.
15. Laurie, D. J., Seeburg, P. H., and Wisden, W. (1992) The distribution of 13 $GABA_A$ receptor subunit mRNAs in the rat brain. II. Olfactory bulb and cerebellum. *J. Neurosci.* **12,** 1063–1076.
16. Calver, A. R., Davies, C. H., and Pangalos, M. (2002) $GABA_B$ receptors: from monogamy to promiscuity. *Neurosignals* **11,** 299–314.
17. Seeburg, P. H., Wisden, W., Verdoorn, T. A., et al. (1990) The $GABA_A$ receptor family: molecular and functional diversity. *Cold Spring Harb. Symp. Quant. Biol.* **55,** 29–40.
18. Roca, D. J., Rozenberg, I., Farrant, M., and Farb, D. H. (1990) Chronic agonist exposure induces down-regulation and allosteric uncoupling of the γ-aminobutyric acid/benzodiazepine receptor complex. *Mol. Pharmacol.* **37,** 37–43.
19. Roca, D. J., Schiller, G. D., Friedman, L., Rozenberg, I., Gibbs, T. T., and Farb, D. H. (1990) γ-aminobutyric $acid_A$ receptor regulation in culture: altered allosteric interactions following prolonged exposure to benzodiazepines, barbiturates, and methylxanthines. *Mol. Pharmacol.* **37,** 710–719.
20. Montpied, P., Ginns, E. I., Martin, B. M., Roca, D., Farb, D. H., and Paul, S. M. (1991) γ-aminobutyric acid (GABA) induces a receptor-mediated reduction in $GABA_A$ receptor α subunit messenger RNAs in embryonic chick neurons in culture. *J. Biol. Chem.* **266,** 6011–6014.

21. Friedman, L. K., Gibbs, T. T., and Farb, D. H. (1996) γ-aminobutyric $acid_A$ receptor regulation: heterologous uncoupling of modulatory site interactions induced by chronic steroid, barbiturate, benzodiazepine, or GABA treatment in culture. *Brain Res.* **707,** 100–109.
22. Lyons, H. R., Land, M. B., Gibbs, T. T., and Farb, D. H. (2001) Distinct signal transduction pathways for GABA-induced $GABA_A$ receptor down-regulation and uncoupling in neuronal culture: a role for voltage-gated calcium channels. *J. Neurochem.* **78,** 1114–1126.
23. Russek, S. J., Bandyopadhyay, S., and Farb, D. H. (2000) An initiator element mediates autologous downregulation of the human type A γ-aminobutyric acid receptor β1 subunit gene. *Proc. Natl. Acad. Sci. USA* **97,** 8600–8605.
24. Kaupmann, K., Huggel, K., Heid, J., et al. (1997) Expression cloning of $GABA_B$ receptors uncovers similarity to metabotropic glutamate receptors. *Nature* **386,** 239–246.
25. Kaupmann, K., Malitschek, B., Schuler, V., et al. (1998) $GABA_B$-receptor subtypes assemble into functional heteromeric complexes. *Nature* **396,** 683–687.
26. Jones, K. A., Borowsky, B., Tamm, J. A., et al. (1998) $GABA_B$ receptors function as a heteromeric assembly of the subunits $GABA_BR1$ and $GABA_BR2$. *Nature* **396,** 674–679.
27. White, J. H., Wise, A., Main, M. J., et al. (1998) Heterodimerization is required for the formation of a functional GABAB receptor. *Nature* **396,** 679–682.
28. Martin, S. C., Russek, S. J., and Farb, D. H. (1999) Molecular identification of the human $GABA_BR2$: cell surface expression and coupling to adenylyl cyclase in the absence of $GABA_BR1$. *Mol. Cell. Neurosci.* **13,** 180–191.
29. Kuner, R., Kohr, G., Grunewald, S., Eisenhardt, G., Bach, A., and Kornau, H. C. (1999) Role of heteromer formation in $GABA_B$ receptor function. *Science* **283,** 74–77.
30. Ng, G. Y., Clark, J., Coulombe, N., et al. (1999) Identification of a $GABA_B$ receptor subunit, gb2, required for functional $GABA_B$ receptor activity. *J. Biol. Chem.* **274,** 7607–7610.
31. Macdonald, R. L. and Olsen, R. W. (1994) $GABA_A$ receptor channels. *Ann. Rev. Neurosci.* **17,** 569–602.
32. Smith, G. B. and Olsen, R.W. (1995) Functional domains of $GABA_A$ receptors. *Trends Pharmacol. Sci.* **16,** 162–168.
33. Sieghart, W. (1995) Structure and pharmacology of γ-aminobutyric $acid_A$ receptor subtypes. *Pharmacol. Rev.* **47,** 181–234.
34. Sieghart, W., Fuchs, K., Tretter, V., et al. (1999) Structure and subunit composition of $GABA_A$ receptors. *Neurochem. Int.* **34,** 379–385.
35. Jechlinger, M., Pelz, R., Tretter, V., Klausberger, T., and Sieghart, W. (1998) Subunit composition and quantitative importance of hetero-oligomeric receptors: $GABA_A$ receptors containing α6 subunits. *J. Neurosci.* **18,** 2449–2457.
36. Chang, Y., Wang, R., Barot, S., and Weiss, D. S. (1996) Stoichiometry of a recombinant $GABA_A$ receptor. *J. Neurosci.* **16,** 5415–5424.
37. Backus, K. H., Arigoni, M., Drescher, U., et al. (1993) Stoichiometry of a recombinant $GABA_A$ receptor deduced from mutation-induced rectification. *Neuroreport* **5,** 285–288.

38. Ranna, M., Sinkkonen, S. T., Moykkynen, T., Uusi-Oukari, M., and Korpi, E. R. (2006) Impact of ε and θ subunits on pharmacological properties of α3β1 $GABA_A$ receptors expressed in *Xenopus* oocytes. *BMC. Pharmacol.* **6,** 1.
39. Darlison, M. G., Pahal, I., and Thode, C. (2005) Consequences of the evolution of the $GABA_A$ receptor gene family. *Cell. Mol. Neurobiol.* **25,** 607–624.
40. Steiger, J. L. and Russek, S. J. (2004) $GABA_A$ receptors: building the bridge between subunit mRNAs, their promoters, and cognate transcription factors. *Pharmacol. Ther.* **101,** 259–281.
41. Bell, M. V., Bloomfield, J., McKinley, M., et al. (1989) Physical linkage of a $GABA_A$ receptor subunit gene to the DXS374 locus in human Xq28. *Am. J. Hum. Genet.* **45,** 883–888.
42. Sinnett, D., Wagstaff, J., Glatt, K., Woolf, E., Kirkness, E. J., and Lalande, M. (1993) High-resolution mapping of the γ-aminobutyric acid receptor subunit β3 and α5 gene cluster on chromosome 15q11-q13, and localization of breakpoints in two Angelman syndrome patients. *Am. J. Hum. Genet.* **52,** 1216–1229.
43. Greger, V., Knoll, J. H., Woolf, E., et al. (1995) The γ-aminobutyric acid receptor γ3 subunit gene (*GABRG3*) is tightly linked to the α5 subunit gene (*GABRA5*) on human chromosome 15q11-q13 and is transcribed in the same orientation. *Genomics* **26,** 258–264.
44. Levin, M. L., Chatterjee, A., Pragliola, A., et al. (1996) A comparative transcription map of the murine bare patches (Bpa) and striated (Str) critical regions and human Xq28. *Genome Res.* **6,** 465–477.
45. Buckle, V. J., Fujita, N., Ryder-Cook, A. S., et al. (1989) Chromosomal localization of $GABA_A$ receptor subunit genes: relationship to human genetic disease. *Neuron* **3,** 647–654.
46. Kirkness, E. F., Kusiak, J. W., Fleming, J. T., et al. (1991) Isolation, characterization, and localization of human genomic DNA encoding the β1 subunit of the $GABA_A$ receptor (*GABRB1*). *Genomics* **10,** 985–995.
47. McLean, P. J., Farb, D. H., and Russek, S. J. (1995) Mapping of the α4 subunit gene (*GABRA4*) to human chromosome 4 defines an α2-α4-β1-γ1 gene cluster: further evidence that modern $GABA_A$ receptor gene clusters are derived from an ancestral cluster. *Genomics* **26,** 580–586.
48. Wilcox, A. S., Warrington, J. A., Gardiner, K., et al. (1992) Human chromosomal localization of genes encoding the γ1 and γ2 subunits of the γ-aminobutyric acid receptor indicates that members of this gene family are often clustered in the genome. *Proc. Natl. Acad. Sci. USA* **89,** 5857–5861.
49. Russek, S. J. and Farb, D. H. (1994) Mapping of the β2 subunit gene (*GABRB2*) to microdissected human chromosome 5q34-q35 defines a gene cluster for the most abundant $GABA_A$ receptor isoform. *Genomics* **23,** 528–533.
50. Kostrzewa, M., Kohler, A., Eppelt, K., et al. (1996) Assignment of genes encoding $GABA_A$ receptor subunits α1, α6, β2, and γ2 to a YAC contig of 5q33. *Eur. J. Hum. Genet.* **4,** 199–204.
51. Johnson, K. J., Sander, T., Hicks, A. A., et al. (1992) Confirmation of the localization of the human $GABA_A$ receptor α1-subunit gene (*GABRA1*) to distal 5q by linkage analysis. *Genomics* **14,** 745–748.

52. Wagstaff, J., Knoll, J. H., Fleming, J., et al. (1991) Localization of the gene encoding the $GABA_A$ receptor β3 subunit to the Angelman/Prader-Willi region of human chromosome 15. *Am. J. Hum. Genet.* **49,** 330–337.
53. Knoll, J. H., Sinnett, D., Wagstaff, J., et al. (1993) FISH ordering of reference markers and of the gene for the α5 subunit of the γ-aminobutyric acid receptor (GABRA5) within the Angelman and Prader-Willi syndrome chromosomal regions. *Hum. Mol. Genet.* **2,** 183–189.
54. Wagstaff, J., Chaillet, J. R., and Lalande, M. (1991) The $GABA_A$ receptor β3 subunit gene: characterization of a human cDNA from chromosome 15q11q13 and mapping to a region of conserved synteny on mouse chromosome 7. *Genomics* **11,** 1071–1078.
55. Nakatsu, Y., Tyndale, R. F., DeLorey, T. M., et al. (1993) A cluster of three $GABA_A$ receptor subunit genes is deleted in a neurological mutant of the mouse p locus. *Nature* **364,** 448–450.
56. Cutting, G. R., Curristin, S., Zoghbi, H., O'Hara, B., Seldin, M. F., and Uhl, G. R. (1992) Identification of a putative γ-aminobutyric acid (GABA) receptor subunit rho_2 cDNA and colocalization of the genes encoding rho_2 (GABRR2) and rho_1 (GABRR1) to human chromosome 6q14-q21 and mouse chromosome 4. *Genomics* **12,** 801–806.
57. Bailey, M. E., Albrecht, B. E., Johnson, K. J., and Darlison, M. G. (1999) Genetic linkage and radiation hybrid mapping of the three human $GABA_C$ receptor ρ subunit genes: *GABRR1*, *GABRR2* and *GABRR3*. *Biochim. Biophys. Acta.* **1447,** 307–312.
58. Sommer, B., Poustka, A., Spurr, N. K., and Seeburg, P. H. (1990) The murine $GABA_A$ receptor δ-subunit gene: structure and assignment to human chromosome 1. *DNA Cell. Biol.* **9,** 561–568.
59. Emberger, W., Windpassinger, C., Petek, E., Kroisel, P. M., and Wagner, K. (2000) Assignment of the human $GABA_A$ receptor δ-subunit gene (GABRD) to chromosome band 1p36.3 distal to marker NIB1364 by radiation hybrid mapping. *Cytogenet. Cell. Genet.* **89,** 281–282.
60. Bailey, M. E., Matthews, D. A., Riley, B. P., et al. (1999) Genomic mapping and evolution of human $GABA_A$ receptor subunit gene clusters. *Mamm. Genome.* **10,** 839–843.
61. Ma, W., Saunders, P. A., Somogyi, R., Poulter, M. O., and Barker, J. L. (1993) Ontogeny of $GABA_A$ receptor subunit mRNAs in rat spinal cord and dorsal root ganglia. *J. Comp. Neurol.* **338,** 337–359.
62. Bank, A. (2006) Regulation of human fetal hemoglobin: new players, new complexities. *Blood* **107,** 435–443.
63. Tian, H., Chen, H. J., Cross, T. H., and Edenberg, H. J. (2005) Alternative splicing and promoter use in the human *GABRA2* gene. *Mol. Brain Res.* **137,** 174–183.
64. Kim, Y., Glatt, H., Xie, W., Sinnett, D., and Lalande, M. (1997) Human γ-aminobutyric acid-type A receptor α5 subunit gene (*GABRA5*): characterization and structural organization of the 5′ flanking region. *Genomics* **42,** 378–387.
65. Kirkness, E. and Fraser, C. (1993) A strong promoter element is located between alternative exons of a gene encoding the human γ-aminobutyric acid-type A receptor β3 subunit (*GABRB3*). *J. Biol. Chem.* **268,** 4420–4428.

66. Fuchs, K. and Celepirovic, N. (2002) The 5′-flanking region of the rat $GABA_A$ receptor α2-subunit gene (*GABRA2*). *J. Neurochem.* **82,** 1512–1523.
67. Edenberg, H. J. (2002) The collaborative study on the genetics of alcoholism: an update. *Alcohol Res. Health* **26,** 214–218.
68. Mu, W., Cheng, Q., Yang, J., and Burt, D. R. (2002) Alternative splicing of the $GABA_A$ receptor α4 subunit creates a severely truncated mRNA. *Brain Res. Bull.* **58,** 447–454.
69. Glatt, K., Glatt, H., and Lalande, M. (1997) Structure and organization of *GABRB3* and *GABRA5*. *Genomics* **41,** 63–69.
70. Davies, P. A., McCartney, M. R., Wang, W., Hales, T. G., and Kirkness, E. F. (2002) Alternative transcripts of the $GABA_A$ receptor ε subunit in human and rat. *Neuropharmacology* **43,** 467–475.
71. Korpi, E. R., Kuner, T., Kristo, P., et al. (1994) Small N-terminal deletion by splicing in cerebellar α6 subunit abolishes $GABA_A$ receptor function. *J. Neurochem.* **63,** 1167–1170.
72. Martinez-Torres, A., Vazquez, A. E., Panicker, M. M., and Miledi, R. (1998) Cloning and functional expression of alternative spliced variants of the ρ1 γ-aminobutyrate receptor. *Proc. Natl. Acad. Sci. USA* **95,** 4019–4022.
73. McKinley, D. D., Lennon, D. J., and Carter, D. B. (1995) Cloning, sequence analysis and expression of two forms of mRNA coding for the human β2 subunit of the $GABA_A$ receptor. *Mol. Brain Res.* **28,** 175–179.
74. Whiting, P., McKernan, R., and Iversen, L. (1990) Another mechanism for creating diversity in γ-aminobutyrate type A receptors: RNA splicing directs expression of two forms of γ2 subunit, one of which contains a protein kinase C phosphorylation site. *Proc. Natl. Acad. Sci. USA* **87,** 9966–9970.
75. Poulsen, C. F., Christjansen, K. N., Hastrup, S., and Hartvig, L. (2000) Identification and cloning of a γ3 subunit splice variant of the human $GABA_A$ receptor. *Mol. Brain Res.* **78,** 201–203.
76. Wang, J. B. and Burt, D. R. (1991) Differential expression of two forms of $GABA_A$ receptor γ2-subunit in mice. *Brain Res. Bull.* **27,** 731–735.
77. Jin, P., Zhang, J., Rowe-Teeter, C., Yang, J., Stuve, L. L., and Fu, G. K. (2004) Cloning and characterization of a $GABA_A$ receptor γ2 subunit variant. *J. Biol. Chem.* **279,** 1408–1414.
78. Fritschy, J. M., Paysan, J., Enna, A., and Mohler, H. (1994) Switch in the expression of rat $GABA_A$-receptor subtypes during postnatal development: an immunohistochemical study. *J. Neurosci.* **14,** 5302–5324.
79. Poulter, M. O., Barker, J. L., O'Carroll, A. M., Lolait, S. J., and Mahan, L. C. (1992) Differential and transient expression of $GABA_A$ receptor α-subunit mRNAs in the developing rat CNS. *J. Neurosci.* **12,** 2888–2900.
80. Roberts, A. A. and Kellogg, C. K. (2000) Synchronous postnatal increase in α1 and γ2L $GABA_A$ receptor mRNAs and high affinity zolpidem binding across three regions of rat brain. *Dev. Brain Res.* **119**, 21–32.
81. Candy, J. M. and Martin, I. L. (1979) The postnatal development of the benzodiazepine receptor in the cerebral cortex and cerebellum of the rat. *J. Neurochem.* **32,** 655–658.

82. Lippa, A. S., Beer, B., Sano, M. C., Vogel, R. A., and Meyerson, L. R. (1981) Differential ontogeny of type 1 and type 2 benzodiazepine receptors. *Life Sci.* **28,** 2343–2347.
83. Chisholm, J., Kellogg, C., and Lippa, A. (1983) Development of benzodiazepine binding subtypes in three regions of rat brain. *Brain Res.* **267,** 388–391.
84. Pirker, S., Schwarzer, C., Wieselthaler, A., Sieghart, W., and Sperk, G. (2000) $GABA_A$ receptors: immunocytochemical distribution of 13 subunits in the adult rat brain. *Neuroscience* **101,** 815–850.
85. Benke, D., Fritschy, J., Trzeciak, A., Bannwarth, W., and Mohler, H. (1994) Distribution, prevalence, and drug binding profile of γ- aminobutyric acid type A receptor subtypes differing in the β-subunit variant. *J. Biol. Chem.* **269,** 27,100–27,107.
86. Li, M. and De Blas, A. L. (1997) Coexistence of two β subunit isoforms in the same γ-aminobutyric acid type A receptor. *J. Biol. Chem.* **272,** 16,564–16,569.
87. Gutierrez, A., Khan, Z. U., and De Blas, A. L. (1994) Immunocytochemical localization of γ2 short and γ2 long subunits of the $GABA_A$ receptor in the rat brain. *J. Neurosci.* **14,** 7168–7179.
88. Moragues, N., Ciofi, P., Lafon, P., Odessa, M. F., Tramu, G., and Garret, M. (2000) cDNA cloning and expression of a γ-aminobutyric acid A receptor ε-subunit in rat brain. *Eur. J. Neurosci.* **12,** 4318–4330.
89. Neelands, T. R. and Macdonald, R. L. (1999) Incorporation of the π subunit into functional γ-aminobutyric $Acid_A$ receptors. *Mol. Pharmacol.* **56,** 598–610.
90. Hedblom, E. and Kirkness, E. F. (1997) A novel class of $GABA_A$ receptor subunit in tissues of the reproductive system. *J. Biol. Chem.* **272,** 15,346–15,350.
91. Boue-Grabot, E., Roudbaraki, M., Bascles, L., Tramu, G., Bloch, B., and Garret, M. (1998) Expression of GABA receptor ρ subunits in rat brain. *J. Neurochem.* **70,** 899–907.
92. Enz, R., Brandstatter, J. H., Hartveit, E., Wassle, H., and Bormann, J. (1995) Expression of GABA receptor ρ1 and ρ2 subunits in the retina and brain of the rat. *Eur. J. Neurosci.* **7,** 1495–1501.
93. Sinkkonen, S. T., Hanna, M. C., Kirkness, E. F., and Korpi, E. R. (2000) $GABA_A$ receptor ε and θ subunits display unusual structural variation between species and are enriched in the rat locus ceruleus. *J. Neurosci.* **20,** 3588–3595.
94. Bonnert, T. P., McKernan, R. M., Farrar, S., et al. (1999) θ, a novel γ-aminobutyric acid type A receptor subunit. *Proc. Natl. Acad. Sci. USA* **96,** 9891–9896.
95. Nusser, Z., Roberts, J. D., Baude, A., Richards, J. G., and Somogyi, P. (1995) Relative densities of synaptic and extrasynaptic $GABA_A$ receptors on cerebellar granule cells as determined by a quantitative immunogold method. *J. Neurosci.* **15,** 2948–2960.
96. Nusser, Z., Sieghart, W., and Somogyi, P. (1998) Segregation of different $GABA_A$ receptors to synaptic and extrasynaptic membranes of cerebellar granule cells. *J. Neurosci.* **18,** 1693–1703.
97. Somogyi, P., Fritschy, J. M., Benke, D., Roberts, J. D., and Sieghart, W. (1996) The γ2 subunit of the $GABA_A$ receptor is concentrated in synaptic junctions

containing the α1 and β2/3 subunits in hippocampus, cerebellum and globus pallidus. *Neuropharmacology* **35,** 1425–1444.
98. Fritschy, J. M., Johnson, D. K., Mohler, H., and Rudolph, U. (1998) Independent assembly and subcellular targeting of $GABA_A$ receptor subtypes demonstrated in mouse hippocampal and olfactory neurons in vivo. *Neurosci. Lett.* **249,** 99–102.
99. Brunig, I., Scotti, E., Sidler, C., and Fritschy, J. M. (2002) Intact sorting, targeting, and clustering of γ-aminobutyric acid A receptor subtypes in hippocampal neurons in vitro. *J. Comp. Neurol.* **443,** 43–55.
100. Wei, W., Zhang, N., Peng, Z., Houser, C. R., and Mody, I. (2003) Perisynaptic localization of δ subunit-containing $GABA_A$ receptors and their activation by GABA spillover in the mouse dentate gyrus. *J. Neurosci.* **23,** 10,650–10,661.
101. Farrant, M. and Nusser, Z. (2005) Variations on an inhibitory theme: phasic and tonic activation of $GABA_A$ receptors. *Nat. Rev. Neurosci.* **6,** 215–229.
102. MacDermott, A. B., Role, L. W., and Siegelbaum, S. A. (1999) Presynaptic ionotropic receptors and the control of transmitter release. *Ann. Rev. Neurosci.* **22,** 443–485.
103. Bohlhalter, S., Weinmann, O., Mohler, H., and Fritschy, J. M. (1996) Laminar compartmentalization of $GABA_A$ receptor subtypes in the spinal cord: an immunohistochemical study. *J. Neurosci.* **16,** 283–297.
104. Sur, C., McKernan, R., and Triller, A. (1995) Subcellular localization of the $GABA_A$ receptor γ2 subunit in the rat spinal cord. *Eur. J. Neurosci.* **7,** 1323–1332.
105. Alvarez, F. J., Taylor-Blake, B., Fyffe, R. E., De Blas, A. L., and Light, A. R. (1996) Distribution of immunoreactivity for the β2 and β3 subunits of the $GABA_A$ receptor in the mammalian spinal cord. *J. Comp. Neurol.* **365,** 392–412.
106. Caillard, O., Ben-Ari, Y., and Gaiarsa, J. L. (1999) Long-term potentiation of GABAergic synaptic transmission in neonatal rat hippocampus. *J. Physiol.* **518,** 109–119.
107. Caillard, O., Ben-Ari, Y., and Gaiarsa, J. L. (1999) Mechanisms of induction and expression of long-term depression at GABAergic synapses in the neonatal rat hippocampus. *J. Neurosci.* **19,** 7568–7577.
108. Gaiarsa, J. L., Caillard, O., and Ben-Ari, Y. (2002) Long-term plasticity at GABAergic and glycinergic synapses: mechanisms and functional significance. *Trends Neurosci.* **25,** 564–570.
109. Poulter, M. O., Ohannesian, L., Larmet, Y., and Feltz, P. (1997) Evidence that $GABA_A$ receptor subunit mRNA expression during development is regulated by $GABA_A$ receptor stimulation. *J. Neurochem.* **68,** 631–639.
110. Clements, J. D. (1996) Transmitter timecourse in the synaptic cleft: its role in central synaptic function. *Trends Neurosci.* **19,** 163–171.
111. Mountcastle, V. B., Talbot, W. H., Sakata, H., and Hyvarinen, J. (1969) Cortical neuronal mechanisms in flutter-vibration studied in unanesthetized monkeys. Neuronal periodicity and frequency discrimination. *J. Neurophysiol.* **32,** 452–484.
112. Ranck, J. B. Jr. (1973) Studies on single neurons in dorsal hippocampal formation and septum in unrestrained rats. I. Behavioral correlates and firing repertoires. *Exp. Neurol.* **41,** 461–531.
113. Hajos, N., Palhalmi, J., Mann, E. O., Nemeth, B., Paulsen, O., and Freund, T. F. (2004) Spike timing of distinct types of GABAergic interneuron during hippocampal gamma oscillations in vitro. *J. Neurosci.* **24,** 9127–9137.

114. Jones, M. V. and Westbrook, G. L. (1995) Desensitized states prolong $GABA_A$ channel responses to brief agonist pulses. *Neuron* **15,** 181–191.
115. Bianchi, M. T. and Macdonald, R. L. (2001) Agonist trapping by $GABA_A$ receptor channels. *J. Neurosci.* **21,** 9083–9091.
116. Timofeev, I., Grenier, F., and Steriade, M. (2002) The role of chloride-dependent inhibition and the activity of fast-spiking neurons during cortical spike-wave electrographic seizures. *Neuroscience* **114,** 1115–1132.
117. Hutchinson, P. J., O'Connell, M. T., Al-Rawi, P. G., et al. (2002) Increases in GABA concentrations during cerebral ischaemia: a microdialysis study of extracellular amino acids. *J. Neurol. Neurosurg. Psychiatry* **72,** 99–105.
118. Swift, C. G., Swift, M. R., Hamley, J., Stevenson, I. H., and Crooks, J. (1984) Side-effect 'tolerance' in elderly long-term recipients of benzodiazepine hypnotics. *Age Ageing* **13,** 335–343.
119. Seidel, W. F., Cohen, S. A., Wilson, L., and Dement, W. C. (1985) Effects of alprazolam and diazepam on the daytime sleepiness of non-anxious subjects. *Psychopharmacology* **87,** 194–197.
120. Haigh, J. R. and Feely, M. (1988) Tolerance to the anticonvulsant effect of benzodiazepines. *Trends Pharmacol. Sci.* **9**, 361–366.
121. Gallager, D. W., Lakoski, J. M., Gonsalves, S. F., and Rauch, S. L. (1984) Chronic benzodiazepine treatment decreases postsynaptic GABA sensitivity. *Nature* **308,** 74–77.
122. Holt, R. A., Bateson, A. N., and Martin, I. L. (1999) Decreased GABA enhancement of benzodiazepine binding after a single dose of diazepam. *J. Neurochem.* **72,** 2219–2222.
123. Lyons, H. R., Gibbs, T. T., and Farb, D. H. (2000) Turnover and down-regulation of $GABA_A$ receptor α1, β2S, and γ1 subunit mRNAs by neurons in culture. *J. Neurochem.* **74,** 1041–1048.
124. Friedman, L., Gibbs, T. T., and Farb, D. H. (1993) γ-aminobutyric $acid_A$ receptor regulation: chronic treatment with pregnanolone uncouples allosteric interactions between steroid and benzodiazepine recognition sites. *Mol. Pharmacol.* **44,** 191–197.
125. Follesa, P., Serra, M., Cagetti, E., et al. (2000) Allopregnanolone synthesis in cerebellar granule cells: roles in regulation of $GABA_A$ receptor expression and function during progesterone treatment and withdrawal. *Mol. Pharmacol.* **57,** 1262–1270.
126. Klein, R. L., Whiting, P. J., and Harris, R. A. (1994) Benzodiazepine treatment causes uncoupling of recombinant $GABA_A$ receptors expressed in stably transfected cells. *J. Neurochem.* **63,** 2349–2352.
127. Wong, G., Lyon, T., and Skolnick, P. (1994) Chronic exposure to benzodiazepine receptor ligands uncouples the γ-aminobutyric acid type A receptor in WSS-1 cells. *Mol. Pharmacol.* **46,** 1056–1062.
128. Ali, N. J. and Olsen, R. W. (2001) Chronic benzodiazepine treatment of cells expressing recombinant $GABA_A$ receptors uncouples allosteric binding: studies on possible mechanisms. *J. Neurochem.* **79,** 1100–1108.
129. Gravielle, M. C., Faris, R., Russek, S. J., and Farb, D. H. (2005) GABA induces activity dependent delayed-onset uncoupling of GABA/benzodiazepine site interactions in neocortical neurons. *J. Biol. Chem.* **280,** 20,954–20,960.

130. Stelzer, A., Kay, A. R., and Wong, R. K. (1988) $GABA_A$-receptor function in hippocampal cells is maintained by phosphorylation factors. *Science* **241,** 339–341.
131. Gyenes, M., Wang, Q., Gibbs, T. T., and Farb, D. H. (1994) Phosphorylation factors control neurotransmitter and neuromodulator actions at the γ-aminobutyric acid type A receptor. *Mol. Pharmacol.* **46,** 542–549.
132. Kanematsu, T., Jang, I. S., Yamaguchi, T., et al. (2002) Role of the PLC-related, catalytically inactive protein p130 in $GABA_A$ receptor function. *EMBO J.* **21,** 1004–1011.
133. Mu, W. and Burt, D. R. (1999) The mouse $GABA_A$ receptor α3 subunit gene and promoter. *Mol. Brain Res.* **73,** 172–180.
134. McLean, P. J., Shpektor, D., Bandyopadhyay, S., Russek, S. J., and Farb, D. H. (2000) A minimal promoter for the $GABA_A$ receptor α6-subunit gene controls tissue specificity. *J. Neurochem.* **74,** 1858–1869.
135. Mu, W. and Burt, D. R. (1999) Transcriptional regulation of $GABA_A$ receptor γ2 subunit gene. *Mol. Brain Res.* **67,** 137–147.
136. Lüscher, B., Hauselmann, R., Leitgeb, S., Rulicke, T., and Fritschy, J. M. (1997) Neuronal subtype-specific expression directed by the $GABA_A$ receptor δ subunit gene promoter/upstream region in transgenic mice and in cultured cells. *Mol. Brain Res.* **51,** 197–211.
137. Bahn, S., Jones, A., and Wisden, W. (1997) Directing gene expression to cerebellar granule cells using γ-aminobutyric acid type A receptor α6 subunit transgenes. *Proc. Natl. Acad. Sci. USA* **94,** 9417–9421.
138. Jones, A., Bahn, S., Grant, A. L., Kohler, M., and Wisden, W. (1996) Characterization of a cerebellar granule cell-specific gene encoding the γ-aminobutyric acid type A receptor α6 subunit. *J. Neurochem.* **67,** 907–916.
139. Bateson, A. N., Ultsch, A., and Darlison, M. G. (1995) Isolation and sequence analysis of the chicken $GABA_A$ receptor α1-subunit gene promoter. *Gene* **153,** 243–247.
140. Chong, J. A., Tapia-Ramirez, J., Kim, S., et al. (1995) REST: a mammalian silencer protein that restricts sodium channel gene expression to neurons. *Cell* **80,** 949–957.
141. Schoenherr, C. J. and Anderson, D. J. (1995) The neuron-restrictive silencer factor (NRSF): a coordinate repressor of multiple neuron-specific genes. *Science* **267,** 1360–1363.
142. Schoenherr, C. J., Paquette, A. J., and Anderson, D. J. (1996) Identification of potential target genes for the neuron-restrictive silencer factor. *Proc. Natl. Acad. Sci. USA* **93,** 9881–9886.
143. Motejlek, K., Hauselmann, R., Leitgeb, S., and Luscher, B. (1994) BSF1, a novel brain-specific DNA-binding protein recognizing a tandemly repeated purine DNA element in the $GABA_A$ receptor δ subunit gene. *J. Biol. Chem.* **269,** 15,265–15,273.
144. Ma, L., Song, L., Radoi, G. E., and Harrison, N. L. (2004) Transcriptional regulation of the mouse gene encoding the α4 subunit of the $GABA_A$ receptor. *J. Biol. Chem.* **279,** 40,451–40,461.
145. Roberts, D. S., Raol, Y. H., Bandyopadhyay, S., et al. (2005) Egr3 stimulation of *GABRA4* promoter activity as a mechanism for seizure-induced up-regulation of

$GABA_A$ receptor α4 subunit expression. *Proc. Natl. Acad. Sci. USA* **102,** 11,894–11,899.
146. Rabow, L. E., Russek, S. J., and Farb, D. H. (1995) From ion currents to genomic analysis: recent advances in $GABA_A$ receptor research. *Synapse* **21,** 189–274.
147. Czajkowski, C. and Farb, D. H. (1986) Transmembrane topology and subcellular distribution of the benzodiazepine receptor. *J. Neurosci.* **6,** 2857–2863.
148. Moss, S. J. and Smart, T. G. (2001) Constructing inhibitory synapses. *Nat. Rev. Neurosci.* **2,** 240–250.
149. Kittler, J. T., McAinsh, K., and Moss, S. J. (2002) Mechanisms of $GABA_A$ receptor assembly and trafficking: implications for the modulation of inhibitory neurotransmission. *Mol. Neurobiol.* **26,** 251–268.
150. Czajkowski, C. and Farb, D. H. (1989) Identification of an intracellular pool of γ-aminobutyric $acid_A$/benzodiazepine receptors en route to the cell surface of brain neurons in culture. *Mol. Pharmacol.* **35,** 183–188.
151. Borden, L. A., Czajkowski, C., Chan, C. Y., and Farb, D. H. (1984) Benzodiazepine receptor synthesis and degradation by neurons in culture. *Science* **226,** 857–860.
152. Borden, L. A. and Farb, D. H. (1988) Mechanism of γ-aminobutyric acid/benzodiazepine receptor turnover in neuronal cells: evidence for nonlysosomal degradation. *Mol. Pharmacol.* **34,** 354–362.
153. Stevens, C. F. (2003) Neurotransmitter release at central synapses. *Neuron* **40,** 381–388.
154. Li, Z. and Sheng, M. (2003) Some assembly required: the development of neuronal synapses. *Nat. Rev. Mol. Cell. Biol.* **4,** 833–841.
155. Lüscher, B. and Keller, C. A. (2004) Regulation of $GABA_A$ receptor trafficking, channel activity, and functional plasticity of inhibitory synapses. *Pharmacol. Ther.* **102,** 195–221.
156. Fritschy, J. M., Schweizer, C., Brunig, I., and Luscher, B. (2003) Pre- and postsynaptic mechanisms regulating the clustering of type A γ-aminobutyric acid receptors ($GABA_A$ receptors). *Biochem. Soc. Trans.* **31,** 889–892.
157. Essrich, C., Lorez, M., Benson, J. A., Fritschy, J. M., and Luscher, B. (1998) Postsynaptic clustering of major $GABA_A$ receptor subtypes requires the γ2 subunit and gephyrin. *Nat. Neurosci.* **1,** 563–571.
158. Schweizer, C., Balsiger, S., Bluethmann, H., et al. (2003) The γ2 subunit of $GABA_A$ receptors is required for maintenance of receptors at mature synapses. *Mol. Cell. Neurosci.* **24,** 442–450.
159. Jacob, T. C., Bogdanov, Y. D., Magnus, C., et al. (2005) Gephyrin regulates the cell surface dynamics of synaptic $GABA_A$ receptors. *J. Neurosci.* **25,** 10,469–10,478.
160. Meyer, G., Kirsch, J., Betz, H., and Langosch, D. (1995) Identification of a gephyrin binding motif on the glycine receptor β subunit. *Neuron* **15,** 563–572.
161. Kneussel, M., Brandstatter, J. H., Gasnier, B., Feng, G., Sanes, J. R., and Betz, H. (2001) Gephyrin-independent clustering of postsynaptic $GABA_A$ receptor subtypes. *Mol. Cell. Neurosci.* **17,** 973–982.

162. Wang, H., Bedford, F. K., Brandon, N. J., Moss, S. J., and Olsen, R. W. (1999) $GABA_A$-receptor-associated protein links $GABA_A$ receptors and the cytoskeleton. *Nature* **397,** 69–72.
163. Coyle, J. E., Qamar, S., Rajashankar, K. R., and Nikolov, D. B. (2002) Structure of GABARAP in two conformations: implications for $GABA_A$ receptor localization and tubulin binding. *Neuron* **33,** 63–74.
164. Kneussel, M., Haverkamp, S., Fuhrmann, J. C., et al. (2000) The γ-aminobutyric acid type A receptor ($GABA_AR$)-associated protein GABARAP interacts with gephyrin but is not involved in receptor anchoring at the synapse. *Proc. Natl. Acad. Sci. USA* **97,** 8594–8599.
165. Kittler, J. T., Rostaing, P., Schiavo, G., et al. (2001) The subcellular distribution of GABARAP and its ability to interact with NSF suggest a role for this protein in the intracellular transport of $GABA_A$ receptors. *Mol. Cell. Neurosci.* **18,** 13–25.
166. Leil, T. A., Chen, Z. W., Chang, C. S. S., and Olsen, R. W. (2004) $GABA_A$ receptor-associated protein traffics $GABA_A$ receptors to the plasma membrane in neurons. *J. Neurosci.* **24,** 11,429–11,438.
167. O'Sullivan, G. A., Kneussel, M., Elazar, Z., and Betz, H. (2005) GABARAP is not essential for GABA receptor targeting to the synapse. *Eur. J. Neurosci.* **22,** 2644–2648.
168. Bedford, F. K., Kittler, J. T., Muller, E., et al. (2001) $GABA_A$ receptor cell surface number and subunit stability are regulated by the ubiquitin-like protein Plic-1. *Nat. Neurosci.* **4,** 908–916.
169. Tehrani, M. H. and Barnes, E. M. Jr. (1993) Identification of $GABA_A$/benzodiazepine receptors on clathrin-coated vesicles from rat brain. *J. Neurochem.* **60,** 1755–1761.
170. Kittler, J. T., Delmas, P., Jovanovic, J. N., Brown, D. A., Smart, T. G., and Moss, S. J. (2000) Constitutive endocytosis of $GABA_A$ receptors by an association with the adaptin AP2 complex modulates inhibitory synaptic currents in hippocampal neurons. *J. Neurosci.* **20,** 7972–7977.
171. Herring, D., Huang, R., Singh, M., Robinson, L. C., Dillon, G. H., and Leidenheimer, N. J. (2003) Constitutive $GABA_A$ receptor endocytosis is dynamin-mediated and dependent on a dileucine AP2 adaptin-binding motif within the β2 subunit of the receptor. *J. Biol. Chem.* **278,** 24,046–24,052.
172. Kittler, J. T., Chen, G., Honing, S., et al. (2005) Phospho-dependent binding of the clathrin AP2 adaptor complex to $GABA_A$ receptors regulates the efficacy of inhibitory synaptic transmission. *Proc. Natl. Acad. Sci. USA* **102,** 14,871–14,876.
173. Charych, E. I., Yu, W., Miralles, C. P., et al. (2004) The brefeldin A-inhibited GDP/GTP exchange factor 2, a protein involved in vesicular trafficking, interacts with the beta subunits of the GABA receptors. *J. Neurochem.* **90,** 173–189.
174. Rathenberg, J., Kittler, J. T., and Moss, S. J. (2004) Palmitoylation regulates the clustering and cell surface stability of $GABA_A$ receptors. *Mol. Cell. Neurosci.* **26,** 251–257.
175. Keller, C. A., Yuan, X., Panzanelli, P., et al. (2004) The γ2 subunit of $GABA_A$ receptors is a substrate for palmitoylation by GODZ. *J. Neurosci.* **24,** 5881–5891.

176. Wan, Q., Xiong, Z. G., Man, H. Y., et al. (1997) Recruitment of functional $GABA_A$ receptors to postsynaptic domains by insulin. *Nature* **388,** 686–690.
177. Wang, Q., Liu, L., Pei, L., et al. (2003) Control of synaptic strength, a novel function of Akt. *Neuron* **38,** 915–928.
178. Chen, G., Kittler, J. T., Moss, S. J., and Yan, Z. (2006) Dopamine D3 receptors regulate $GABA_A$ receptor function through a phospho-dependent endocytosis mechanism in nucleus accumbens. *J. Neurosci.* **26,** 2513–2521.
179. Mielke, J. G. and Wang, Y. T. (2005) Insulin exerts neuroprotection by counteracting the decrease in cell-surface GABA receptors following oxygen-glucose deprivation in cultured cortical neurons. *J. Neurochem.* **92,** 103–113.
180. Brunig, I., Penschuck, S., Berninger, B., Benson, J., and Fritschy, J. M. (2001) BDNF reduces miniature inhibitory postsynaptic currents by rapid downregulation of $GABA_A$ receptor surface expression. *Eur. J. Neurosci.* **13,** 1320–1328.
181. Cheng, Q. and Yeh, H. H. (2003) Brain-derived neurotrophic factor attenuates mouse cerebellar granule cell $GABA_A$ receptor-mediated responses via postsynaptic mechanisms. *J. Physiol.* **548,** 711–721.
182. Henneberger, C., Juttner, R., Rothe, T., and Grantyn, R. (2002) Postsynaptic action of BDNF on GABAergic synaptic transmission in the superficial layers of the mouse superior colliculus. *J. Neurophysiol.* **88,** 595–603.
183. Tanaka, T., Saito, H., and Matsuki, N. (1997) Inhibition of $GABA_A$ synaptic responses by brain-derived neurotrophic factor (BDNF) in rat hippocampus. *J. Neurosci.* **17,** 2959–2966.
184. Hewitt, S. A. and Bains, J. S. (2006) Brain-derived neurotrophic factor silences GABA synapses onto hypothalamic neuroendocrine cells through a postsynaptic dynamin-mediated mechanism. *J. Neurophysiol.* **95,** 2193–2198.
185. Mizoguchi, Y., Kanematsu, T., Hirata, M., and Nabekura, J. (2003) A rapid increase in the total number of cell surface functional $GABA_A$ receptors induced by brain-derived neurotrophic factor in rat visual cortex. *J. Biol. Chem.* **278,** 44,097–44,102.
186. Swanwick, C. C., Murthy, N. R., and Kapur, J. (2006) Activity-dependent scaling of GABAergic synapse strength is regulated by brain-derived neurotrophic factor. *Mol. Cell. Neurosci.* **31,** 481–492.
187. Elmariah, S. B., Oh, E. J., Hughes, E. G., and Balice-Gordon, R. J. (2005) Astrocytes regulate inhibitory synapse formation via Trk-mediated modulation of postsynaptic $GABA_A$ receptors. *J. Neurosci.* **25,** 3638–3650.
188. Jovanovic, J. N., Thomas, P., Kittler, J. T., Smart, T. G., and Moss, S. J. (2004) Brain-derived neurotrophic factor modulates fast synaptic inhibition by regulating $GABA_A$ receptor phosphorylation, activity, and cell-surface stability. *J. Neurosci.* **24,** 522–530.
189. Palma, E., Torchia, G., Limatola, C., et al. (2005) BDNF modulates $GABA_A$ receptors microtransplanted from the human epileptic brain to *Xenopus* oocytes. *Proc. Natl. Acad. Sci. USA* **102,** 1667–1672.
190. Chapell, R., Bueno, O. F., Alvarez-Hernandez, X., Robinson, L. C., and Leidenheimer, N. J. (1998) Activation of protein kinase C induces γ-aminobutyric acid type A receptor internalization in *Xenopus* oocytes. *J. Biol. Chem.* **273,** 32,595–32,601.

191. Connolly, C. N., Kittler, J. T., Thomas, P., et al. (1999) Cell surface stability of γ-aminobutyric acid type A receptors. Dependence on protein kinase C activity and subunit composition. *J. Biol. Chem.* **274,** 36,565–36,572.
192. Cinar, H. and Barnes, E. M. Jr. (2001) Clathrin-independent endocytosis of $GABA_A$ receptors in HEK 293 cells. *Biochemistry* **40,** 14,030–14,036.
193. Herring, D., Huang, R., Singh, M., Dillon, G. H., and Leidenheimer, N. J. (2005) PKC modulation of $GABA_A$ receptor endocytosis and function is inhibited by mutation of a dileucine motif within the receptor β2 subunit. *Neuropharmacology* **48,** 181–194.
194. Thomas, P., Mortensen, M., Hosie, A. M., and Smart, T. G. (2005) Dynamic mobility of functional $GABA_A$ receptors at inhibitory synapses. *Nat. Neurosci.* **8,** 889–897.
195. Balasubramanian, S., Teissere, J. A., Raju, D. V., and Hall, R. A. (2004) Hetero-oligomerization between $GABA_A$ and $GABA_B$ receptors regulates $GABA_B$ receptor trafficking. *J. Biol. Chem.* **279,** 18,840–18,850.
196. Marini, C., Harkin, L. A., Wallace, R. H., Mulley, J. C., Scheffer, I. E., and Berkovic, S. F. (2003) Childhood absence epilepsy and febrile seizures: a family with a $GABA_A$ receptor mutation. *Brain* **126,** 230–240.
197. Harkin, L. A., Bowser, D. N., Dibbens, L. M., et al. (2002) Truncation of the $GABA_A$ receptor γ2 subunit in a family with generalized epilepsy with febrile seizures plus. *Am. J. Hum. Genet.* **70,** 530–536.
198. Cossette, P., Liu, L., Brisebois, K., et al. (2002) Mutation of *GABRA1* in an autosomal dominant form of juvenile myoclonic epilepsy. *Nat. Genet.* **31,** 184–189.
199. Baulac, S., Huberfeld, G., Gourfinkel-An, I., et al. (2001) First genetic evidence of $GABA_A$ receptor dysfunction in epilepsy: a mutation in the γ2-subunit gene. *Nat. Genet.* **28,** 46–48.
200. Gallagher, M. J., Shen, W., Song, L., and Macdonald, R. L. (2005) Endoplasmic reticulum retention and associated degradation of a $GABA_A$ receptor epilepsy mutation that inserts an aspartate in the M3 transmembrane segment of the α1 subunit. *J. Biol. Chem.* **280,** 37,995–38,004.
201. Gallagher, M. J., Song, L., Arain, F., and Macdonald, R. L. (2004) The juvenile myoclonic epilepsy $GABA_A$ receptor α1 subunit mutation A322D produces asymmetrical, subunit position-dependent reduction of heterozygous receptor currents and α1 subunit protein expression. *J. Neurosci.* **24,** 5570–5578.
202. Kang, J. Q., Shen, W., and Macdonald, R. L. (2006) Why does fever trigger febrile seizures? $GABA_A$ receptor γ2 subunit mutations associated with idiopathic generalized epilepsies have temperature-dependent trafficking deficiencies. *J. Neurosci.* **26,** 2590–2597.
203. Zhang, G., Raol, Y. H., Hsu, F. C., Coulter, D. A., and Brooks-Kayal, A. R. (2004) Effects of status epilepticus on hippocampal $GABA_A$ receptors are age-dependent. *Neuroscience* **125,** 299–303.
204. Ni, H., Jiang, Y. W., Bo, T., Wang, J. M., Pan, H., and Wu, X. R. (2004) Long-term effects of neonatal seizures on subsequent N-methyl-D-aspartate receptor-1 and γ-aminobutyric acid receptor A α1 receptor expression in hippocampus of the Wistar rat. *Neurosci. Lett.* **368,** 254–257.

205. Ni, H., Jiang, Y. W., Bo, T., Wang, J. M., and Wu, X. R. (2005) c-Fos, N-methyl-d-aspartate receptor 2C, $GABA_A$ $\alpha 1$ immonoreactivity, seizure latency and neuronal injury following single or recurrent neonatal seizures in hippocampus of Wistar rat. *Neurosci. Lett.* **380,** 149–154.
206. Meguro, R., Lu, J., Gavrilovici, C., and Poulter, M. O. (2004) Static, transient and permanent organization of GABA receptor expression in calbindin-positive interneurons in response to amygdala kindled seizures. *J. Neurochem.* **91,** 144–154.
207. Houser, C. R. and Esclapez, M. (2003) Downregulation of the $\alpha 5$ subunit of the $GABA_A$ receptor in the pilocarpine model of temporal lobe epilepsy. *Hippocampus* **13,** 633–645.
208. Kapur, J. and Coulter, D. A. (1995) Experimental status epilepticus alters γ-aminobutyric acid type A receptor function in CA1 pyramidal neurons. *Ann. Neurol.* **38,** 893–900.
209. Kapur, J., Lothman, E. W., and DeLorenzo, R. J. (1994) Loss of $GABA_A$ receptors during partial status epilepticus. *Neurology* **44,** 2407–2408.
210. Kokaia, M., Pratt, G. D., Elmer, E., et al. (1994) Biphasic differential changes of $GABA_A$ receptor subunit mRNA levels in dentate gyrus granule cells following recurrent kindling-induced seizures. *Mol. Brain. Res.* **23,** 323–332.
211. Kamphuis, W., De Rijk, T. C., and Lopes da Silva, F. H. (1995) Expression of $GABA_A$ receptor subunit mRNAs in hippocampal pyramidal and granular neurons in the kindling model of epileptogenesis: an in situ hybridization study. *Mol. Brain. Res.* **31,** 33–47.
212. Brooks-Kayal, A. R., Shumate, M. D., Jin, H., Rikhter, T. Y., and Coulter, D. A. (1998) Selective changes in single cell $GABA_A$ receptor subunit expression and function in temporal lobe epilepsy. *Nat. Med.* **4,** 1166–1172.
213. Poulter, M. O., Brown, L. A., Tynan, S., Willick, G., William, R., and McIntyre, D. C. (1999) Differential expression of $\alpha 1$, $\alpha 2$, $\alpha 3$, and $\alpha 5$ $GABA_A$ receptor subunits in seizure-prone and seizure-resistant rat models of temporal lobe epilepsy. *J. Neurosci.* **19,** 4654–4661.
214. Schwarzer, C., Tsunashima, K., Wanzenbock, C., Fuchs, K., Sieghart, W., and Sperk, G. (1997) $GABA_A$ receptor subunits in the rat hippocampus II: altered distribution in kainic acid-induced temporal lobe epilepsy. *Neuroscience* **80,** 1001–1017.
215. Tsunashima, K., Schwarzer, C., Kirchmair, E., Sieghart, W., and Sperk, G. (1997) $GABA_A$ receptor subunits in the rat hippocampus III: altered messenger RNA expression in kainic acid-induced epilepsy. *Neuroscience* **80,** 1019–1032.
216. Nusser, Z., Hajos, N., Somogyi, P., and Mody, I. (1998) Increased number of synaptic $GABA_A$ receptors underlies potentiation at hippocampal inhibitory synapses. *Nature* **395,** 172–177.
217. Kumar, S., Fleming, R. L., and Morrow, A. L. (2004) Ethanol regulation of γ-aminobutyric acid A receptors: genomic and nongenomic mechanisms. *Pharmacol. Ther.* **101,** 211–226.
218. Ticku, M. K. and Burch, T. (1980) Alterations in γ-aminobutyric acid receptor sensitivity following acute and chronic ethanol treatments. *J. Neurochem.* **34,** 417–423.

219. Kumar, S., Sieghart, W., and Morrow, A. L. (2002) Association of protein kinase C with $GABA_A$ receptors containing α1 and α4 subunits in the cerebral cortex: selective effects of chronic ethanol consumption. *J. Neurochem.* **82,** 110–117.
220. Kumar, S., Kralic, J. E., O'Buckley, T. K., Grobin, A. C., and Morrow, A. L. (2003) Chronic ethanol consumption enhances internalization of α1 subunit-containing $GABA_A$ receptors in cerebral cortex. *J. Neurochem.* **86,** 700–708.
221. Follesa, P., Mancuso, L., Biggio, F., et al. (2003) γ-hydroxybutyric acid and diazepam antagonize a rapid increase in $GABA_A$ receptors α4 subunit mRNA abundance induced by ethanol withdrawal in cerebellar granule cells. *Mol. Pharmacol.* **63,** 896–907.
222. Devaud, L. L., Smith, F. D., Grayson, D. R., and Morrow, A. L. (1995) Chronic ethanol consumption differentially alters the expression of γ-aminobutyric $acid_A$ receptor subunit mRNAs in rat cerebral cortex: competitive, quantitative reverse transcriptase-polymerase chain reaction analysis. *Mol. Pharmacol.* **48,** 861–868.
223. Devaud, L. L., Fritschy, J. M., Sieghart, W., and Morrow, A. L. (1997) Bidirectional alterations of $GABA_A$ receptor subunit peptide levels in rat cortex during chronic ethanol consumption and withdrawal. *J. Neurochem.* **69,** 126–130.
224. Grobin, A. C., Papadeas, S. T., and Morrow, A. L. (2000) Regional variations in the effects of chronic ethanol administration on $GABA_A$ receptor expression: potential mechanisms. *Neurochem. Int.* **37,** 453–461.
225. Mhatre, M. C. and Ticku, M. K. (1992) Chronic ethanol administration alters γ-aminobutyric $acid_A$ receptor gene expression. *Mol. Pharmacol.* **42,** 415–422.
226. Mhatre, M. C., Pena, G., Sieghart, W., and Ticku, M. K. (1993) Antibodies specific for $GABA_A$ receptor α subunits reveal that chronic alcohol treatment down-regulates α-subunit expression in rat brain regions. *J. Neurochem.* **61,** 1620–1625.
227. Matthews, D. B., Devaud, L. L., Fritschy, J. M., Sieghart, W., and Morrow, A. L. (1998) Differential regulation of $GABA_A$ receptor gene expression by ethanol in the rat hippocampus versus cerebral cortex. *J. Neurochem.* **70,** 1160–1166.
228. Follesa, P., Biggio, F., Caria, S., Gorini, G., and Biggio, G. (2004) Modulation of $GABA_A$ receptor gene expression by allopregnanolone and ethanol. *Eur. J. Pharmacol.* **500,** 413–425.
229. Follesa, P., Mostallino, M. C., Biggio, F., et al. (2005) Distinct patterns of expression and regulation of GABA receptors containing the δ subunit in cerebellar granule and hippocampal neurons. *J. Neurochem.* **94,** 659–671.
230. Sanna, E., Mostallino, M. C., Busonero, F., et al. (2003) Changes in $GABA_A$ receptor gene expression associated with selective alterations in receptor function and pharmacology after ethanol withdrawal. *J. Neurosci.* **23,** 11,711–11,724.
231. Mahmoudi, M., Kang, M. H., Tillakaratne, N., Tobin, A. J., and Olsen, R. W. (1997) Chronic intermittent ethanol treatment in rats increases $GABA_A$ receptor α4 subunit expression: possible relevance to alcohol dependence. *J. Neurochem.* **68,** 2485–2492.
232. Cagetti, E., Liang, J., Spigelman, I., and Olsen, R. W. (2003) Withdrawal from chronic intermittent ethanol treatment changes subunit composition, reduces

synaptic function, and decreases behavioral responses to positive allosteric modulators of $GABA_A$ receptors. *Mol. Pharmacol.* **63,** 53–64.
233. Follesa, P., Biggio, F., Mancuso, L., et al. (2004) Ethanol withdrawal-induced up-regulation of the α2 subunit of the $GABA_A$ receptor and its prevention by diazepam or γ-hydroxybutyric acid. *Mol. Brain. Res.* **120,** 130–137.
234. Saba, L., Porcella, A., Sanna, A., et al. (2005) Five mutations in the $GABA_A$ α6 gene 5' flanking region are associated with a reduced basal and ethanol-induced α6 upregulation in mutated Sardinian alcohol non-preferring rats. *Mol. Brain Res.* **137,** 252–257.
235. Wallner, M., Hanchar, H. J., and Olsen, R. W. (2003) Ethanol enhances α4β3δ and α6β3δ γ-aminobutyric acid type A receptors at low concentrations known to affect humans. *Proc. Natl. Acad. Sci. USA* **100,** 15,218–15,223.
236. Wei, W., Faria, L. C., and Mody, I. (2004) Low ethanol concentrations selectively augment the tonic inhibition mediated by δ subunit-containing $GABA_A$ receptors in hippocampal neurons. *J. Neurosci.* **24,** 8379–8382.
237. Lewis, D. A., Hashimoto, T., and Volk, D. W. (2005) Cortical inhibitory neurons and schizophrenia. *Nat. Rev. Neurosci.* **6,** 312–324.
238. Ohnuma, T., Augood, S. J., Arai, H., McKenna, P. J., and Emson, P. C. (1999) Measurement of GABAergic parameters in the prefrontal cortex in schizophrenia: focus on GABA content, $GABA_A$ receptor α1 subunit messenger RNA and human GABA transporter-1 (HGAT-1) messenger RNA expression. *Neuroscience.* **93,** 441–448.
239. Benes, F. M., Vincent, S. L., Marie, A., and Khan, Y. (1996) Up-regulation of $GABA_A$ receptor binding on neurons of the prefrontal cortex in schizophrenic subjects. *Neuroscience.* **75,** 1021–1031.
240. Deng, C. and Huang, X. F. (2006) Increased density of $GABA_A$ receptors in the superior temporal gyrus in schizophrenia. *Exp. Brain Res.* **168,** 587–590.
241. Perlstein, W. M., Carter, C. S., Noll, D. C., and Cohen, J. D. (2001) Relation of prefrontal cortex dysfunction to working memory and symptoms in schizophrenia. *Am. J. Psychiatry.* **158,** 1105–1113.
242. Wassef, A., Baker, J., and Kochan, L. D. (2003) GABA and schizophrenia: a review of basic science and clinical studies. *J. Clin. Psychopharmacol.* **23,** 601–640.
243. Cryan, J. F. and Kaupmann, K. (2005) Don't worry 'B' happy!: a role for $GABA_B$ receptors in anxiety and depression. *Trends Pharmacol. Sci.* **26,** 36–43.
244. Gassmann, M., Shaban, H., Vigot, R., et al. (2004) Redistribution of $GABA_{B(1)}$ protein and atypical $GABA_B$ responses in $GABA_{B(2)}$-deficient mice. *J. Neurosci.* **24,** 6086–6097.
245. Galvez, T., Parmentier, M. L., Joly, C., et al. (1999) Mutagenesis and modeling of the $GABA_B$ receptor extracellular domain support a venus flytrap mechanism for ligand binding. *J. Biol. Chem.* **274,** 13,362–13,369.
246. Malitschek, B., Schweizer, C., Keir, M., et al. (1999) The N-terminal domain of γ-aminobutyric $acid_B$ receptors is sufficient to specify agonist and antagonist binding. *Mol. Pharmacol.* **56,** 448–454.
247. Pfaff, T., Malitschek, B., Kaupmann, K., et al. (1999) Alternative splicing generates a novel isoform of the rat metabotropic $GABA_BR1$ receptor. *Eur. J. Neurosci.* **11,** 2874–2882.

248. Green, A., Walls, S., Wise, A., Green, R. H., Martin, A. K., and Marshall, F. H. (2000) Characterization of [^{3}H]-CGP54626A binding to heterodimeric $GABA_B$ receptors stably expressed in mammalian cells. *Br. J. Pharmacol.* **131,** 1766–1774.
249. Margeta-Mitrovic, M., Jan, Y. N., and Jan, L. Y. (2000) A trafficking checkpoint controls $GABA_B$ receptor heterodimerization. *Neuron.* **27,** 97–106.
250. Calver, A. R., Robbins, M. J., Cosio, C., et al. (2001) The C-terminal domains of the $GABA_B$ receptor subunits mediate intracellular trafficking but are not required for receptor signaling. *J. Neurosci.* **21,** 1203–1210.
251. Pagano, A., Rovelli, G., Mosbacher, J., et al. (2001) C-terminal interaction is essential for surface trafficking but not for heteromeric assembly of $GABA_B$ receptors. *J. Neurosci.* **21,** 1189–1202.
252. Grunewald, S., Schupp, B. J., Ikeda, S. R., et al. (2002) Importance of the γ-aminobutyric $acid_B$ receptor C-termini for G-protein coupling. *Mol. Pharmacol.* **61,** 1070–1080.
253. Goei, V. L., Choi, J., Ahn, J., Bowlus, C. L., Raha-Chowdhury, R., and Gruen, J. R. (1998) Human γ-aminobutyric acid B receptor gene: complementary DNA cloning, expression, chromosomal location, and genomic organization. *Biol. Psychiatry* **44,** 659–666.
254. Grifa, A., Totaro, A., Rommens, J. M., et al. (1998) GABA (γ-amino-butyric acid) neurotransmission: identification and fine mapping of the human $GABA_B$ receptor gene. *Biochem. Biophys. Res. Commun.* **250,** 240–245.
255. Peters, H. C., Kammer, G., Volz, A., et al. (1998) Mapping, genomic structure, and polymorphisms of the human $GABA_BR1$ receptor gene: evaluation of its involvement in idiopathic generalized epilepsy. *Neurogenetics.* **2,** 47–54.
256. Lander, E. S., Linton, L. M., Birren, B., et al. (2001) Initial sequencing and analysis of the human genome. *Nature.* **409,** 860–921.
257. Modrek, B. and Lee, C. (2002) A genomic view of alternative splicing. *Nat. Genet.* **30,** 13–19.
258. Trinklein, N. D., Aldred, S. J., Saldanha, A. J., and Myers, R. M. (2003) Identification and functional analysis of human transcriptional promoters. *Genome Res.* **13,** 308–312.
259. Landry, J. R., Mager, D. L., and Wilhelm, B. T. (2003) Complex controls: the role of alternative promoters in mammalian genomes. *Trends Genet.* **19,** 640–648.
260. Ayoubi, T. A. and Van De Ven, W. J. (1996) Regulation of gene expression by alternative promoters. *FASEB J.* **10,** 453–460.
261. Blein, S., Hawrot, E., and Barlow, P. (2000) The metabotropic GABA receptor: molecular insights and their functional consequences. *Cell. Mol. Life Sci.* **57,** 635–650.
262. Schwarz, D. A., Barry, G., Eliasof, S. D., Petroski, R. E., Conlon, P. J., and Maki, R. A. (2000) Characterization of γ-aminobutyric acid receptor $GABA_{B(1e)}$, a $GABA_{B(1)}$ splice variant encoding a truncated receptor. *J. Biol. Chem.* **275,** 32174–32181.
263. Isomoto, S., Kaibara, M., Sakurai-Yamashita, Y., et al. (1998) Cloning and tissue distribution of novel splice variants of the rat $GABA_B$ receptor. *Biochem. Biophys. Res. Commun.* **253,** 10–15.

264. Wei, K., Eubanks, J. H., Francis, J., Jia, Z., and Snead, O. C. 3rd (2001) Cloning and tissue distribution of a novel isoform of the rat $GABA_BR1$ receptor subunit. *Neuroreport.* **12,** 833–837.
265. Wei, K., Jia, Z., Wang, Y. T., Yang, J., Liu, C. C., and Snead, O. C. 3rd (2001) Cloning and characterization of a novel variant of rat $GABA_BR1$ with a truncated C-terminus. *Mol. Brain Res.* **89,** 103–110.
266. Clark, J. A., Mezey, E., Lam, A. S., and Bonner, T. I. (2000) Distribution of the $GABA_B$ receptor subunit gb2 in rat CNS. *Brain Res. Rev.* **860,** 41–52.
267. Owens, D. F. and Kriegstein, A. R. (2002) Is there more to GABA than synaptic inhibition? *Nat. Rev. Neurosci.* **3,** 715–727.
268. Ben-Ari, Y. (2002) Excitatory actions of GABA during development: the nature of the nurture. *Nat. Rev. Neurosci.* **3,** 728–739.
269. Gaiarsa, J. L., McLean, H., Congar, P., et al. (1995) Postnatal maturation of γ-aminobutyric acidA and B-mediated inhibition in the CA3 hippocampal region of the rat. *J. Neurobiol.* **26,** 339–349.
270. McLean, H. A., Caillard, O., Khazipov, R., Ben-Ari, Y., and Gaiarsa, J. L. (1996) Spontaneous release of GABA activates $GABA_B$ receptors and controls network activity in the neonatal rat hippocampus. *J. Neurophysiol.* **76,** 1036–1046.
271. Lopez-Bendito, G., Shigemoto, R., Kulik, A., Vida, I., Fairen, A., and Lujan, R. (2004) Distribution of metabotropic GABA receptor subunits $GABA_{B1a/b}$ and $GABA_B2$ in the rat hippocampus during prenatal and postnatal development. *Hippocampus* **14,** 836–848.
272. Côrrea, S. A., Munton, R., Nishimune, A., Fitzjohn, S., and Henley, J. M. (2004) Development of $GABA_B$ subunits and functional $GABA_B$ receptors in rat cultured hippocampal neurons. *Neuropharmacology* **47,** 475–484.
273. Fritschy, J. M., Meskenaite, V., Weinmann, O., Honer, M., Benke, D., and Mohler, H. (1999) $GABA_B$ receptor splice variants GB1a and GB1b in rat brain: developmental regulation, cellular distribution and extrasynaptic localization. *Eur. J. Neurosci.* **11,** 761–768.
274. Fritschy, J. M., Sidler, C., Parpan, F., et al. (2004) Independent maturation of the $GABA_B$ receptor subunits $GABA_{B1}$ and $GABA_{B2}$ during postnatal development in rodent brain. *J. Comp. Neurol.* **477,** 235–252.
275. Malitschek, B., Ruegg, D., Heid, J., et al. (1998) Developmental changes of agonist affinity at $GABA_BR1$ receptor variants in rat brain. *Mol. Cell. Neurosci.* **12,** 56–64.
276. Kulik, A., Vida, I., Lujan, R., et al. (2003) Subcellular localization of metabotropic $GABA_B$ receptor subunits $GABA_{B1a/b}$ and $GABA_{B2}$ in the rat hippocampus. *J. Neurosci.* **23,** 11,026–11,035.
277. Liang, F., Hatanaka, Y., Saito, H., Yamamori, T., and Hashikawa, T. (2000) Differential expression of γ-aminobutyric acid type B receptor-1a and -1b mRNA variants in GABA and non-GABAergic neurons of the rat brain. *J. Comp. Neurol.* **416,** 475–495.
278. Martin, S. C., Russek, S. J., and Farb, D. H. (1999) Molecular identification of the human GABABR2: cell surface expression and coupling to adenylyl cyclase in the absence of GABABR1. *Mol. Cell Neurosci.* **13,** 180–191.

279. Ng, T. K. and Yung, K. K. (2001) Differential expression of $GABA_BR1$ and $GABA_BR2$ receptor immunoreactivity in neurochemically identified neurons of the rat neostriatum. *J. Comp. Neurol.* **433,** 458–470.
280. Waldvogel, H. J., Billinton, A., White, J. H., Emson, P. C., and Faull, R. L. (2004) Comparative cellular distribution of $GABA_A$ and $GABA_B$ receptors in the human basal ganglia: immunohistochemical colocalization of the α1 subunit of the $GABA_A$ receptor, and the $GABA_BR1$ and $GABA_BR2$ receptor subunits. *J. Comp. Neurol.* **470,** 339–356.
281. Panzanelli, P., Lopez-Bendito, G., Lujan, R., and Sassoe-Pognetto, M. (2004) Localization and developmental expression of $GABA_B$ receptors in the rat olfactory bulb. *J. Neurocytol.* **33,** 87–99.
282. Yung, K. K., Ng, T. K., and Wong, C. K. (1999) Subpopulations of neurons in the rat neostriatum display $GABA_BR1$ receptor immunoreactivity. *Brain Res.* **830,** 345–352.
283. Charara, A., Heilman, T. C., Levey, A. I., and Smith, Y. (2000) Pre- and postsynaptic localization of $GABA_B$ receptors in the basal ganglia in monkeys. *Neuroscience.* **95,** 127–140.
284. Knight, A. R. and Bowery, N. G. (1996) The pharmacology of adenylyl cyclase modulation by $GABA_B$ receptors in rat brain slices. *Neuropharmacology.* **35,** 703–712.
285. Wise, A., Green, A., Main, M. J., Wilson, R., Fraser, N., and Marshall, F. H. (1999) Calcium sensing properties of the $GABA_B$ receptor. *Neuropharmacology.* **38,** 1647–1656.
286. Zhu, X. Z. and Chuang, D. M. (1987) Modulation of calcium uptake and D-aspartate release by $GABA_B$ receptors in cultured cerebellar granule cells. *Eur. J. Pharmacol.* **141,** 401–408.
287. De Erausquin, G., Brooker, G., Costa, E., and Wojcik, W. J. (1992) Stimulation of high affinity γ-aminobutyric $acid_B$ receptors potentiates the depolarization-induced increase of intraneuronal ionized calcium content in cerebellar granule neurons. *Mol. Pharmacol.* **42,** 407–414.
288. Park, S. K., An, S. J., Hwang, I. K., et al. (2004) Altered $GABA_B$ receptor immunoreactivity in the gerbil hippocampus induced by baclofen and phaclofen, not seizure activity. *Neurosci. Res.* **49,** 405–416.
289. Lehmann, A., Mattsson, J. P., Edlund, A., Johansson, T., and Ekstrand, A. J. (2003) Effects of repeated administration of baclofen to rats on $GABA_B$ receptor binding sites and subunit expression in the brain. *Neurochem. Res.* **28,** 387–393.
290. Sands, S. A., McCarson, K. E., and Enna, S. J. (2003) Differential regulation of $GABA_B$ receptor subunit expression and function. *J. Pharmacol. Exp. Ther.* **305,** 191–196.
291. Ghorbel, M. T., Becker, K. G., and Henley, J. M. (2005) Profile of changes in gene expression in cultured hippocampal neurones evoked by the $GABA_B$ receptor agonist baclofen. *Physiol. Genomics.* **22,** 93–98.
292. Zhou, W., Mailloux, A. W., and McGinty, J. F. (2005) Intracerebral baclofen administration decreases amphetamine-induced behavior and neuropeptide gene expression in the striatum. *Neuropsychopharmacology* **30,** 880–890.
293. Zhou, W., Mailloux, A. W., Jung, B. J., Edmunds, H. S. Jr., and McGinty, J. F. (2004) $GABA_B$ receptor stimulation decreases amphetamine-induced behavior and neuropeptide gene expression in the striatum. *Brain Res.* **1004,** 18–28.

294. Herdegen, T. and Leah, J. D. (1998) Inducible and constitutive transcription factors in the mammalian nervous system: control of gene expression by Jun, Fos and Krox, and CREB/ATF proteins. *Brain Res. Rev.* **28,** 370–490.
295. Hai, T. and Hartman, M. G. (2001) The molecular biology and nomenclature of the activating transcription factor/cAMP responsive element binding family of transcription factors: activating transcription factor proteins and homeostasis. *Gene.* **273,** 1–11.
296. De Cesare, D. and Sassone-Corsi, P. (2000) Transcriptional regulation by cyclic AMP-responsive factors. *Prog. Nucleic Acid Res. Mol. Biology* **64,** 343–369.
297. Barthel, F., Kienlen Campard, P., Demeneix, B. A., Feltz, P., and Loeffler, J. P. (1996) $GABA_B$ receptors negatively regulate transcription in cerebellar granular neurons through cyclic AMP responsive element binding protein-dependent mechanisms. *Neuroscience.* **70,** 417–427.
298. Ito, Y., Ishige, K., Zaitsu, E., Anzai, K., and Fukuda, H. (1995) γ-Hydroxybutyric acid increases intracellular $Ca^{2(}$ concentration and nuclear cyclic AMP-responsive element- and activator protein 1 DNA-binding activities through $GABA_B$ receptor in cultured cerebellar granule cells. *J. Neurochem.* **65,** 75–83.
299. Helm, K. A., Haberman, R. P., Dean, S. L., et al. (2005) $GABA_B$ receptor antagonist SGS742 improves spatial memory and reduces protein binding to the cAMP response element (CRE) in the hippocampus. *Neuropharmacology* **48,** 956–964.
300. Nehring, R. B., Horikawa, H. P., El Far, O., et al. (2000) The metabotropic $GABA_B$ receptor directly interacts with the activating transcription factor 4 (ATF-4). *J. Biol. Chem.* **275,** 35,185–35,191.
301. White, J. H., McIllhinney, R. A., Wise, A., et al. (2000) The $GABA_B$ receptor interacts directly with the related transcription factors CREB2 and ATFx. *Proc. Natl. Acad. Sci. USA* **97,** 13,967–13,972.
302. Vernon, E., Meyer, G., Pickard, L., et al. (2001) $GABA_B$ receptors couple directly to the transcription factor ATF4. *Mol. Cell Neurosci.* **17,** 637–645.
303. Sauter, K., Grampp, T., Fritschy, J. M., et al. (2005) Subtype-selective interaction with the transcription factor CCAAT/enhancer-binding protein (C/EBP) homologous protein (CHOP) regulates cell surface expression of $GABA_B$ receptors. *J. Biol. Chem.* **280,** 33,566–33,572.
304. Bartsch, D., Ghirardi, M., Skehel, P. A., et al. (1995) Aplysia CREB2 represses long-term facilitation: relief of repression converts transient facilitation into long-term functional and structural change. *Cell.* **83,** 979–992.
305. Chen, A., Muzzio, I. A., Malleret, G., et al. (2003) Inducible enhancement of memory storage and synaptic plasticity in transgenic mice expressing an inhibitor of ATF4 (CREB-2) and C/EBP proteins. *Neuron* **39,** 655–669.
306. Yin, J. C., Wallach, J. S., Del Vecchio, M., et al. (1994) Induction of a dominant negative CREB transgene specifically blocks long-term memory in Drosophila. *Cell* **79,** 49–58.
307. Ishige, K., Aizawa, M., Ito, Y., and Fukuda, H. (1996) γ-Butyrolactone-induced absence-like seizures increase nuclear CRE- and AP-1 DNA-binding activities in mouse brain. *Neuropharmacology* **35,** 45–55.
308. Ishige, K., Endo, H., Saito, H., and Ito, Y. (2001) Repeated administration of CGP 46381, a γ-aminobutyric $acid_B$ antagonist, and ethosuximide suppresses seizure-associated cyclic adenosine 3'5' monophosphate response element- and activator

protein-1 DNA-binding activities in lethargic (*lh/lh*) mice. *Neurosci. Lett.* **297,** 207–210.
309. Ishige, K., Ito, Y., and Fukuda, H. (1999) Characterization of absence seizure-dependent cyclic AMP responsive element-and activator protein 1 DNA-binding activities in lethargic (lh/lh) mice. *Neurosci. Lett.* **262,** 53–56.
310. Wetherington, J. P. and Lambert, N. A. (2002) $GABA_B$ receptor activation desensitizes postsynaptic $GABA_B$ and A_1 adenosine responses in rat hippocampal neurones. *J. Physiol.* **544,** 459–467.
311. Fairfax, B. P., Pitcher, J. A., Scott, M. G., et al. (2004) Phosphorylation and chronic agonist treatment atypically modulate $GABA_B$ receptor cell surface stability. *J. Biol. Chem.* **279,** 12,565–12,573.
312. Dario, A., Scamoni, C., Picano, M., Casagrande, F., and Tomei, G. (2004) Pharmacological complications of the chronic baclofen infusion in the severe spinal spasticity. Personal experience and review of the literature. *J. Neurosurg. Sci.* **48,** 177–181.
313. Coffey, J. R., Cahill, D., Steers, W., et al. (1993) Intrathecal baclofen for intractable spasticity of spinal origin: results of a long-term multicenter study. *J. Neurosurg.* **78,** 226–232.
314. Penn, R. D. and Kroin, J. S. (1987) Long-term intrathecal baclofen infusion for treatment of spasticity. *J. Neurosurg.* **66,** 181–185.
315. Nielsen, J. F., Hansen, H. J., Sunde, N., and Christensen, J. J. (2002) Evidence of tolerance to baclofen in treatment of severe spasticity with intrathecal baclofen. *Clin. Neurol. Neurosurg.* **104,** 142–145.
316. McCarson, K. E. and Enna, S. J. (1999) Nociceptive regulation of $GABA_B$ receptor gene expression in rat spinal cord. *Neuropharmacology* **38,** 1767–1773.
317. Sands, S. A., Reisman, S. A., and Enna, S. J. (2004) Effect of antidepressants on $GABA_B$ receptor function and subunit expression in rat hippocampus. *Biochem. Pharmacol.* **68,** 1489–1495.
318. McCarson, K. E., Duric, V., Reisman, S. A., Winter, M., and Enna, S. J. (2006) $GABA_B$ receptor function and subunit expression in the rat spinal cord as indicators of stress and the antinociceptive response to antidepressants. *Brain Res.* **1068,** 109–117.
319. Sands, S. A., McCarson, K. E., and Enna, S. J. (2004) Relationship between the antinociceptive response to desipramine and changes in $GABA_B$ receptor function and subunit expression in the dorsal horn of the rat spinal cord. *Biochem. Pharmacol.* **67,** 743–749.
320. Francis, J., Jung, B. P., Zhang, G., et al. (2001) Perforant pathway kindling transiently induces the mRNA expression of $GABA_B$ receptor subtypes R1A and R2 in the adult rat hippocampus. *Mol. Brain Res.* **91,** 159–162.
321. Furtinger, S., Bettler, B., and Sperk, G. (2003) Altered expression of $GABA_B$ receptors in the hippocampus after kainic-acid-induced seizures in rats. *Mol. Brain Res.* **113,** 107–115.
322. Straessle, A., Loup, F., Arabadzisz, D., Ohning, G. V., and Fritschy, J. M. (2003) Rapid and long-term alterations of hippocampal $GABA_B$ receptors in a mouse model of temporal lobe epilepsy. *Eur. J. Neurosci.* **18,** 2213–2226.

323. Billinton, A., Baird, V. H., Thom, M., Duncan, J. S., Upton, N., and Bowery, N. G. (2001) $GABA_{B1}$ mRNA expression in hippocampal sclerosis associated with human temporal lobe epilepsy. *Mol. Brain Res.* **86,** 84–89.
324. Princivalle, A. P., Duncan, J. S., Thom, M., and Bowery, N. G. (2003) $GABA_{B1a}$, $GABA_{B1b}$ and $GABA_{B2}$ mRNA variants expression in hippocampus resected from patients with temporal lobe epilepsy. *Neuroscience* **122,** 975–984.
325. Bading, H. (1999) Nuclear calcium-activated gene expression: possible roles in neuronal plasticity and epileptogenesis. *Epilepsy Res.* **36,** 225–231.
326. Park, S. A., Kim, T. S., Choi, K. S., Park, H. J., Heo, K., and Lee, B. I. (2003) Chronic activation of CREB and p90RSK in human epileptic hippocampus. *Exp. Mol. Med.* **35,** 365–370.
327. Cousins, M. S., Roberts, D. C., and de Wit, H. (2002) $GABA_B$ receptor agonists for the treatment of drug addiction: a review of recent findings. *Drug Alcohol Depend.* **65,** 209–220.
328. Brebner, K., Childress, A. R., and Roberts, D. C. (2002) A potential role for $GABA_B$ agonists in the treatment of psychostimulant addiction. *Alcohol Alcohol* **37,** 478–484.
329. Ranaldi, R. and Poeggel, K. (2002) Baclofen decreases methamphetamine self-administration in rats. *Neuroreport.* **13,** 1107–1110.
330. Shoaib, M., Swanner, L. S., Beyer, C. E., Goldberg, S. R., and Schindler, C. W. (1998) The $GABA_B$ agonist baclofen modifies cocaine self-administration in rats. *Behav. Pharmacol.* **9,** 195–206.
331. Akhondzadeh, S., Ahmadi-Abhari, S. A., Assadi, S. M., Shabestari, O. L., Kashani, A. R., and Farzanehgan, Z. M. (2000) Double-blind randomized controlled trial of baclofen vs. clonidine in the treatment of opiates withdrawal. *J. Clin. Pharm. Ther.* **25,** 347–353.
332. Assadi, S. M., Radgoodarzi, R., and Ahmadi-Abhari, S. A. (2003) Baclofen for maintenance treatment of opioid dependence: a randomized double-blind placebo-controlled clinical trial. *BMC. Psychiatry* **3,** 16.
333. Ling, W., Shoptaw, S., and Majewska, D. (1998) Baclofen as a cocaine anti-craving medication: a preliminary clinical study. *Neuropsychopharmacology* **18,** 403–404.
334. Shoptaw, S., Yang, X., Rotheram-Fuller, E. J., et al. (2003) Randomized placebo-controlled trial of baclofen for cocaine dependence: preliminary effects for individuals with chronic patterns of cocaine use. *J. Clin. Psychiatry* **64,** 1440–1448.
335. Li, J., Olinger, A. B., Dassow, M. S., and Abel, M. S. (2003) Up-regulation of $GABA_B$ receptor mRNA and protein in the hippocampus of cocaine- and lidocaine-kindled rats. *Neuroscience* **118,** 451–462.
336. Li, J., Olinger, A. B., Dassow, M. S., and Abel, M. S. (2002) $GABA_B$ receptor gene expression is not altered in cocaine-sensitized rats. *J. Neurosci. Res.* **68,** 241–247.
337. Yamaguchi, M., Suzuki, T., Abe, S., Baba, A., Hori, T., and Okado, N. (2002) Repeated cocaine administration increases $GABA_{B(1)}$ subunit mRNA in rat brain. *Synapse.* **43,** 175–180.
338. Graybiel, A. M., Moratalla, R., and Robertson, H. A. (1990) Amphetamine and cocaine induce drug-specific activation of the c-fos gene in striosome-matrix

compartments and limbic subdivisions of the striatum. *Proc. Natl. Acad. Sci. USA*. **87,** 6912–6916.
339. Nguyen, T. V., Kosofsky, B. E., Birnbaum, R., Cohen, B. M., and Hyman, S. E. (1992) Differential expression of c-fos and zif268 in rat striatum after haloperidol, clozapine, and amphetamine. *Proc. Natl. Acad. Sci. USA* **89,** 4270–4274.
340. Moratalla, R., Robertson, H. A., and Graybiel, A. M. (1992) Dynamic regulation of NGFI-A (zif268, egr1) gene expression in the striatum. *J. Neurosci.* **12,** 2609–2622.
341. Konradi, C., Cole, R. L., Heckers, S., and Hyman, S. E. (1994) Amphetamine regulates gene expression in rat striatum via transcription factor CREB. *J. Neurosci.* **14,** 5623–5634.
342. Cole, R. L., Konradi, C., Douglass, J., and Hyman, S. E. (1995) Neuronal adaptation to amphetamine and dopamine: molecular mechanisms of prodynorphin gene regulation in rat striatum. *Neuron* **14,** 813–823.
343. Konradi, C., Leveque, J. C., and Hyman, S. E. (1996) Amphetamine and dopamine-induced immediate early gene expression in striatal neurons depends on postsynaptic NMDA receptors and calcium. *J. Neurosci.* **16,** 4231–4239.
344. Shaw-Lutchman, T. Z., Impey, S., Storm, D., and Nestler, E. J. (2003) Regulation of CRE-mediated transcription in mouse brain by amphetamine. *Synapse* **48,** 10–17.
345. Terwilliger, R. Z., Beitner-Johnson, D., Sevarino, K. A., Crain, S. M., and Nestler, E. J. (1991) A general role for adaptations in G-proteins and the cyclic AMP system in mediating the chronic actions of morphine and cocaine on neuronal function. *Brain Res.* **548,** 100–110.
346. Kano, T., Suzuki, Y., Shibuya, M., Kiuchi, K., and Hagiwara, M. (1995) Cocaine-induced CREB phosphorylation and c-Fos expression are suppressed in Parkinsonism model mice. *Neuroreport* **6,** 2197–2200.
347. Pliakas, A. M., Carlson, R. R., Neve, R. L., Konradi, C., Nestler, E. J., and Carlezon, W. A. Jr. (2001) Altered responsiveness to cocaine and increased immobility in the forced swim test associated with elevated cAMP response element-binding protein expression in nucleus accumbens. *J. Neurosci.* **21,** 7397–7403.
348. Carlezon, W. A. Jr., Thome, J., Olson, V. G., et al. (1998) Regulation of cocaine reward by CREB. *Science* **282,** 2272–2275.
349. Walters, C. L. and Blendy, J. A. (2001) Different requirements for cAMP response element binding protein in positive and negative reinforcing properties of drugs of abuse. *J. Neurosci.* **21,** 9438–9444.
350. Quandt, K., Frech, K., Karas, H., Wingender, E., and Werner, T. (1995) MatInd and MatInspector: new fast and versatile tools for detection of consensus matches in nucleotide sequence data. *Nucleic Acids Res.* **23,** 4878–4884.

9

Chemistry of $GABA_B$ Modulators

Wolfgang Froestl, Nigel G. Cooke, and Stuart J. Mickel

Summary

Nearly 10 yr have passed since the authors review "Chemistry of $GABA_B$ Modulators" appeared in the book "The GABA Receptors," 2nd edition, edited by S. J. Enna and N. G. Bowery, Humana Press, Totowa, 1997 *(1)*. In this update the authors wish to outline only new developments not covered in the 1997 paper. Baclofen, synthesized for the first time in September 1962, is still the only γ-aminobutyric acid ($GABA_B$)-receptor agonist marketed for the treatment of spasticity and trigeminal neuralgia. It is fascinating to learn how many highly competent chemists devised manifold and elegant synthetic procedures for either racemic or (*R*)-(–)-baclofen and the structurally closely related potent antidepressant (*R*)-(–)-Rolipram, a selective phosphodiesterase-4 inhibitor. The new syntheses published after 1996 are listed in alphabetical order *(2–16)*. Very recently a prodrug of (*R*)-(–)-baclofen (Fig. 1) was described which enhanced the oral bioavailability in Cynomolgus monkeys to more than 80% *(17)*. It is planned to test this prodrug in clinical trials for the treatment of spasticity and gastroesophageal reflux disease.

Key Words: $GABA_B$ receptor agonists; positive modulators; antagonists; radioactive and photoaffinity ligands.

1. $GABA_B$-Receptor Agonists

A new series of potent and selective $GABA_B$-receptor agonists was discovered by replacing the carboxylic acid groups of GABA or γ-amino-β-hydroxybutyric acid by phosphinic acid residues leading to the much investigated compounds γ-aminopropyl-phosphinic acid (APPA) (or CGP27492), CGP35024

From: *The Receptors: The GABA Receptors, Third Edition*
Edited by: S. J. Enna and H. Möhler © Humana Press Inc., Totowa, NJ

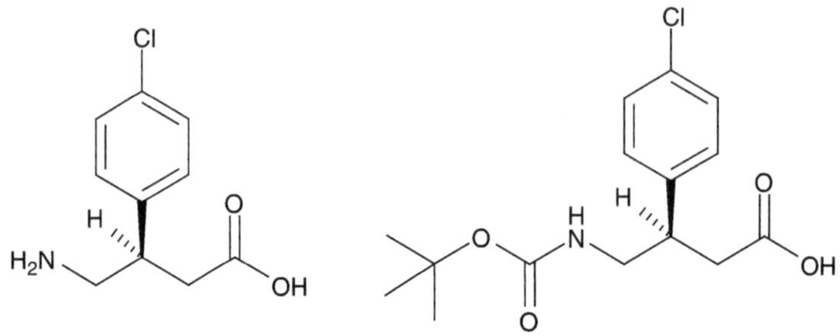

(R)-(–)-Baclofen   Prodrug of (R)-(–)-Baclofen

Fig. 1. Structures of (*R*)-(–)-baclofen and its prodrug.

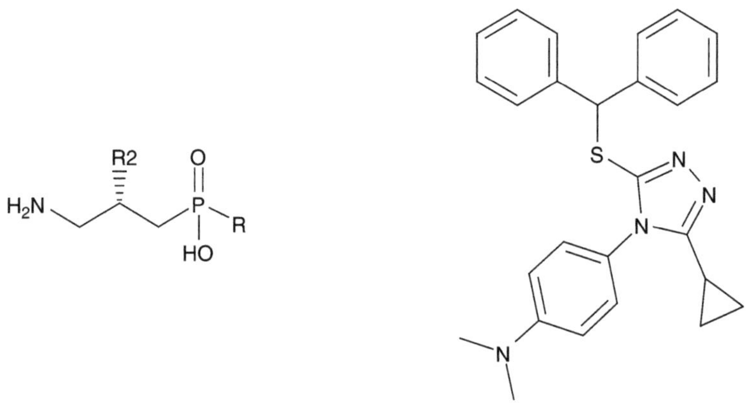

R = H R_2 = H APPA = CGP27492

R = Me R_2 = H SK&F97541 = CGP35024

R = Me R_2 = OH CGP44532

Bayer Healthcare AG

Fig. 2. Structures of novel $GABA_B$-receptor agonists.

(or SK&F97541), and CGP44532 (Fig. 2). CGPs 35024 and 44532 were extensively evaluated in models of neuropathic pain *(18)*. In addition, CGP44532 was tested in several animal models of suppression of craving for diverse drugs of abuse, such as alcohol *(19)*, cocaine *(20,21)*, and nicotine *(22–24)*.

In 2005 a new heterocyclic class of $GABA_B$-receptor agonists, probably discovered through high-throughput screening, was described in the patent literature by chemists of Bayer Healthcare AG in Germany and Japan, i.e., 5-substituted-2-phenylmethyl-thio-4-phenyl-4H-1,2,4-triazoles (Fig. 2) for the treatment of urinary incontinence and pain *(25)*. Other companies did not succeed to identify novel $GABA_B$-receptor agonists through high throughput screening, but described the lessons learned in this process in a very instructive poster *(26)*.

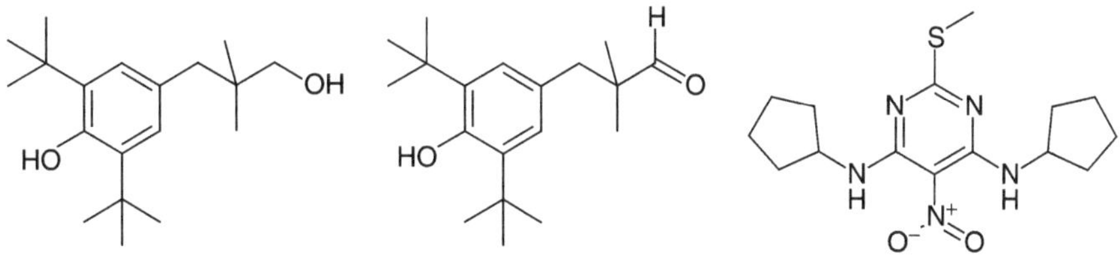

CGP7930 CGP13501 GS39783

Fig. 3. Structures of positive modulators of GABA$_B$ receptors.

2. Positive Modulators of GABA$_B$ Receptors

Allosteric modulators are molecules that bind to a site on a neurotransmitter or hormone receptor which is topographically distinct from the orthosteric-binding pocket for agonists or competitive antagonists *(27)*. Allosteric agents have little or no intrinsic agonistic activity of their own, but induce conformational changes in the receptor protein, which affects its interaction with allosteric ligands. These compounds may present distinct advantages over full agonists *(28)*.

In a high throughput screen using a GTPγ[^{35}S] assay in membranes from Chinese hamster ovary K1 cells stably transfected with human GABA$_{B1b}$ and rat GABA$_{B2}$ complementary DNAs, two compounds CGP7930 and its oxidation product the aldehyde CGP13501 (Fig. 3) were identified as positive modulators of GABA$_B$ receptor function *(29)*. They potentiate GABA-stimulated GTPγ[^{35}S] binding at low micromolar concentrations and are inactive in the absence of GABA. They increase both agonist potency and maximal efficacy: the effective concentration to reach a 50% effect (EC$_{50}$) for CGP7930 in presence of 1 μ*M* GABA in recombinant GABA$_B$-receptor heterodimer expressed in CHO cells is 4.6 μ*M*, in the presence of 20 μ*M* GABA 1.87 μ*M*. The maximal effect is dose dependent: GABA at 20 μ*M* stimulates basal activity to 301%, in presence of 1 μ*M* CGP7930 to 328%, in presence of 3 μ*M* CGP7930 to 377%, in presence of 10 μ*M* CGP7930 to 394% and in presence of 30 μ*M* CGP7930 to 427%. For the in vivo potentiating effects of CGP7930 of the GABA$_B$-receptor agonist baclofen (*see* ref. *30*). It was shown that CGP7930 interacts with the heptahelical domain of the GABA$_{B2}$-receptor subunit thus being the first described agonist of GABA$_{B2}$ *(31)*.

Later a more potent compound was identified in GS39783 increasing the potency of GABA about eightfold: EC$_{50}$ of GABA in absence of GS39783 was 3.59 μ*M* and in presence of 30 μ*M* GS39783: EC$_{50}$ = 0.45 μ*M*. The maximal intrinsic efficacy increased from 100 to 217% *(32)*. For a full behavioral characterization of GS39783 (*see* refs. *33,34*). GS39783 also attenuates the reward-facilitating effects of cocaine *(35,36)*.

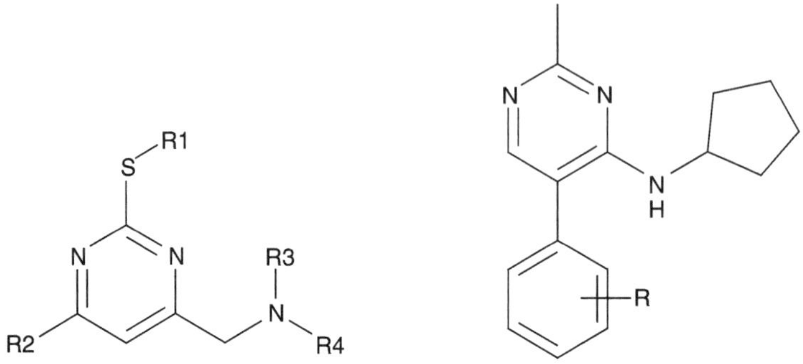

Roche NIH

Fig. 4. Novel positive modulators of $GABA_B$ receptors.

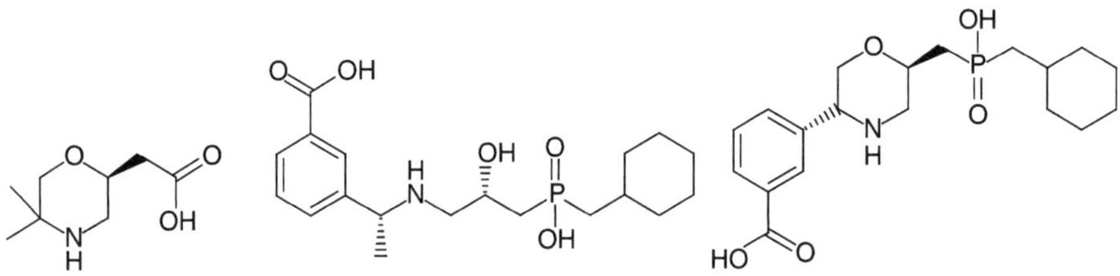

SCH50911 CGP56999 CGP76290

Fig. 5. Combination of Schering and Novartis $GABA_B$-receptor antagonists.

In 2005, novel structures, both based on the lead of GS39783 (Fig. 4), were disclosed by chemists from Roche *(37)* and Novartis, the latter in collaboration with National Institutes of Health, in particular National Institute of Drug Abuse and National Institute of Mental Health (Grant U01 MH069062) to obtain compounds for the treatment of craving after smoking cessation *(38)*.

3. $GABA_B$-Receptor Antagonists

Potent $GABA_B$-receptor antagonists originating from Novartis and Schering have been described in exhaustive detail in the review in the book "The GABA Receptors," 2nd edition, in 1997 *(1)*. Since then hybrid structures between the Schering and Novartis compounds, such as CGP76290 have been prepared and extensively tested (Fig. 5). The synthesis of the (3*S*, 6*R*) enantiomer CGP76290 is outlined in Scheme 1 using an intramolecular Michael reaction for the formation of the morpholine ring *(39)*.

The IC_{50} of CGP76290 was determined to be 1.85 n*M* (inhibition of binding of [^{3}H]CGP27492 to membranes from rat cerebral cortex). The IC_{50} values of the (3*R*, 6*S*) enantiomer CGP76291 was measured as 69 n*M*, the racemic

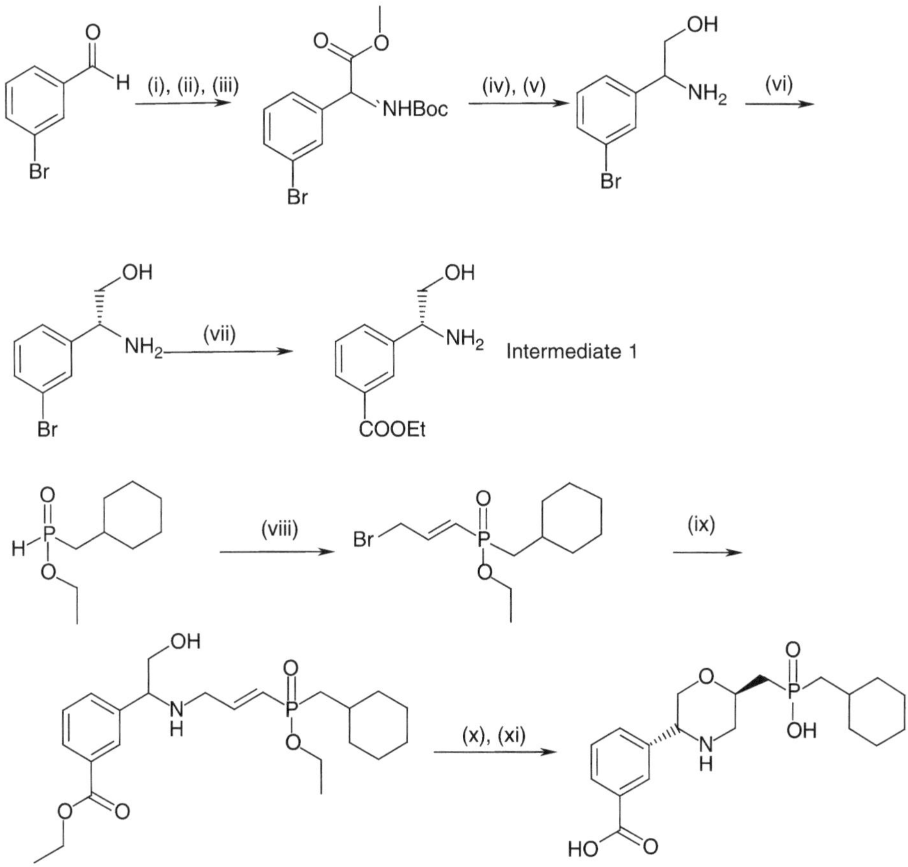

Scheme 1. Reagents and conditions: (i) NaCN, NH_4Cl, NH_4OH, MeOH, H_2O, rt, 24 h; (ii) MeOH, HCl (g), –20°C, 2 d; (iii) $(Boc)_2O$, MeOH, Et_3N, 60°C, 30 min; (iv) $NaBH_4$, EtOH, rt, 6 h; (v) TFA, CH_2Cl_2, 0°C, 30 min, rt, 3 h; (vi) D-tartaric acid, recrystallize from H_2O; (vii) 10 mol% $(Ph_3P)_2PdCl_2$, EtOH, CO (50 psi), EtOH, Et_3N, 100°C; (viii) 1,3-dibromopropene, *bis*-(trimethylsilyl)-acetamide, $(MeO)_3PO$, CH_2Cl_2, rt, 18 h; (ix) intermediate 1, DBU, toluene/THF 1:1, 80°C, 5 h; (x) NaH, toluene, rt, 20 h; (xi) TMSBr, CH_2Cl_2, rt, 24 h.

compound CGP71982 as 8 n*M* *(40)*. The benzoic acid derivative is by far the best compound: unsubstituted phenyl-, p-nitro-, m, p-dichloro, p-iodo, or p-cyano compounds showed inferior affinities *(40)*. Substitution by m-carboxybenzyl instead of m-carboxyphenyl led to a loss of affinity: IC_{50} = 140 nm for CGP81009. Methylation of the nitrogen of the morpholine leads to a massive loss of affinity: IC_{50} = 3.92 μ*M* for CGP77322. The corresponding racemic thiomorpholine derivative CGP79856 (S instead of O in the six-membered ring) was marginally more active than CGP71982 with IC_{50} of 5 n*M*. The corresponding

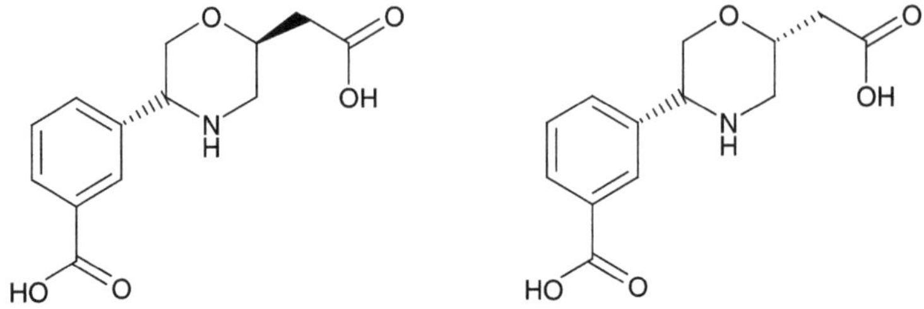

trans (2S, 5R) CGP83713 *cis*-(2R, 5R) CGP83721

Fig. 6. *Trans*- and *cis*-5-aryl-morpholino-2-acetic acids.

racemic piperazine derivative CGP77315 (NH instead of O in the six-membered ring) displayed an IC_{50} of 18 n*M*, as did the racemic piperidine derivative CGP77328 (CH_2 instead of O in the six-membered ring).

The corresponding morpholine acetic acids have been prepared (Fig. 6) *(41)*. However, their affinities to $GABA_B$ receptors is about one order of magnitude lower, i.e., racemic trans CGP79648 with IC_{50} of 51 n*M*, the more potent *trans* (2*S*, 5*R*)-enantiomer CGP83713 (Fig. 6) with IC_{50} of 23 n*M*, the less potent *trans* (2*S*, 5*S*)-enantiomer CGP83719 with IC_{50} of 2.95 µ*M*. The syntheses proceed through a similar intramolecular Michael reaction using ethyl 4-bromocrotonate instead of ethyl 3-bromo-propenyl-cyclohexylmethyl-phosphinate. However, contrary to the phosphinic acid derivatives the cyclization yielded *cis*- and *trans*-derivatives of which the *cis*-compounds displayed inferior affinities to $GABA_B$ receptors: *cis* (2*R*, 5*R*)-enantiomer CGP83721 (Fig. 6) with IC_{50} of 527 n*M*, and *cis* (2*S*, 5*S*)-enantiomer CGP 83720 with IC_{50} of 2.33 µ*M*. The corresponding racemic *trans*-thiomorpholine derivative was again marginally more active, i.e., CGP82066 with IC_{50} of 30 n*M*, and the more potent (2*S*, 5*R*) enantiomer, NVP-AAR108 with IC_{50} of 17 n*M*. In this series the racemic piperazine derivative CGP81007 (NH instead of O in the six-membered ring) showed an IC_{50} of only 262 n*M*.

Surprisingly, in the series of the oxazepines, i.e., homomorpholines, the *cis*-compounds displayed higher affinities than the trans compounds and the *p*-benzoic acids were more active than the meta-substituted compounds: racemic *m*-benzoic acid, *cis*, CGP83727, IC_{50} = 180 n*M*, *trans*: CGP83724; IC_{50} = 598 n*M*, racemic *p*-benzoic acid, *cis*, CGP83728, IC_{50} = 79 n*M*, *trans*: CGP83726; IC_{50} = 1.2 µ*M*. (Fig. 7).

4. Radioactive $GABA_B$-Receptor Antagonist Ligands

Several radioactive antagonist ligands have proven to be very useful for in vitro elucidations of different aspects of $GABA_B$ receptor research. Well known

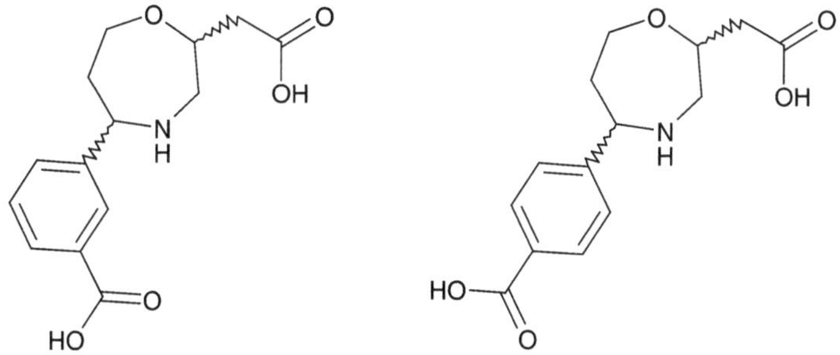

m-benzoic acid, *cis:* CGP83727
m-benzoic acid, *trans:* CGP 83724

p-benzoic acid, *cis:* CGP83728
p-benzoic acid, *trans:* CGP83726

Fig. 7. 5-Aryl-homomorpholine-2-acetic acids.

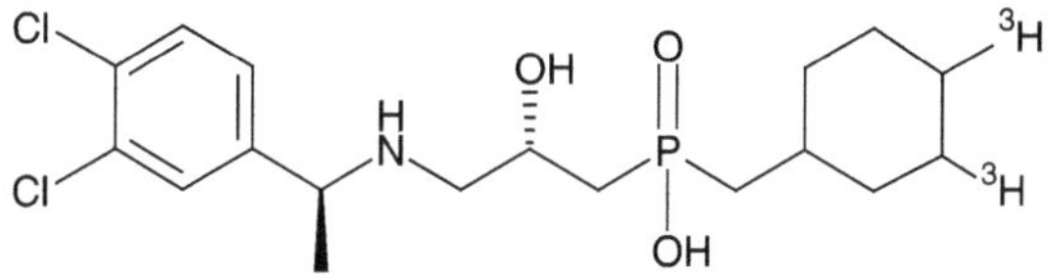

[^{3}H]CGP54626

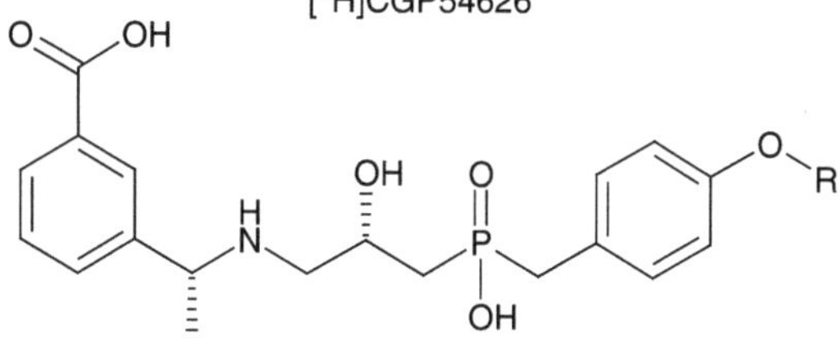

R = C^{3}H$_3$: [^{3}H]CGP62349
R = ^{11}CH$_3$: [^{11}C]CGP62349

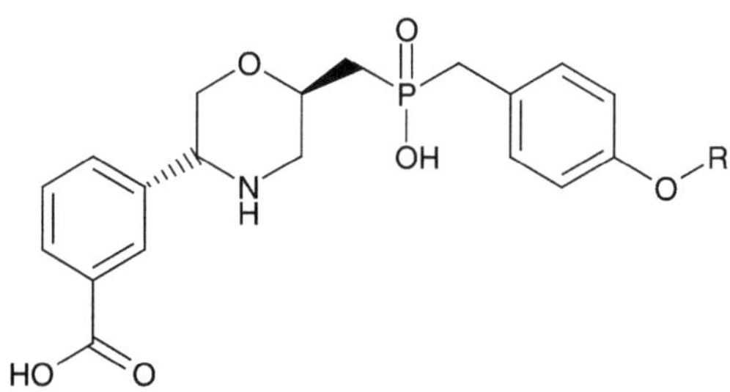

R = ^{11}CH$_3$: [^{11}C]NVP-AAV516

Fig. 8. Radioactive GABA$_B$-receptor antagonist ligands.

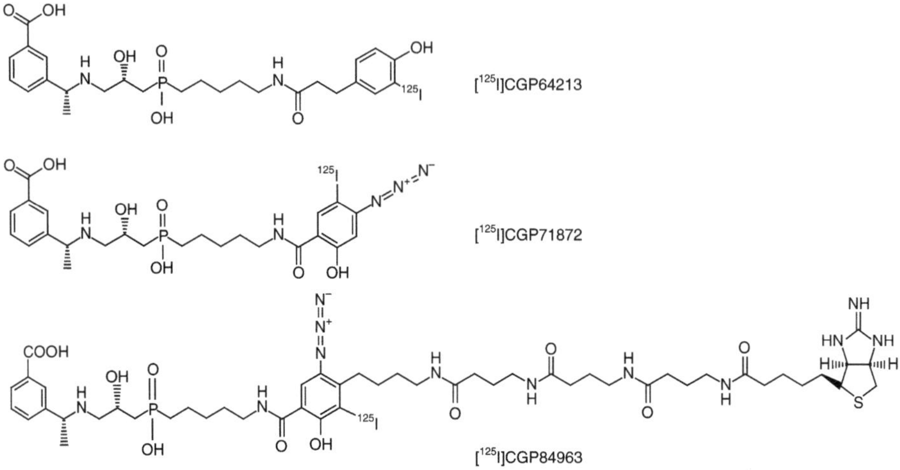

Fig. 9. Ligands for $GABA_B$ receptors with very high specific radioactivity.

are [^{3}H]CGP54626 (IC_{50} = 4 n*M*) *(42)* and [^{3}H]CGP62349 (IC_{50} = 2 n*M*) *(43–45)*. [^{11}C]CGP62349 containing the positron emitting isotope ^{11}C was prepared as a potential PET ligand *(46,47)*. A second potential PET ligand, i.e., [^{11}C]NVP-AAV516, was prepared by Dr. Catherine Aubert under Prof. Victor W. Pike at the Hammersmith Hospital in London (Fig. 8).

For the expression cloning experiments the iodinated ligands [^{125}I]CGP64213 (IC_{50} = 1.6 n*M*, i.e., inhibition of binding of [^{125}I]CGP64213 to $GABA_B$ receptors on rat cerebral cortex membranes), and the photoaffinity ligand [^{125}I]CGP71872 (IC_{50} = 2.4 n*M*), both with high specific radioactivity of >2000 Ci/mmol were developed, and used for the structure elucidation of the $GABA_{B1}$-receptor subunit (Fig. 9) *(48)*. The photoaffinity ligand [^{125}I]CGP71872 allowed the determination of the molecular weight of two $GABA_{B1a}$ and $GABA_{B1b}$ receptor isoforms, i.e., 130 and 100 kD, whereas [^{125}I]CGP64213 served as a means to identify the clone with the $GABA_B$ receptor after transfection into COS cells finally allowing the hitherto elusive structure elucidation of $GABA_B$ receptors. The syntheses of both ligands are described with full experimental details in the authors patent *(49)*. The synthesis of [^{125}I]CGP71872 has been described years later also by chemists of Merck *(50)*.

The ligand [^{125}I]CGP84963 (IC_{50} = 6 n*M*, i.e., inhibition of binding of [^{125}I]CGP64213 to $GABA_B$ receptors on rat cerebral cortex membranes) (Fig. 9), combines in one molecule a $GABA_B$ receptor-binding part, an azidosalicylic acid as photoaffinity moiety separated by a spacer of three GABA molecules from 2-iminobiotin, which binds to avidine in a reversible, pH-dependent fashion.

This compound was prepared to facilitate isolation and purification of the extracellular N-terminal GABA$_{B1}$-receptor fragment for crystallization and X-ray studies of the GABA$_{B1}$ receptor-binding site *(51)*. The 31-step synthesis of this compound is described in detail *(52)*.

Acknowledgments

We wish to express our gratitude to Drs. H. Allgeier, C. Angst, J. Carey, L. Diorazio, P. Gull, E. Irving, K. Kaupmann, M. Koller, K. Lingenhöhl, Ph. Stampf, S. Urwyler, Anne R. B., Williams and K. Zimmermann for the design of new compounds and their biological testing, respectively, and to S. W. L. Bennett, Herta Bieler, Nicola Blair, R. Bösch, M. Erb, M. Gunzenhauser, W. Gunzenhauser, J. Heid, C. Holder, Monika Horvath, Agnes Jeker, M. Kessler, Corinne Marx, Pascal Merklen, G. Pilgrim, Chantal Portet, and D. Strub for their skilful experimental work.

References

1. Froestl, W. and Mickel, S. J. (1997) Chemistry of GABA$_B$ modulators, in *The GABA Receptors* (Enna, S. J. and Bowery, N. G., eds.) Humana Press, Totowa, NJ, 271–296.
2. Baldoli, C., Maiorana, S., Licandro, E., Perdicchia, D., and Vandoni, B. (2000) Michael addition of nitromethane to non-racemic chiral $Cr(CO)_3$ complexes of ethyl cinnamate derivatives: stereoselective synthesis of (***R***)-(−)-baclofen. *Tetrahedron Asymmetry* **11,** 2007–2014.
3. Barnes, D. M., Ji, J., Fickes, M. G., et al. (2002) Development of a catalytic enantioselective conjugate addition of 1,3-dicarbonyl compounds to nitroalkenes for the synthesis of endothelin-A antagonist ABT-546. Scope, mechanism, and further application to the synthesis of the antidepressant Rolipram. *J. Am. Chem. Soc.* **124,** 13,097–13,105.
4. Brenna, E., Caraccia, N., Fuganti, C., Fuganti, D., and Grasselli, P. (2003) Enantioselective synthesis of β-substituted butyric acid derivatives via orthoester Claisen rearrangement of enzymatically resolved allylic alcohols: application to the synthesis of (*R*)-(−)-baclofen. *Tetrahedron Asymmetry* **8,** 3801–3805.
5. Chang, M. Y., Sun, P. P., Chen, S. T., and Chang, N. C. (2003) Synthesis of (±)-Rolipram. *Heterocycles* **60,** 1865–1872.
6. Corey, E. J. and Zhang, F. Y. (2000) Enantioselective Michael addition of nitromethane to α,β-enones catalyzed by chiral quarternary ammonium salts A simple synthesis of (*R*)-baclofen. *Org. Lett.* **2,** 4257–4259.
7. Doyle, M. P. and Hu, W. (2002) Enantioselective carbon-hydrogen insertion is an effective and efficient methodology for the synthesis of (*R*)-(−)-baclofen. *Chirality* **14,** 169–172.
8. Garcia, A. L., Carpes, M. J., de Oca, A. C., dos Santos, M. A., Santana, C. C., and Correia, C. R. (2005) Synthesis of 4-aryl-2-pyrrolidones and β-aryl-γ-amino-butyric acid (GABA) analogues by Heck arylation of 3-pyrrolines with arenediazonium

tetrafluoroborates. Synthesis of (±)-Rolipram on a multigram scale and chromatographic resolution by semipreparative chiral simulated moving bed chromatography. *J. Org. Chem.* **70,** 1050–1053.

9. Hayashi, M. and Ogasawara, K. A. (2003) diastereocontrolled route to (*R*)-(–)-baclofen using a cyclopentanoid chiral building block. *Heterocycles* **59**, 785–791.
10. Langlois, N. and Wang, H. S. (1997) Synthesis of the novel antidepressant (*R*)-(–)-Rolipram. *Synth. Commun.* **27,** 3133–3144.
11. Langlois, N., Dahuron, N., and Wang, H. S. (1996) Enantioselective syntheses of (*R*)-3-phenyl GABA, (*R*)-baclofen and 4-arylpyrrolidin-2-ones. *Tetrahedron* **52,** 15,117–15,126.
12. Licandro, E., Maiorana, S., Baldoli, C., Capella, L., and Perdicchia, D. (2000) Enantioselective synthesis of (*R*)-(–)-baclofen using Fischer-type carbene anions. *Tetrahedron Asymmetry* **11,** 975–980.
13. Mazzini, C., Lebreton, J., Alphand, V., and Furstoss, R. A chemoselective strategy for the synthesis of enantiopure (*R*)-(–)-baclofen. *Tetrahedron Lett.* **38,** 1195–1196.
14. Resende, P., Almeida, W. P., and Coelho, F. (1999) An efficient synthesis of (*R*)-(–)-baclofen. *Tetrahedron Asymmetry* **10,** 2113–2118.
15. Tehrani, M. H. H., Farnia, M., and Nazer, M. S. (2003) Synthesis of baclofen; an alternative approach. *Iranian J. Pharm. Res.* **2,** 1–3.
16. Thakur, V. V., Nikalje, M. D., and Sudalai, S. (2003) Enantioselective synthesis of (*R*)-(–)-baclofen via Ru(II)-BINAP catalyzed asymmetric hydrogenation. *Tetrahedron Asymmetry* **14,** 581–586.
17. Gallop, M. A., Yao, F., Ludwikow, M. J., Phan, T., and Peng, G. (2005) New acyloxyalkyl carbamate prodrugs of baclofen—useful for the prevention and treatment of e.g., spasticity, gastro-esophageal reflux disease, drug addiction, alcohol abuse, nicotine addiction, cough and emesis. WO2005/019163-A2, 2005; Publ. date: 20050303, Priority: US-932374, 20040813 for Xenoport Inc.
18. Patel, S., Naeem, S., Kesingland, A., et al. (2001) The effects of $GABA_B$ agonists and gabapentin on mechanical hyperalgesia in models of neuropathic and inflammatory pain in rats. *Pain* **90,** 217–226.
19. Colombo, G., Serra, S., Brunetti, G., et al. (2002) The $GABA_B$ receptor agonists baclofen and CGP 44532 prevent acquisition of alcohol drinking behaviour in alcohol-preferring rats. *Alcohol Alcoholism* **37,** 499–503.
20. Brebner, K., Froestl, W., Andrews, M., Phelan, R., and Roberts, D. C. (1999) The $GABA_B$ agonist CGP 44532 decreases cocaine self-administration in rats: demonstration using a progressive ratio and a discrete trials procedure. *Neuropharmacology* **38,** 1797–1804.
21. Dobrovitsky, V., Pimentel, P., Duarte, A., Froestl, W., Stellar, J. R., and Trzcinska, M. (2002) CGP44532, a $GABA_B$ receptor agonist, is hedonically neutral and reduces cocaine-induced enhancement of reward. *Neuropharmacology* **42,** 626–632.
22. Paterson, N. E., Froestl, W., and Markou, A. (2004) The $GABA_B$ receptor agonists baclofen and CGP44532 decrease nicotine self-administration in the rat. *Psychopharmacology* **172,** 179–186.
23. Paterson, N. E., Bruijnzeel, A. W., Kenny, P. J., Wright, C. D., Froestl, W., and Markou, A. (2005a) Prolonged nicotine exposure does not alter $GABA_B$ receptor-mediated regulation of brain reward function. *Neuropharmacology* **49,** 953–962.

24. Paterson, N. E., Froestl, W., and Markou, A. (2005b) Repeated administration of the $GABA_B$ receptor agonist CGP44532 decreased nicotine self-administration, and acute administration decreased cue-induced reinstatement of nicotine-seeking in rats. *Neuropsychopharmacology* **30,** 119–128.
25. Bauser, M., Krueger, J., Meier, H., et al. (2005) 5-substituted 2-(phenylmethyl) thio-4-phenyl-4H-1,2,4-triazole derivatives and related compounds as GABA-agonists for the treatment of urinary incontinence and related diseases. WO2005/039569 A1. 2005; Publ. date: 20050506, Priority: EP-23701, 20031018 for Bayer Healthcare AG.
26. Gijsen, H. J. M., Berthelot, D., De Cleyn, M., Surkyn, M., De Breucker, S., and Smans, K. (2004) Lessons learned from a HTS campaign on $GABA_B$ agonism. *Drugs Future* **29(Suppl. A),** 307.
27. Urwyler, S., Gjoni, T., Koljatic, J., and Dupuis, D. S. (2005) Mechanisms of allosteric modulation at $GABA_B$ receptors by CGP7930 and GS39783: effects on affinities and efficacies or orthosteric ligands with distinct intrinsic properties. *Neuropharmacology* **48,** 343–353.
28. Marshall, F. H. (2005) Is the $GABA_B$ heterodimer a good drug target? *J. Mol. Neurosci.* **26,** 169–176.
29. Urwyler, S., Mosbacher, J., Lingenhoehl, K., et al. (2001)Positive allosteric modulation of native and recombinant $GABA_B$ receptors by 2,6-di-tert.-butyl-4-(3-hydroxy-2,2-di-methyl-propyl)-phenol (CGP7930) and its aldehyde analogue CGP13501. *Mol. Pharmacol.* **60,** 963–971.
30. Carai, M. A. M., Colombo, G., Froestl, W., and Gessa, G. L. (2004) In vivo effectiveness of CGP7930, a positive allosteric modulator of the $GABA_B$ receptor. *Eur. J. Pharmacol.* **504,** 213–216.
31. Binet, V., Brajon, C., Le Corre, L., Acher, F., Pin, J. P., and Prézeau, L. (2004) The heptahelical domain of $GABA_{B2}$ is activated directly by CGP7930, a positive allosteric modulator of the $GABA_B$ receptor. *J. Biol. Chem.* **279,** 29,085–29,091.
32. Urwyler, S., Pozza, M. F., Lingenhoehl, K., et al. (2003) *N,N'*-dicyclopentyl-2-methylsulfanyl-5-nitro-pyrimidine-4,6-diamine (GS39783) and structurally related compounds: novel allosteric enhancers of γ-aminobutyric $acid_B$ receptor function. *J. Pharmacol. Exp. Ther.* **307,** 322–330.
33. Cryan, J. F., Kelly, P. H., Chaperon, F., et al. (2004) Behavioral characterization of the novel $GABA_B$ receptor-positive modulator GS39783 (*N,N'*-dicyclopentyl-2-methylsulfanyl-5-nitro-pyrimidine-4,6-diamine): anxiolytic-like activity without side effects associated with baclofen or benzodiazepines. *J. Pharmacol. Exp. Ther.* **310,** 952–963.
34. Mombereau, C., Kaupmann, K., Froestl, W., Sansig, G., van der Putten, H., and Cryan, J. F. (2004) Genetic and pharmacological evidence of a role for $GABA_B$ receptors in the modulation of anxiety- and antidepressant-like behavior. *Neuropsychopharmacology* **29,** 1050–1062.
35. Smith, M. A., Yancey, D. L., Morgan, D., Liu, Y., Froestl, W., and Roberts, D. C. (2004) Effects of positive allosteric modulators of the $GABA_B$ receptor on cocaine self-administration in rats. *Psychopharmacology* **173,** 105–111.
36. Slattery, D. A., Markou, A., Froestl, W., and Cryan, J. F. (2005) The $GABA_B$ receptor-positive modulator GS39783 and the $GABA_B$ receptor agonist baclofen

attenuate the reward-facilitating effects of cocaine: intercranial self-stimulation studies in the rat. *Neuropsychopharmacology* **30,** 2065–2072.

37. Malherbe, P., Masciadri, R., Prinssen, E., Spooren, W., and Thomas, A. W. (2005) Aminomethylpyrimidines as allosteric enhancers of the $GABA_B$ receptors. US 2005/0197337 A1 2005; Publ. date: 20050908, Priority: EP2004-100830, 20040302, for Hoffmann-La Roche Inc.
38. Guery, S., Floersheim, P., and Froestl, W. (2005) Syntheses and identification of new GS39783 analogs as positive allosteric modulators of $GABA_B$ receptors. Second Italian-Swiss Meeting on Medicinal Chemistry, Modena, Poster 104, Sept. 12–16, 2005.
39. Froestl, W., Mickel, S. J., Cooke, N. G., Bennett, S. N. L., and Williams, A. R. B. (1997) Substituted phosphinic compounds and their use as pharmaceuticals. WO 97/09335 A1 1997; Publ. date: 19970313, Priority: GB-013047, 19960621 for Ciba-Geigy AG.
40. Ong, J., Kerr, D. I. B., Bittiger, H., et al. (1998) Morpholin-2-yl-phosphinic acids are potent $GABA_B$ receptor antagonists in rat brain. *Eur. J. Pharmacol.* **362,** 27–34.
41. Cooke, N. G., Irving, E., and Froestl, W. (1998) (Thio)morpholine-substituted carboxylic and phosphinic acids. WO 98/28313, 1998; Publ. date: 19980702. Priority: 19961224 for Novartis AG.
42. Bittiger, H., Reymann, N., Froestl, W., and Mickel, S. J. (1992) ^{3}H-CGP54626: A potent antagonist radioligand for $GABA_B$ receptors. *Pharmacol. Commun.* **2,** 23.
43. Bittiger, H., Bellouin, C., Froestl, W., Heid, J., Schmutz, M., and Stampf, P. (1996) [^{3}H]CGP 62349: A new potent $GABA_B$ receptor antagonist radioligand. *Pharmacol. Rev. Commun.* **8,** 97–98.
44. Stampf, P., Froestl, W., Mickel, S. J., and Zimmermann, K. (1997) Novel hydroxylated and alkoxylated N- and/or P-aralkylphosphinic acid derivatives. WO 97/11953 A1 1997; Publ. date: 19970403, Priority: EP96/04062, 19950928 for Ciba-Geigy AG.
45. Ambardekar, A. V., Ilinsky, I. A., Froestl, W., Bowery, N. G., and Kultas-Ilinsky, K. (1999) Distribution and properties of $GABA_B$ antagonist [^{3}H]CGP62349 binding in the Rhesus monkey thalamus and basal ganglia and the influence of lesions in the reticular thalamic nucleus. *Neuroscience* **93,** 1339–1347.
46. Todde, S., Froestl, W., Stampf, P., et al. (1997) [^{11}C]CGP62349, a $GABA_B$ receptor antagonist ligand: radiosynthesis and PET studies in monkeys. *J. Labelled Comp. Radiopharm.* **40,** 651–652.
47. Todde, S., Moresco, R. M.,'Froestl, W., et al. (2000) Synthesis and in vivo evaluation of [^{11}C]CGP62349, a new $GABA_B$ receptor antagonist. *Nucl. Med. Biol.* **27,** 565–569.
48. Kaupmann, K., Huggel, K., Heid, J., et al. (1997a) Expression cloning of $GABA_B$ receptors uncovers similarity to metabotropic glutamate receptors. *Nature* **386,** 239–246.
49. Kaupmann, K., Bettler, B., Bittiger, H., Froestl, W., and Mickel, S. J. (1997b) Metabotropic $GABA_B$ receptors, receptor-specific ligands and their uses. WO 97/46675 1997b; Publ. date: 19971211, Priority: 19960530 for Novartis AG.
50. Belley, M., Sullivan, R., Reeves, A., Evans, J., O'Neill, G., and Ng, G. Y. K. (1999) Synthesis of the nanomolar photoaffinity GABAB receptor ligand CGP71872

reveals diversity in the tissue distribution of GABAB receptor forms. *Bioorg. Med. Chem.* **7,** 2697–2704.

51. Froestl, W., Bettler, B., Bittiger, H., et al. (1999) Ligands for the isolation of GABA$_B$ receptors. *Neuropharmacology* **38,** 1641–1646.
52. Froestl, W., Bettler, B., Bittiger, H., et al. Ligands for expression cloning and isolation of GABA$_B$ receptors. *Farmaco* **58,** 173–183.

10

The Unusual Functioning of the $GABA_B$-Receptor Heterodimer

An Old Receptor Teaching New Functional Tricks?

Andrés Couve, Stephen J. Moss, and Menelas N. Pangalos

Summary

Functional γ-aminobutyric acid ($GABA_B$) receptors are formed by the association of $GABA_BR1$ and $GABA_BR2$, two closely related subunits belonging to the G protein coupled receptor superfamily. An increasing number of reports have now uncovered that $GABA_B$ receptors display unique functional properties, including unusual activation and inactivation mechanisms following binding of agonists. In this chapter the focus will be on nonclassical aspects of $GABA_B$ receptor phosphorylation, desensitization, downregulation, signaling, and association with a variety of protein partners. In addition, the extent to which the heterodimeric structure of the receptor contributes to establish these unusual properties will be discussed.

Key Words: $GABA_B$; receptor signaling; phosphorylation; GPCR; heterodimer; signalosome.

1. Introduction

γ-Aminobutyric acid (GABA) receptors are the major inhibitory receptors in the central nervous system (CNS). Whereas $GABA_A$- and $GABA_C$ ionotropic receptors mediate fast synaptic inhibition, metabotropic $GABA_B$ receptors induce slower, more prolonged effects through the activation of an array of downstream effector systems, such as inhibition of adenylyl-cyclase, inhibition of presynaptic calcium channels, and activation of postsynaptic potassium channels. The far-reaching effects of $GABA_B$ receptors coupled with their widespread

From: *The Receptors: The GABA Receptors, Third Edition*
Edited by: S. J. Enna and H. Möhler © Humana Press Inc., Totowa, NJ

distribution in the CNS have raised expectations that modulation of this receptor class will be of clinical importance for pharmaceutical intervention in a number of disease areas such as epilepsy, depression, anxiety, drug addiction, and cognitive enhancement for patients affected with neurodegenerative conditions such as Alzheimer's disease (for review *see* refs. *1,2*). Since the original discovery of the $GABA_B$-binding site over 20 yr back *(3)*, much work has gone into identifying the molecular components mediating $GABA_B$ receptor-mediated function and physiology. In this chapter the most recent work investigating the molecular underpinnings of $GABA_B$ receptor function will be discussed and the unusual nature of this receptor complex will be highlighted.

2. $GABA_B$ Receptor Phosphorylation

2.1. *$GABA_B$ Receptors Exist as Phosphoproteins in the CNS*

Phosphorylation of neurotransmitter receptors constitutes a widely used post-translational modification that produces rapid changes in receptor activity. In this regard, it is extensively used to promote long-term changes in the CNS, many of which are associated with modulation of synaptic plasticity *(4)*. Even before the cloning of the individual $GABA_B$-receptor subunits $GABA_BR1$ and $GABA_BR2$ protein kinases (PK), such as PKC and cyclic AMP (cAMP)-dependent PKA, were shown to modulate the activity of $GABA_B$ receptors *(5–7)*. Following the molecular identification of the $GABA_B$-receptor heterodimer it was confirmed that, like many other G protein-coupled receptors (GPCRs), the two subunits of the receptor are phosphorylated.

Both $GABA_BR1$- and $GABA_BR2$-subunits are phosphorylated basally when overexpressed in HEK-293 or COS cells *(8,9)*. No reports on the details of $GABA_BR1$ phosphorylation have yet been published, but the $GABA_BR2$-subunit has been investigated in depth. Studies have conclusively shown that the carboxyl terminal domain of $GABA_BR2$ is phosphorylated in recombinant systems, preparations of brain membranes, and in primary cultures of cortical and hippocampal neurons *(10)*. The principal phosphorylated residue appears to be serine 892 (S892) and PKA is responsible for phosphorylation of this site both in vitro and in vivo. Phosphorylation is increased by agents that stimulate cAMP production, such as forskolin and β-adrenergic agonists, and is blocked by specific inhibitors of PKA *(10)*. Results suggest that additional phosphorylated residues are also located on the carboxyl terminal domains of $GABA_BR1$ and $GABA_BR2$. In addition, recent evidence suggests that phosphorylation sites may be shared with other kinases (Couve, A, Moss, SJ, Pangalos, MN, personal communication).

To date, however, $GABA_BR2$-S892 remains the best characterized site for $GABA_B$ receptor phosphorylation. GPCR phosphorylation is generally regarded as an early requirement for receptor desensitization and internalization during the inactivation process *(11)*. In the next section, the unusual desensitization properties of $GABA_B$ receptors in relation to the functional consequences of receptor phosphorylation will be examined.

3. Desensitization, Endocytosis, and Downregulation of $GABA_B$ Receptors

3.1. *The Inactivation of GPCRs is a Conserved Process*

The majority of GPCRs respond to agonists in the same manner. The cycle of activation was first characterized for the β-adrenergic receptor, but can now be applied to GPCRs of all three families (A–C) *(11)*. Briefly, agonist binding produces a conformational change in the receptor that promotes rapid phosphorylation by second messenger kinases such as PKA, PKC, or specific G protein receptor kinases (GRKs). Phosphorylation occurs in either the carboxyl terminal domain or the intracellular loops of the receptor and generally results in uncoupling of the receptor from the heterotrimeric G protein. This relatively fast process is generally referred to as desensitization and occurs in a matter of seconds. Receptors are then recruited to clathrin-coated vesicles in a phosphorylation-dependent or -independent manner, and internalized in the order of minutes in a phenomenon known as receptor endocytosis (Fig. 1). Finally, receptors are sorted for recycling or degraded in a process that may take hours or even days called receptor downregulation. Recently, the classical hypothesis of desensitization and endocytosis of GPCRs has been expanded to accommodate a growing body of evidence that suggests a variety of alternative inactivation responses, including phosphorylation independent mechanisms, proteolytic cleavage, and agonist specific desensitization *(11–13)*.

3.2. *The "Unusual" Desensitization of $GABA_B$ Receptors*

3.2.1. *$GABA_B$ Receptors Acutely Desensitize After Agonist Exposure*

To address whether $GABA_B$ receptors follow the classical desensitization response on agonist exposure, the coupling of the receptor to inwardly rectifying K^+ channels (GIRKs) has been studied during continuous exposure to the $GABA_B$ agonist baclofen *(14)*. As occurs with other GPCRs, the activity of the receptor is attenuated in the order of seconds, thus providing strong evidence that $GABA_B$ receptors desensitize normally. These findings are supported by the desensitization of recombinant $GABA_B$ receptors in transfected cells following pretreatment with GABA *(15)*. Desensitization of native receptors has also been observed in neurons. Several reports have suggested that pre- and postsynaptic $GABA_B$ receptors display different desensitization properties. For example, whereas postsynaptic $GABA_B$ receptors that couple to GIRKs in hippocampal neurons desensitize rapidly after agonist treatment, presynaptic receptors appear more resistant and fail to desensitize even after 24 h of agonist pretreatment *(16)*.

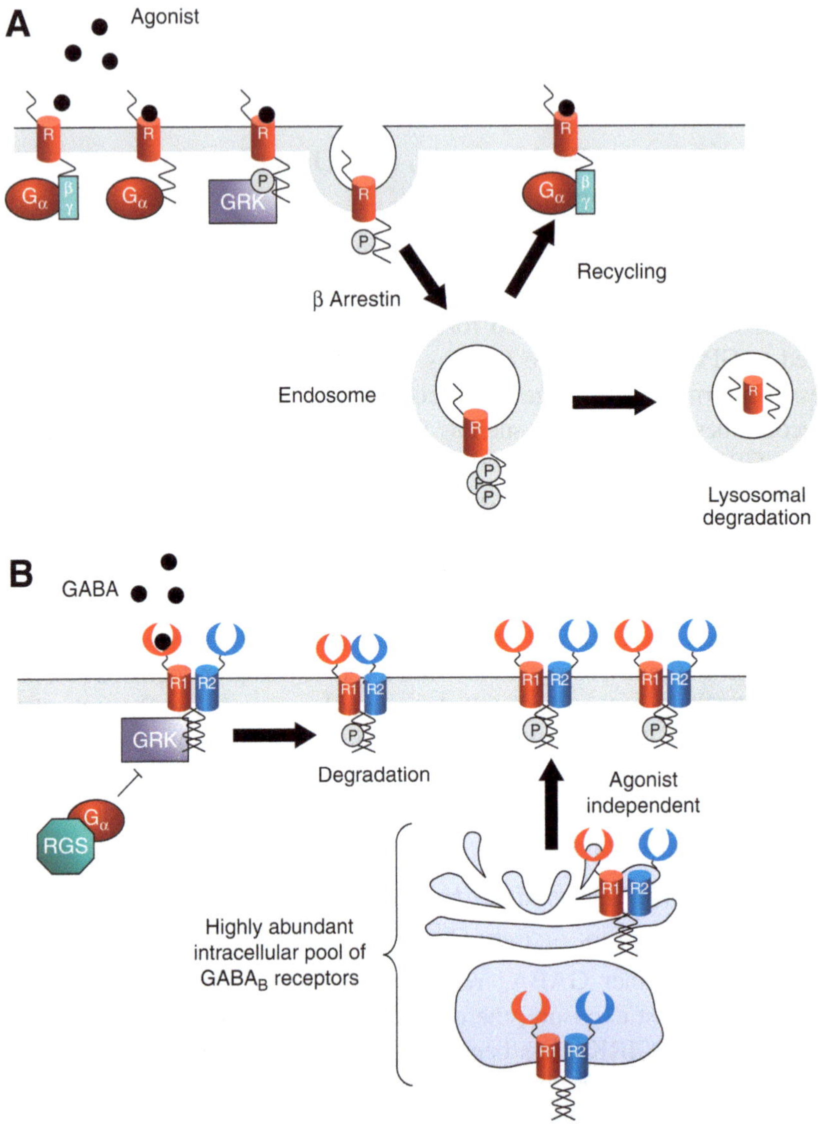

Fig. 1. The mechanism of $GABA_B$-receptor inactivation is highly unusual. **(A)** *The classical model of GPCR-inactivation*: Agonist binding produces a conformational change in the receptor that promotes rapid phosphorylation by PK. Phosphorylation results in the uncoupling of the receptor from the heterotrimeric G protein. Receptors are then recruited to clathrin-coated pits in a phosphorylation dependent or independent manner and internalized through clathrin-coated vesicles. Finally, receptors are sorted for recycling to the plasma membrane or degradation in lysosomes. **(B)** *$GABA_B$-receptor inactivation diverts from the classical model*: Agonist binding produces rapid desensitization of $GABA_B$ receptors but no detectable phosphorylation or internalization

Similarly, slow (0–15 min) desensitization of GABA$_B$ receptors in response to continuous baclofen exposure has been observed in dopaminergic neurons of the ventral tegmental area, but fast desensitization (≤5 min) has been registered in the soma of pyramidal cells of the neocortex *(17,18)*.

Further evidence supporting the existence of a variety of desensitization responses, but conflicting with the results described earlier, comes from a study suggesting the existence of acute presynaptic GABA$_B$ desensitization but a complete lack of postsynaptic desensitization in neonatal rat hippocampus *(19)*. Moreover, this study reports presynaptic desensitization to be more pronounced for autoreceptors compared with heteroreceptors. The desensitization kinetics of the GABA$_B$ receptors coupled with GIRK channels has also been studied in rat pyramidal neurons and appears to be made up of multiple phases *(17)*. The mechanisms controlling each phase of desensitization appear related to the cycle of G protein coupling, nucleotide exchange, or GTP hydrolysis, and not to intrinsic GPCR, or GIRK channel properties. Nevertheless, additional studies will be needed to fully elucidate this phenomenon.

The existing evidence indicates that GABA$_B$ receptor desensitization depends not only on the subcellular localization of the receptor subunits but also on the effector systems being used. It also implies that a unique combination of accessory molecules, associated with the receptor heterodimer is responsible for modulating receptor desensitization in specific cellular and subcellular domains.

3.2.2. Phosphorylation Does Not Contribute to Acute Desensitization

Given the clear existence of GABA$_B$ receptor desensitization, a number of recent studies have addressed the functional consequences of receptor phosphorylation in this regard. Unexpectedly, two reports found that neither GABA$_B$R1- nor GABA$_B$R2-subunits are phosphorylated in response to agonist *(8,9)*. These findings have been further explored by analyzing the effect of PK during desensitization of GABA$_B$ receptors after continuous exposure to baclofen *(14)*. As mentioned previously, coupling of GABA$_B$ receptors to GIRKs is rapidly desensitized in transfected cell systems. However, the kinetics and degree of GABA$_B$ receptor desensitization were unchanged by PKA activation, the best-characterized GABA$_B$ receptor kinase *(14)*. These results provide preliminary evidence that phosphorylation of GABA$_B$ receptors is not a regulator of acute receptor desensitization. However, the influence of other kinases on receptor

---

Fig. 1. *(Continued)* is observed. Desensitization may be mediated by GRK4 or RGS proteins, which may interfere with the function of G proteins. Slow degradation of plasma membrane receptors occurs in response to agonist but the total pool of receptors remains constant, probably through constitutive synthesis and a large intracellular receptor reserve pool. Contrary to other GPCRs, phosphorylation of GABA$_B$ receptors results in potentiation of receptor function.

desensitization need to be further investigated before the effects of receptor phosphorylation in this particular phenomenon can be ruled out.

3.2.3. *GRK4 Promotes Phosphorylation-Independent Desensitization*

The lack of receptor phosphorylation in response to agonists has been reported for recombinant $GABA_BR1$ and $GABA_BR2$ in COS and HEK-293 cells *(9,20)*. Taking advantage of the presence of $GABA_B$ receptors in cerebellar granule cells, a recent study demonstrated that $GABA_B$ receptors also desensitize rapidly in this cell type *(20)*. By depleting GRK4, through small interfering RNA, the authors have shown that the presence of GRK4 is *necessary* to promote agonist-dependent desensitization *(20)*. Similarly, they have presented a *sufficiency* argument by demonstrating that in HEK-293 cells addition of exogenous GRK4 produces desensitization properties that are remarkably similar to the ones observed in cerebellar granule neurons *(20)*. These findings provide a plausible desensitization mechanism for $GABA_B$ receptors. However, two points must be carefully considered and further explored. First, $GABA_B$ receptors are widely expressed in the CNS, whereas GRK4 has a limited distribution in the brain, potentially limiting the physiological relevance of this interaction. Second, the effect of GRK4 does not require its catalytic domain to promote the desensitization response *(20)*. At first glance this latter finding seems puzzling. However, $GABA_B$ receptors are now placed in a continuously growing list of GPCRs that are modulated by GRKs in a phosphorylation-independent manner *(12)*. Although more evidence is required to confirm these findings, the authors have identified the protein–protein interaction as the key element for the GRK4-mediated receptor desensitization, likely through blockade of the receptor and G protein interaction *(12,20)*.

3.2.4. *Regulators of G Protein Signaling Also Promotes $GABA_B$-Receptor Desensitization*

From the data reviewed earlier, it is clear that novel regulatory mechanisms are responsible for controlling $GABA_B$ receptor desensitization. One further report has provided evidence of an alternative mechanism. Working on $GABA_BR1$/$GABA_BR2$-transfected cells, Mutneja and coworkers have reported that acute $GABA_B$ receptor desensitization depends on the G_α protein used in the assays. A G_α-subunit with an altered regulators of G protein signaling (RGS)-binding motif was able to modify the kinetics of receptor desensitization, suggesting that endogenous RGS proteins may be able to modulate $GABA_B$ receptor desensitization *(14)*. Although intriguing, direct evidence for an interaction between RGS proteins $GABA_B$ receptors remains to be demonstrated and warrants further investigation.

3.2.5. *Receptor Phosphorylation Prevents Inactivation*

The role of second messenger kinases has been explored relative to other aspects of the attenuation response. First, the effect of PKA phosphorylation has been studied during GIRK activation after repetitive pulses of agonist. In these studies, the coupling of GABA$_B$ receptors to GIRKs decreases gradually during the experimental time period *(10,17)*. After monitoring GIRK function for 30 min in the absence or presence of cAMP, it was demonstrated that the decrease in receptor activity was significantly inhibited by cAMP, in a manner dependent on the S892 residue of GABA$_B$R2. The same effect was obtained when the experiment was performed in cultured hippocampal neurons or HEK cells transfected with GABA$_B$R1, GABA$_B$R2, and GIRKs *(10)*. These findings suggest that although receptor phosphorylation does not affect acute desensitization, the activity of GABA$_B$ receptors is stimulated by PKA phosphorylation during the later phase of receptor inactivation. Similar stimulatory rather than inhibitory effects of PKA have been reported for metabotropic glutamate receptors *(21)*, suggesting this related family of GPCRs display a conserved mechanism of potentiation following PKA phosphorylation.

Whereas cAMP potentiation during repetitive agonist application was observed after 20 min *(10)*, no potentiation by cAMP was obtained during continuous exposure to agonist (single pulse of 60 s) *(14)*. Although these observations seem conflicting, they relate to very different phases of the response and suggest that GABA$_B$ receptors have early and late components of functional attenuation. Even though much progress has been made, these studies also highlight the need for a systematic and extensive analysis of the effects of kinases and G protein modulators on the desensitization of GABA$_B$ receptors. Analyses should consider the cell types (cell lines, neuronal cultures, acute slices), exposure times to drugs (s/min/h), and the mode of drug application (repetitive/continuous). Data generated from such studies can be directly compared and contrasted, and will allow us to better understand the mechanisms involved in regulating GABA$_B$ receptor desensitization.

3.2.6. *GABA$_B$ Receptors Do Not Internalize in Response to Agonist*

In addition to the unusual findings that agonists fail to stimulate GABA$_B$ receptor phosphorylation, it has recently been shown that GABA$_B$ receptors fail to undergo agonist-mediated endocytosis (Fig. 1). These results have been obtained using antibody internalization and enzyme-linked immunosorbent assay-type assays in cells expressing recombinant receptors *(8,9,14)*. Similar results have been obtained with endogenous receptors in cultured cortical and hippocampal neurons employing antibody internalization and cell-surface biotinylation assays *(9)*. Results from these studies converge on the fact that although desensitization of GABA$_B$ receptors occurs normally, receptor internalization does not contribute to the attenuation of signaling. Whether endocytosis takes place in cerebellar granule cells where GRK4-dependent desensitization is evident remains to be determined. However, the lack of endocytosis in differentiated cortical and hippocampal neurons in culture, two cell types that express functionally competent $GABA_B$ receptors, make this seem unlikely.

Despite the apparent lack of receptor endocytosis following agonist stimulation, decreased levels of cell surface receptors have been reported *(10,15)*. Measurements in these studies have used live cells before fixation and point a note of caution to any general conclusions suggesting that $GABA_B$ receptors lack the ability to internalize.

3.2.7. $GABA_B$-Receptor Downregulation

The process of downregulation is defined as the loss of total number of receptors induced by agonist after hours or even days of exposure *(11)*. Interestingly, long-term treatment of cortical neurons with baclofen resulted in a significant decrease of cell surface receptors *(9)*. Given the lack of agonist-induced phosphorylation and internalization, it is possible that $GABA_B$ receptors may be directly degraded at the plasma membrane. In contrast, long-term treatments with agonist had little effect on the total number of receptors. The simplest explanation for these observations is that the population of intracellular receptors is much more abundant than the cell surface pool and, additionally, not directly sensitive to agonist (Fig. 1). Thus, the results are consistent with agonist dependent removal and degradation of surface receptors, but not with agonist-independent recycling of the entire receptor population. Therefore, they suggest an independence of $GABA_B$ receptor activity and overall abundance. These findings are also compatible with an accelerated rate of synthesis in the presence of agonist, but experiments are needed to explore this possibility. Combined, these experiments indicate that $GABA_B$ receptors are resilient to chronic downregulation *(9)*.

The presence of an unusually stable population of $GABA_B$ receptors is supported by clinical and experimental evidence that demonstrates the existence of a functionally stable pool of $GABA_B$ receptors. For example, the response to baclofen over long periods of time after intrathecal application is stable in patients treated for spasticity and dystonia, and there is no continual escalation in dose required to maintain the response *(22–25)*. Furthermore, although rats develop tolerance after repetitive injections of baclofen, the total messenger RNA and protein levels for $GABA_B$ receptors remain unchanged, suggesting that tolerance is not produced by downregulation *(26,27)*.

The experiments reported so far establish a new receptor model in which the levels of $GABA_B$ receptors are not correlated with receptor activity. Combined, they rule out the role of phosphorylation in agonist-dependent receptor desensitization and internalization, two trademarks of classical GPCR signaling. In this respect, it remains to be established how receptors that fail to internalize, are removed from the plasma membrane and replaced by *de novo* synthesized or recycled receptors and are amenable to new rounds of activation.

4. $GABA_B$-Associated Proteins and Their Influence on Receptor Function

Intense efforts have been directed toward elucidating the role of associated proteins in the biosynthesis and function of the $GABA_B$ receptor. Among the many proteins described to date the concentration will be only on those molecular interactions whose initial characterization have been further explored in terms of receptor function (Fig. 2). For additional information on protein–protein associations a series of recent reviews provide a comprehensive view on the subject *(1,2,28–30)*.

4.1. *Interaction Between $GABA_BR1$ and $GABA_BR2$ Regulates the Subcellular Distribution of the Receptor*

The interaction between the two subunits of the $GABA_B$ heterodimer is essential to constitute a fully functional receptor. In this regard, the molecular interactions between the domains of the two subunits play a crucial role in cellular signaling and receptor activation. A general model has been proposed in which the $GABA_BR1$-subunit binds a single agonist molecule whereas the $GABA_BR2$-subunit mediates G protein coupling *(31,32)*. Both intra- and intermolecular interactions among the subunits are necessary to mediate this highly unusual mechanism of transactivation *(28,32,33)*.

Apart from the direct implications in receptor activation, dimerization between the two subunits seems to have profound consequences in the biosynthesis and subcellular distribution of $GABA_B$ receptors. It was noted early that knockout mice lacking $GABA_BR1$ show a substantial decrease in the levels of $GABA_BR2$ protein, but retain normal levels of the $GABA_BR2$ transcript *(34,35)*. Similarly, two reports now show that $GABA_BR2$ mutant animals show a marked decrease in the protein levels of $GABA_BR1$ *(36,37)*. Two interpretations of these findings are possible. Either the synthesis of the two subunits is intrinsically bound, or alternatively, the stability and degradation of the receptor subunits are strongly dependent on the formation of the heterodimer. This latter alternative seems more plausible in view of the tight control of misfolded proteins in the endoplasmic reticulum (ER) and the extensive interactions between the subunits needed to constitute a fully functional, and therefore a correctly folded, receptor heterodimer.

Together with the substantial changes in protein expression, one report suggests that $GABA_BR1$ is incorrectly localized to the soma and proximal neurites in the absence of the $GABA_BR2$-subunit *(37)*. However, given the dramatic decrease in $GABA_BR1$ protein, this conclusion seems somewhat premature. Nevertheless, it opens the interesting possibility that targeting signals exist within the $GABA_BR2$ sequence and implies that the correct placement of functional $GABA_B$ receptors along with distal projections may depend on retention sequences in one subunit and targeting sequences in the other.

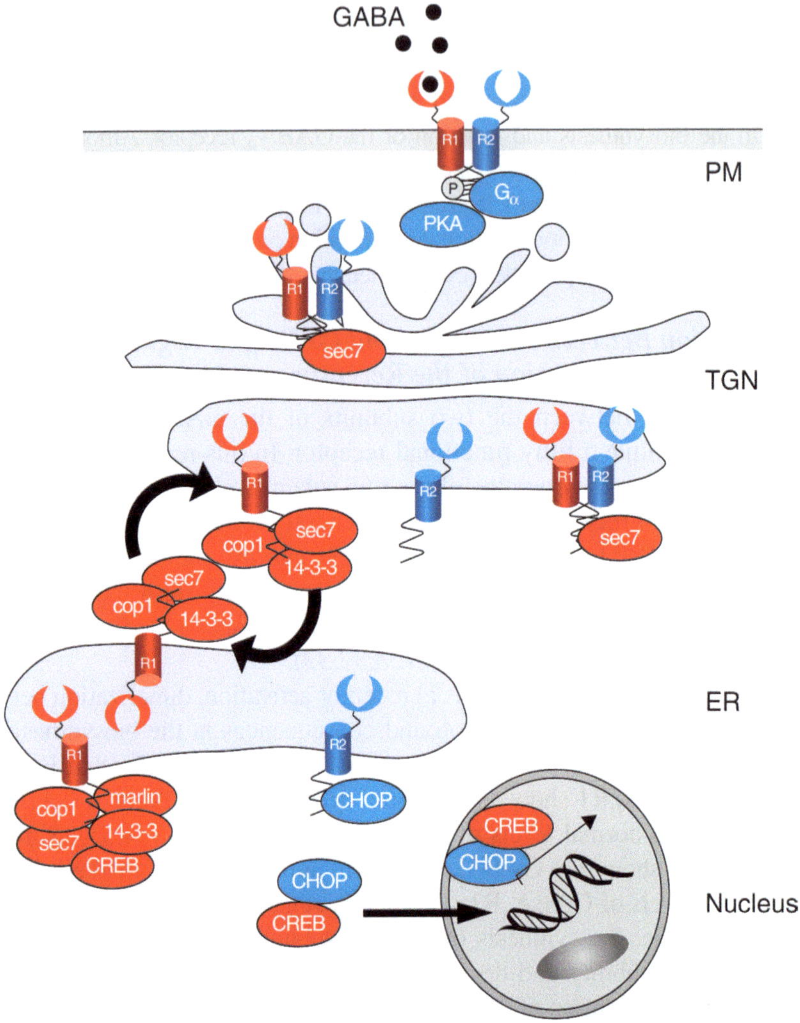

Fig. 2. $GABA_B$ receptors are modulated by a variety of associated proteins. $GABA_BR1$ interacts with CREB2/ATF4, 14-3-3, Marlin-1, COPI and msec7 whereas $GABA_BR2$ interacts with CHOP and PKA. Many of these associations are displaced by the formation of the $GABA_B$ heterodimer. This indicates that many of the interactions probably occur in the ER where receptors are found in monomeric form. It is possible that CREB2/ATF4 and CHOP are released from individual subunits and shuttle to the nucleus to participate in the modulation of the expression of $GABA_B$-subunit isoforms. COPI and 14-3-3 have been implicated in the retention or retrieval of $GABA_BR1$-subunits from ER. The interaction between $GABA_BR1$, Marlin-1, and later $GABA_BR2$ may ensure the correct assembly and stability of the heterodimer as it is transported through the secretory pathway. Finally, msec7 modulates the forward transport of the receptor heterodimer and at the plasma membrane the $GABA_BR2$-subunit associates with $G_{\alpha i}$ to constitute a signaling receptor.

4.2. Interaction With G Proteins

4.2.1. $GABA_BR2$ Mediates the Interaction to G Proteins

Several reports have demonstrated that the two subunits of the $GABA_B$ heterodimer are not functionally equivalent *(31,38–40)*. Based on these studies a general model of activation had been proposed in which the $GABA_BR1$-subunit was responsible for agonist binding, whereas the second and third intracellular loops of $GABA_BR2$ were directly responsible for G protein coupling *(31,32)*. A strong argument in support of this hypothesis has now been provided in using receptor-G protein fusions. The recombinant tool of G proteins fused to their corresponding receptors has been used widely to explore the signaling and coupling properties of specific GPCR pathways *(41)*. Using this methodology it has been demonstrated that G protein activation occurs only when it is directly fused to the carboxyl terminus of the $GABA_BR2$-subunit *(42)*. These results were obtained with $G_{\alpha i2}$ and chimeric $G_{\alpha qi5}$ proteins, confirming the role of $G_{\alpha i2}$ in coupling with $GABA_B$ receptors *(43)*.

4.3. Interaction With Multiprotein Complexes

4.3.1. The Identification of Multiprotein Complexes Containing $GABA_B$ Receptors Has Remained Elusive

Large multiprotein complexes have been shown to mediate signaling through GPCRs *(44)*. Unexpectedly, there is no evidence to date of large protein complexes containing $GABA_B$ receptors. This lack of information raises questions as to the existence of the macromolecular complex as an essential component of $GABA_B$ receptor signaling. This is particularly surprising when one considers that macromolecular complexes have been described for classic effector systems of $GABA_B$ receptors such as the voltage-dependent Ca^{2+} channel. In this particular instance, Ca^{2+} channels have been shown to function as adaptor proteins for multiple molecules that bind the channel and signal to downstream components following $GABA_B$-receptor activation *(45)*.

Possible reasons for the difficulty in identifying macromolecular complexes associated with $GABA_B$ receptors may lie within the receptor heterodimer itself. Most likely, the formation of a compact dimer between $GABA_BR1$ and $GABA_BR2$ results in the exposure of novel molecular surfaces for protein interactions, probably absent in the individual subunits and, therefore, likely to be missed in yeast two-hybrid screens (or similar screens) probing for interactions with individual $GABA_B$-subunits. In addition, only a limited number of receptor domains have been used to search for specific interacting proteins. Therefore, the search for components of a macromolecular complex involved in the direct modulation of $GABA_B$ receptor signaling may require the use of new biochemical approaches using native $GABA_B$-receptor heterodimers.

4.3.2. $GABA_BR1$ Associates With a Variety of Transcription Factors

4.3.2.1. ATF4/CREB2

Although no macromolecular $GABA_B$ containing complex has yet been described, a number of studies have identified proteins interacting specifically with $GABA_B$-receptor subunits (Fig. 2). Of particular interest are the findings that at least two transcription factors have been shown to interact with both $GABA_BR1$ and $GABA_BR2$. Several reports have shown that the $GABA_BR1$-subunit associates with the transcription factor ATF4/CREB2 *(46–48)*. These studies have highlighted a direct interaction between the $GABA_B$ receptor and the transcription factor, mediated by the coiled-coil region of the receptor and the leucine-zipper domains present in the carboxyl terminus of ATF4/CREB2. It remains unclear whether a tripartite interaction between $GABA_BR1$, $GABA_BR2$, and ATF4/CREB2 is possible. However, recent studies using immunoprecipitation from mouse brain extracts, have demonstrated that ATF4/CREB2 is able to associate with both $GABA_BR1$- and $GABA_BR2$-subunits. These findings suggest that the heterodimeric receptor interacts with the transcription factor, providing evidence for an interaction with a physiologically functional $GABA_B$ receptor *(49)*.

The interaction between ATF4/CREB2 and the $GABA_BR$ does not affect activity of the receptor, although reports do suggest that translocation of the transcription factor to and from the nucleus can be regulated by $GABA_B$ agonists *(46,47)*. A recent study has shed further light onto the physiological implications of this protein–protein interaction. The two principal isoforms of $GABA_BR1$ ($GABA_BR1a$ and $GABA_BR1b$) are transcribed from alternative promoter regions *(50)*, rather than by alternative splicing, as has been previously suggested *(51)*. The authors argue that expression of the two $GABA_BR$ splice variants, from distinct transcriptional start sites is controlled by binding of ATF4/CREB and upstream stimulatory factor to the specific $GABA_BR1a$ or $GABA_BR1b$ promoter regions. Given that expression of specific transcription factors varies between cell types, the authors propose that expression of $GABA_BR1a$ and $GABA_BR1b$ in post- or presynaptic sites, may be controlled by the interaction of cell-type specific transcription factors with the appropriate $GABA_BR1$ subtype specific promoter *(50)*. Under such circumstances it is conceivable that functional $GABA_B$ receptors are able to modulate their cell specific gene expression by interacting with specific transcription factors.

4.3.3. $GABA_BR2$ Associates With a Variety of Transcription Factors: CCAAT/Enhancer-Binding Protein Homologous Protein

Another transcription factor, namely CCAAT/enhancer-binding protein (C/EBP) homologous protein (CHOP), has been found to associate with $GABA_BR2$ *(52)*. As with the $GABA_BR1$-ATF4/CREB2 interaction, $GABA_BR2$ and CHOP interact through leucine-zipper domains located within the carboxyl termini of each protein. Immunoprecipitation experiments have also shown that CHOP is capable of binding to the GABA$_B$R1a-subunit, however, in contrast to GABA$_B$R2 this interaction does not require the leucine-zipper domain. Surprisingly, although binding to CHOP occurs in a region conserved between GABA$_B$R1a and GABA$_B$R1b, GABA$_B$R1b is unable to bind CHOP. The significance of the GABA$_B$R interaction with CHOP is not fully understood. Recombinant expression of GABA$_B$R1, GABA$_B$R2, and CHOP in culture systems results in a reduced expression of GABA$_B$ receptors at the cell surface. However, further work is clearly required to delineate the role of this interaction in neurons and other physiologically expressing GABA$_B$ cells.

4.3.4. 14-3-3, COPI, and msec7 Regulate GABA$_B$ Receptor Trafficking

The first proteins identified as GABA$_B$ receptor interactors belong to the 14-3-3 family of proteins *(53)*. Recently, these associations have been explored further and, in what is becoming a general trend for GABA$_B$ interacting proteins, it appears that they do not affect the functional coupling of GABA$_B$ receptors to their G proteins *(54)*. However, 14-3-3s do bind within the coiled-coil region of GABA$_B$R1 and binding is abolished when a mutant subunit is used in which the RXR-type retention sequence in the carboxyl terminal domain of GABA$_B$R1 is mutated (<u>R</u>S<u>R</u>R to <u>A</u>S<u>A</u>R). Interestingly, the same sequence in GABA$_B$R1 binds αCOP and βCOP, two protein components of the COPI complex involved in the retrieval of molecules from the trans-Golgi network (TGN) to the ER. These experiments suggest that by binding to the retention/retrieval sequence in GABA$_B$R1, 14-3-3, and COPI proteins may control the retention of GABA$_B$R1 in the ER or in the very early stages of the Golgi apparatus. It is important to note that experimental data to date suggests that these proteins do not regulate the kinetics of GABA$_B$ receptor "heterodimer" cell surface trafficking *(54)*.

Further evidence for the carefully controlled trafficking of GABA$_B$ receptors has been obtained following the recent uncovering of an association between GABA$_B$R1 and msec7. Msec7 is the mammalian ortholog of the yeast sec7, a family of small GTPase implicated in vesicular membrane transport between the ER, TGN, and endosomes *(55)*. Msec7 contains a conserved coiled-coil domain at the amino terminus and modulates the level of cell surface receptor in a manner that is dependent on a dileucine motif found in the carboxyl terminal domain of GABA$_B$R1. These results suggest that, contrary to 14-3-3 and COPI proteins, msec7 modulates the forward trafficking of GABA$_B$R from the ER to the TGN and on to the cell surface *(56)*.

4.3.5. A GABA$_B$/GABA$_A$ Interaction Regulates GABA$_B$ Trafficking

Unexpectedly, a further protein interaction with a GABA$_A$-receptor subunit, GABA$_A$ γ_2S, has been reported, which results in a 10–20-fold increase in the surface expression of $GABA_BR1$. Not surprisingly, the $GABA_B/GABA_A$ multimer does not couple with known effectors of the $GABA_B$ receptor and the presence of the γ_2S-subunit does not affect the potency of GABA at $GABA_BR1/GABA_BR2$ receptors *(57)*. The trafficking of the $GABA_BR1$-subunit by unrelated proteins with no functional consequences had been reported previously *(58)*, however, there is an aspect of this $GABA_B/GABA_A$ association that makes this a potentially intriguing interaction. Although it is now generally believed that $GABA_B$ receptors fail to internalize in response to agonists, in the presence of the $GABA_A$ γ_2S-subunit, there is a clear internalization of the $GABA_BR1$-subunit after 30 min of $GABA_B$ agonist exposure *(57)*. In view of these findings it is surprising that no internalization of $GABA_B$ receptors is observed in more physiological experimental systems such as cortical neurons, where the expression of the γ_2S-subunit is abundant. As such, the influence of ionotropic receptors on the function of the corresponding metabotropic receptors remains far from clear and will require further investigation and confirmation.

4.3.6. Marlin-1 and Receptor Assembly

Using a yeast two-hybrid screen the authors have recently identified Marlin-1, a brain specific protein that associates specifically with the $GABA_BR1$-subunit. Marlin-1 belongs to a new protein family represented by three different genes, highly conserved in vertebrates. It encodes a 626 amino acid protein with a predicted MW of about 75 kDa and displays a granular distribution in neurons, with abundant expression in the soma and proximal dendrites. Marlin-1 binds different RNA species in vitro, including the 3′ untranslated region of the messenger RNAs for $GABA_BR1$ and $GABA_BR2$. It sediments with ribonuclear markers in brain preparations and forms discrete granules that stain with RNA dyes in the soma and dendrites of cultured primary neurons. Depletion of Marlin-1 through small interfering RNA technology increases the abundance of $GABA_BR2$, providing evidence that a receptor-associated protein can modulate $GABA_B$ receptor biogenesis at the level of RNA metabolism *(1,2)*. In addition to its role in binding $GABA_B$ receptors, a complementary study has shown that recombinant Marlin-1 associates with Janus kinases and polymerized microtubules when transfected into Jurkat cells *(59)*. Combined, these results suggest that Marlin-1 may associate with cytoskeletal elements and participate in the transport of cargo vesicles or cargo particles affecting the abundance of proteins at their target destination *(59)*.

5. Conclusions

When compared with the majority of GPCRs, the $GABA_B$ receptor displays unique properties of activation and inactivation. For example, the transactivation of the receptor following agonist binding is a direct result of the heterodimeric structure of the receptor. Whether this structure also influences the lack of receptor phosphorylation, the lack of receptor endocytosis, and the lack of receptor downregulation remain to be established. However, it is clear that the unusual sequence of events that occur during receptor activation is mirrored during receptor inactivation, and can be described, not without some controversy, as unusual.

All screens to identify interacting proteins have thus far failed to isolate proteins that directly modulate the activity of the receptor, and have instead identified proteins involved in the regulation of receptor biosynthesis, trafficking, and localization. The majority of interactions involving GABA$_B$ receptors occur through coiled-coil and leucine-zipper domains that appear to regulate binding to the carboxyl terminal domains of the receptor. Nevertheless, it is still necessary to elucidate if the proteins identified thus far bind simultaneously to the receptor, whether there is a cellular or subcellular preference for specific associations, or whether low affinity interactions can facilitate a sequential set of interactions facilitating the progress of the receptor through the secretory pathway on its way to the neuronal plasma membrane.

The existence of numerous protein interactions modulating GABA$_B$R trafficking, suggest an ER to membrane pathway with multiple and strict checkpoints along the route, a finding not dissimilar to the protein–protein interactions reported for the ionotropic glutamate receptors *(60)*. Ultimately, the complex and delicate control of cell surface expression of this multisubunit GABA$_B$ receptor, by a variety of protein–protein interactions, is not surprising given the essential role these receptors play in synaptic neurotransmission *(61,62)*.

Acknowledgments

AC funded by FONDECYT 1040083, Iniciativa Científico Milenio P04-068-F, FONDAP 15010006.

References

1. Couve, A., Calver, A. R., Fairfax, B., Moss, S. J., and Pangalos, M. N. (2004) Unravelling the unusual signalling properties of the GABA(B) receptor. *Biochem. Pharmacol.* **68(8),** 1527–1536.
2. Couve, A., Restituito, S., Brandon, J. M., et al. (2004) Marlin-1, a novel RNA-binding protein associates with GABA receptors. *J. Biol. Chem.* **279(14),** 13,934–13,943.
3. Bowery, N. G., Price, G. W., Hudson, A. L., Hill, D. R., Wilkin, G. P., and Turnbull, M. J. (1984) GABA receptor multiplicity. Visualization of different receptor types in the mammalian CNS. *Neuropharmacology* **23(2B),** 219–231.
4. Sweatt, J. D. (2001) Memory mechanisms: the yin and yang of protein phosphorylation. *Curr. Biol.* **11(10),** R391–R394.
5. Taniyama, K., Niwa, M., Kataoka, Y., and Yamashita, K. (1992) Activation of protein kinase C suppresses the gamma-aminobutyric acidB receptor-mediated

inhibition of the vesicular release of noradrenaline and acetylcholine. *J. Neurochem.* **58(4),** 1239–1245.

6. Yoshimura, M., Yoshida, S., and Taniyama, K. (1995) Desensitization by cyclic AMP-dependent protein kinase of $GABA_B$ receptor expressed in Xenopus oocytes. *Life Sci.* **57,** 2397–2401.
7. Kamatchi, G. L. and Ticku, M. J. (1990) Functional coupling of presynaptic $GABA_B$ receptors with voltage-gated Ca^{2+} channel: regulation by protein kinase A and C in cultured spinal neurons. *Mol. Pharmacol.* **38,** 342–347.
8. Perroy, J., Adam, L., Qanbar, R., Chenier, S., and Bouvier, M. (2003) Phosphorylation-independent desensitization of GABA(B) receptor by GRK4. *EMBO J.* **22(15),** 3816–3824.
9. Fairfax, B. P., Pitcher, J. A., Scott, M. G., et al. (2004) Phosphorylation and chronic agonist treatment atypically modulate GABAB receptor cell surface stability. *J. Biol. Chem.* **279(13),** 12,565–12,573.
10. Couve, A., Thomas, P., Calver, A. R., et al. (2002) Cyclic AMP-dependent protein kinase phosphorylation facilitates GABA(B) receptor-effector coupling. *Nat. Neurosci.* **5(5),** 415–424.
11. Tsao, P. I. and von Zastrow, M. (2001) Diversity and specificity in the regulated endocytic membrane trafficking of G-protein-coupled receptors. *Pharmacol. Ther.* **89(2),** 139–147.
12. Pao, C. S. and Benovic, J. L. (2002) Phosphorylation-independent desensitization of G protein-coupled receptors? *Sci. STKE* **2002(153),** PE42.
13. Perez, D. M. and Karnik, S. S. (2005) Multiple signaling states of G-protein-coupled receptors. *Pharmacol. Rev.* **57(2),** 147–161.
14. Mutneja, M., Berton, F., Suen, K. F., Luscher, C., and Slesinger, P. A. (2005) Endogenous RGS proteins enhance acute desensitization of GABA(B) receptor-activated GIRK currents in HEK-293T cells. *Pflugers Arch.* **450(1),** 61–73.
15. Gonzalez-Maeso, J., Wise, A., Green, A., and Koenig, J. A. (2003) Agonist-induced desensitization and endocytosis of heterodimeric GABAB receptors in CHO-K1 cells. *Eur. J. Pharmacol.* **481(1),** 15–23.
16. Wetherington, J. P. and Lambert, N. A. (2002) Differential desensitization of responses mediated by presynaptic and postsynaptic A1 adenosine receptors. *J. Neurosci.* **22(4),** 1248–1255.
17. Sickmann, T. and Alzheimer, C. (2003) Short-term desensitization of G-protein-activated, inwardly rectifying K+ (GIRK) currents in pyramidal neurons of rat neocortex. *J. Neurophysiol.* **90(4),** 2494–2503.
18. Cruz, H. G., Ivanova, T., Lunn, M. L., Stoffel, M., Slesinger, P. A., and Luscher, C. (2004) Bi-directional effects of GABA(B) receptor agonists on the mesolimbic dopamine system. *Nat. Neurosci.* **7(2),** 153–159.
19. Tosetti, P., Bakels, R., Colin-Le Brun, I., Ferrand, N., Gaiarsa, J. L., and Caillard, O. (2004) Acute desensitization of presynaptic GABAB-mediated inhibition and induction of epileptiform discharges in the neonatal rat hippocampus. *Eur. J. Neurosci.* **19(12),** 3227–3234.
20. Perroy, J., Adam, L., Qanbar, R., Chenier, S., and Bouvier, M. (2003) Phosphorylation-independent desensitization of $GABA_B$ receptor by GRK4. *EMBO J.* **22(15),** 3816–3824.

21. Mundell, S. J., Pula, G., More, J. C., Jane, D. E., Roberts, P. J., and Kelly, E. (2004) Activation of cyclic AMP-dependent protein kinase inhibits the desensitization and internalization of metabotropic glutamate receptors 1a and 1b. *Mol. Pharmacol.* **65(6),** 1507–1516.
22. Meythaler, J. M. (1999) Use of intrathecally delivered medications for spasticity and dystonia in acquired brain injury, in *Spinal drug delivery* (Yaksh, T. L., eds.), Elsevier, New York, NY, 513–554.
23. Meythaler, J. M., Guin-Renfroe, S., Law, C., Grabb, P., and Elsevier Hadley, M. N. (2001) Continuously infused intrathecal baclofen over 12 months for spastic hypertonia in adolescents and adults with cerebral palsy. *Arch. Phys. Med. Rehabil.* **82,** 155–161.
24. Coffey, R. J., Cahill, D., Steers, W., et al. (1993) Intrathecal baclofen for intractable spasticity of spinal origin: results of a long-term multicenter study. *J. Neurosurg.* **78,** 226–232.
25. Armstrong, R. W., Steinbok, P., Cochrane, D. D., Kube, S. D., Fife, S. E., and Farrell, K. (1997) Intrathecally administered baclofen for treatment of children with spasticity of cerebral origin. *J. Neurosurg.* **87,** 409–414.
26. Lehmann, A., Mattsson, J. P., Edlund, A., Johansson, T., and Ekstrand, A. J. (2003) Effects of repeated administration of baclofen to rats on GABAB receptor binding sites and subunit expression in the brain. *Neurochem. Res.* **28,** 387–393.
27. Sands, S. A., McCarson, K. E., and Enna, S. J. (2003) Differential regulation of $GABA_B$ receptor subunit expression and function. *J. Pharmacol. Exp. Ther.* **305,** 191–196.
28. Bettler, B., Kaupmann, K., Mosbacher. J., and Gassmann, M. (2004) Molecular structure and physiological functions of GABA(B) receptors. *Physiol Rev.* **84(3),** 835–867.
29. El Far, O. and Betz, H. (2002) G-protein-coupled receptors for neurotransmitter amino acids: C-terminal tails, crowded signalosomes. *Biochem. J.* **365(Pt 2),** 329–336.
30. Calver, A. R., Davies, C. H., and Pangalos, M. (2002) GABA(B) receptors: from monogamy to promiscuity. *Neurosignals* **11(6),** 299–314.
31. Robbins, M. J., Calver, A. R., Filippov, A. K., et al. (2001) GABA(B2) is essential for g-protein coupling of the GABA(B) receptor heterodimer. *J. Neurosci.* **21(20),** 8043–8052.
32. Pin, J. P., Kniazeff, J., Binet, V., et al. (2004b) Activation mechanism of the heterodimeric GABA(B) receptor. *Biochem. Pharmacol.* **68(8),** 1565–1572.
33. Pin, J. P., Kniazeff, J., Goudet, C., et al. (2004a) The activation mechanism of class-C G-protein coupled receptors. *Biol. Cell.* **96(5),** 335–342.
34. Prosser, H. M., Gill, C. H., Hirst, W. D., et al. (2001) Epileptogenesis and enhanced prepulse inhibition in GABA(B1)-deficient mice. *Mol. Cell. Neurosci.* **17(6),** 1059–1070.
35. Schuler, V., Lüscher, C., Blanchet, C., et al. (2001) Epilepsy, hyperalgesia, impaired memory, and loss of pre- and postsynaptic GABA(B) responses in mice lacking GABA(B1). *Neuron* **31,** 47–58.
36. Thuault, S. J., Brown, J. T., Sheardown, S. A., et al. (2004). The $GABA_B2$ subunit is critical for the trafficking and function of native $GABA_B$ receptors. *Biochem. Pharmacol.* **68(8),** 1655–1666.

37. Gassmann, M., Shaban, H., Vigot, R., et al. (2004) Redistribution of GABAB(1) protein and atypical GABAB responses in GABAB(2)-deficient mice. *J. Neurosci.* **24(27),** 6086–6097.
38. Galvez, T., Prezeau, L., Milioti, G., et al. (2000) Mapping the agonist-binding site of GABAB type 1 subunit sheds light on the activation process of GABAB receptors. *J. Biol. Chem.* **275(52),** 41,166–41,174.
39. Kniazeff, J., Galvez, T., Labesse, G., and Pin, J. P. (2002) No ligand binding in the GB2 subunit of the GABA(B) receptor is required for activation and allosteric interaction between the subunits. *J. Neurosci.* **22(17),** 7352–7361.
40. Havlickova, M., Prezeau, L., Duthey, B., Bettler, B., Pin, J. P., and Blahos, J. (2002) The intracellular loops of the GB2 subunit are crucial for G-protein coupling of the heteromeric gamma-aminobutyrate B receptor. *Mol. Pharmacol.* **62(2),** 343–350.
41. Milligan, G. (2002) The use of receptor G-protein fusion proteins for the study of ligand activity. *Receptors Channels* **8(5–6),** 309–317.
42. Uezono, Y., Kanaide, M., Kaibara, M., et al. (2006) Coupling of the GABAB receptor GABAB2 subunit to G proteins: Evidence from the Xenopus oocyte and baby kidney hamster cell expression system. *Am. J. Physiol. Cell. Physiol.* **290,** 200–207.
43. Franek, M., Pagano, A., Kaupmann, K., Bettler, B., Pin, J. P., and Blahos, J. (1999) The heteromeric GABA-B receptor recognizes G-protein alpha subunit C-termini. *Neuropharmacology* **38(11),** 1657–1666.
44. Ciruela, F., Canela, L., Burgueno, J., et al. (2005) Heptaspanning membrane receptors and cytoskeletal/scaffolding proteins: focus on adenosine, dopamine, and metabotropic glutamate receptor function. *J. Mol. Neurosci.* **26(2-3),** 277–292.
45. Richman, R. W., Tombler, E., Lau, K. K., et al. (2004) N-type $Ca2^+$ channels as scaffold proteins in the assembly of signaling molecules for GABAB receptor effects. *J. Biol. Chem.* **279(23),** 24,649–24,658.
46. Vernon, E., Meyer, G., Pickard. L., et al. (2001) GABA(B) receptors couple directly to the transcription factor ATF4. *Mol. Cell. Neurosci.* **17,** 637–645.
47. White, J. H., McIllhinney, R. A., Wise, A., et al. (2000) The $GABA_B$ receptor interacts directly with the related transcription factors CREB2 and ATFx. *Proc. Natl. Acad. Sci. USA.* **97,** 13,967–13,972.
48. Nehring, R. B., Horikawa, H. P., El Far, O., et al. (2000) The metabotropic $GABA_B$ receptor directly interacts with the activating transcription factor 4. *J. Biol. Chem.* **275,** 35,185–35,191.
49. Ritter, B., Zschuntsch, J., Kvachnina, E., Zhang, W., and Ponimaskin, E. G. (2004) The GABA(B) receptor subunits R1 and R2 interact differentially with the activation transcription factor ATF4 in mouse brain during the postnatal development. *Brain Res. Dev. Brain Res.* **149(1),** 73–77.
50. Steiger, J. L., Bandyopadhyay, S., Farb, D. H., and Russek, S. J. (2004) cAMP response element-binding protein, activating transcription factor-4, and upstream stimulatory factor differentially control hippocampal GABABR1a and GABABR1b subunit gene expression through alternative promoters. *J. Neurosci.* **24(27),** 6115–6126.
51. Kaupmann, K., Huggel, K., Heid, J., et al. (1997) Expression cloning of $GABA_B$ receptors uncovers similarity to metabotropic glutamate receptors. *Nature* **386,** 239–246.

52. Sauter, K., Grampp, T., Fritschy, J. M., et al. (2005) Subtype-selective interaction with the transcription factor CCAAT/enhancer-binding protein (C/EBP) homologous protein (CHOP) regulates cell surface expression of GABA(B) receptors. *J. Biol. Chem.* **280(39),** 33,566–33,572.
53. Couve, A., Kittler, J. T., Uren, J. M., et al. (2001) Association of GABA(B) receptors and members of the 14-3-3 family of signaling proteins. *Mol. Cell. Neurosci.* **17(2),** 317–328.
54. Brock, C., Boudier, L., Maurel, D., Blahos, J., and Pin, J. P. (2005) Assembly-dependent Surface Targeting of the Heterodimeric GABAB Receptor Is Controlled by COPI but Not 14-3-3. *Mol. Biol. Cell.* **16(12),** 5572–5578.
55. Jackson, C. L. and Casanova, J. E. (2000) Turning on ARF: the Sec7 family of guanine-nucleotide-exchange factors. *Trends. Cell. Biol.* **10,** 60–67.
56. Restituito, S., Couve, A., Bawagan, H., et al. (2005) Multiple motifs regulate the trafficking of GABA(B) receptors at distinct checkpoints within the secretory pathway. *Mol. Cell. Neurosci.* **28(4),** 747–756.
57. Balasubramanian, S., Teissere, J. A., Raju, D. V., and Hall, R. A. (2004) Hetero-oligomerization between GABAA and GABAB receptors regulates GABAB receptor trafficking. *J. Biol. Chem.* **279(18),** 18,840–18,850.
58. Sullivan, R., Chateauneuf, A., Coulombe, N., et al. (2000) Co-expression of full-length gamma-aminobutyric acid(B) (GABAB) receptors with truncated receptors and metabotropic glutamate receptor 4 supports the GABA(B) heterodimer as the functional receptor. *J. Pharmacol. Exp. Ther.* **293(2),** 460–467.
59. Steindler, C., Li, Z., Algarte, M., et al. (2004) Jamip1 (marlin-1) defines a family of proteins interacting with janus kinases and microtubules. *J. Biol. Chem.* **279(41),** 43,168–43,177.
60. Henley, J. M. (2003) Proteins interactions implicated in AMPA receptor trafficking: a clear destination and an improving route map. *Neurosci. Res.* **45(3),** 243–254.
61. Ferguson, S. S. (2001) Evolving concepts in G protein-coupled receptor endocytosis: the role in receptor desensitization and signaling. *Pharmacol. Rev.* **53(1),** 1–24.
62. Vazquez-Prado, J., Casas-Gonzalez, P., and Garcia-Sainz, J. A. (2003) G protein-coupled receptor cross-talk: pivotal roles of protein phosphorylation and protein-protein interactions. *Cell Signal* **15(6),** 549–557.

11

Characteristics of $GABA_B$ Receptor Mutant Mice

Jim Yu-Hsiang Tiao and Bernhard Bettler

Summary

γ-Aminobutyric acid B ($GABA_B$) receptors are the G protein-coupled receptors for the inhibitory neurotransmitter GABA. Cloning efforts revealed that functional $GABA_B$ receptors can be assembled from the products of the $GABA_{B1}$ and $GABA_{B2}$ genes. However, cloned receptors do not exhibit the reported functional and pharmacological diversity of native receptors, which led to speculations about the existence of additional $GABA_B$ receptor genes. Several groups now have generated mice with ablations in the $GABA_{B1}$ and $GABA_{B2}$ genes, which allowed studying the contributions of cloned gene products to pre- and postsynaptic $GABA_B$ functions. In this chapter, we review the phenotypes of $GABA_B$ knockout mice and discuss how genetic experiments have contributed to our current understanding of the $GABA_B$ receptor system.

Key Words: $GABA_B$; knockout; epilepsy; hyperalgesia; hyperlocomotor activity; anxiety; γ-hydroxy butyrate, GHB.

1. Introduction

Metabotropic γ-aminobutyric acid ($GABA_B$) receptors were first identified based on the receptor's distinct pharmacological profile compared with ionotropic $GABA_{A/C}$ receptors *(1)*. It was subsequently shown that $GABA_B$ receptors are G protein coupled receptors (GPCRs) that inhibit adenylyl cyclase through the $G_{\alpha i/o}$-subunits of the activated G protein *(2)*. It is now well-established that $GABA_B$ receptors also regulate Ca^{2+}- and K^+-channels at pre- and postsynaptic sites through the $G_{\beta\gamma}$-subunits *(3–5)*. Presynaptic $GABA_B$ receptors are subdivided into auto- and heteroreceptors that control the release of GABA and other neurotransmitters,

From: *The Receptors: The GABA Receptors, Third Edition*
Edited by: S. J. Enna and H. Möhler © Humana Press Inc., Totowa, NJ

respectively. Stimulation of presynaptic $GABA_B$ receptors suppresses neurotransmitter release by inhibition of voltage-sensitive Ca^{2+} channels or by a direct modulation of synaptic vesicle priming *(6,7)*. Postsynaptic $GABA_B$ receptors induce a late inhibitory postsynaptic current by activating Kir3-type K^+-channels, which hyperpolarizes the neurons and shunts excitatory neurotransmission *(8)*. The physiological consequences of inhibiting adenylyl cyclase activity through $GABA_B$ receptors are still poorly understood, but include effects on transcription *(9)* and phosphorylation *(10)*. Considerable evidence has accumulated over the years, using a variety of preparations and techniques, to support the notion that pharmacologically distinct subtypes of $GABA_B$ receptors exist *(3,11–16)*.

Molecular identification of $GABA_B$ receptors did not occur until the high-affinity antagonist [^{125}I]CGP64213 allowed expression cloning of the $GABA_{B1a}$-subunit using a radioligand-binding assay *(17)*. Subsequently, $GABA_{B1b}$, a subunit variant with an alternative transcription initiation site was identified. $GABA_{B1b}$ differs from $GABA_{B1a}$ by the absence of a tandem pair of sushi-repeats at the extreme amino-terminus. Using the $GABA_{B1a}$ sequence, a third subunit, $GABA_{B2}$, which derives from a separate gene was identified *(18–21)*. $GABA_{B1}$- and $GABA_{B2}$-subunits need to heterodimerize in order to form a fully functional receptor in heterologous cells. Dimerization of the two subunits masks an endoplasmatic reticulum retention signal in the $GABA_{B1}$-subunit, which triggers surface trafficking of the assembled receptor complex *(22,23)*. Recombinant experiments support that interaction of $GABA_{B1}$ with $GABA_{B2}$ is not only mandatory for cell surface expression but also for G protein coupling *(22–29)*. The demonstration that $GABA_B$ receptors exist as heterodimers triggered the identification of additional heterodimeric GPCRs *(30)*. Although not generally accepted yet, recent data on rhodopsin imply that all GPCRs need to homo- or heterodimerize for function *(31)*.

The search for $GABA_B$-receptor subtypes did not lead to the expected identification of $GABA_B$ genes other than $GABA_{B1}$ and $GABA_{B2}$. Therefore, $GABA_{B1a}$, $GABA_{B1b}$, and $GABA_{B2}$ remain the only abundant receptor subunits, with additional complementary DNA isoforms reported for $GABA_{B1}$ but not $GABA_{B2}$ *(4,32)*. Importantly, however, the existence of a stable $GABA_{B1}$ protein product has only been demonstrated for $GABA_{B1a}$ and $GABA_{B1b}$ but not for the other reported $GABA_{B1}$ isoforms. Recombinant heteromeric $GABA_{B(1a,2)}$ and $GABA_{B(1b,2)}$ receptors couple to all prominent effector systems of native $GABA_B$ receptors—that is adenylyl cyclase, Kir3-type K^+-channels, and P/Q and N-type Ca^{2+}-channels *(21,33,34)*. However, no unique pharmacological or functional properties could be assigned to $GABA_{B1a}$ or $GABA_{B1b}$. The apparent lack of molecular and pharmacological heterogeneity in the $GABA_B$ system was a surprise to many in the field who expected a variety of pharmacologically distinct subunits, as predicted from work on native receptors *(3,11–16)*. Given that cloning efforts did not substantiate the claim for receptor heterogeneity, it became important to understand to which $GABA_B$ functions the cloned receptors can contribute in vivo. This prompted several groups to generate $GABA_{B1}$- and $GABA_{B2}$-deficient mice, referred to as $GABA_{B1}^{-/-}$ and $GABA_{B2}^{-/-}$ mice. The diverse biochemical, pharmacological, electrophysiological, and behavioral abnormalities observed in $GABA_B$ knockout mice are discussed later. Furthermore, we cover pharmacological experiments with $GABA_B$ knockout mice that helped clarify the relationship between $GABA_B$ receptors and the putative receptors for γ-hydroxy butyrate (GHB), an emerging drug of abuse *(35)*.

2. $GABA_{B1}^{-/-}$ Mice

Several mouse strains with a loss of function mutation in the gene encoding $GABA_{B1}$-subunits have been reported *(36–38)*. These mouse strains differ both in the genetic background and the introduced gene deletions (targeted exons: 7–10 *[36]*, 2–6 *[37]*, 1a1–1a/b *[38]*). Only the mice generated in the Balb/c genetic background are viable and show a normal life expectancy *(36)*. $GABA_{B1}^{-/-}$ mice generated in the C57B1/6J or C57B16/129Sv genetic background do not survive past 4–8 wk after birth, presumably because of generalized epileptic seizures *(37,38)*. Consequently, behavioral data are only available for the mice generated in the Balb/c genetic background.

Biochemical and pharmacological analysis of $GABA_{B1}^{-/-}$ mice emphasizes the importance of the $GABA_{B1}$-subunit in the formation of functional $GABA_B$ receptors. $GABA_{B2}$ protein in $GABA_{B1}^{-/-}$ mice is almost undetectable in immunoblots of brain membrane extracts, suggesting that the $GABA_{B1}$ protein stabilizes $GABA_{B2}$ protein and prevents its degradation. In agreement with previous biochemical studies *(39)*, this indicates that virtually all $GABA_{B2}$ protein is associated with $GABA_{B1}$ protein. It is now well established that the $GABA_{B1}$-subunit contains the binding site for GABA *(40)*. Consistent with this, competitive $GABA_B$ radioligand antagonists ($[^{125}I]$CGP71872, $[^{125}I]$CGP64213, $[^{3}H]$CGP54626) and agonists ($[^{3}H]$APPA, $[^{3}H]$baclofen) fail to detect binding sites in $GABA_{B1}^{-/-}$ tissue samples *(36)*. $[^{35}S]$GTPγS binding assesses the activation of $G_{\alpha i/o}$-type G proteins, the main effectors of native $GABA_B$ receptors, and therefore provides a biochemical assay to determine functional receptor levels *(41)*. $GABA_{B1}^{-/-}$ brain membranes show no $[^{35}S]$GTPγS binding, regardless of whether baclofen or the endogenous agonist GABA is used for stimulation *(36)*. This indicates that most if not all $GABA_B$ receptors in the brain incorporate a $GABA_{B1}$-subunit.

Electrophysiological analysis of $GABA_{B1}^{-/-}$ mice revealed a loss of typical baclofen-induced $GABA_B$ responses in the hippocampus *(36,37)*. At presynaptic sites, both $GABA_B$ hetero- and autoreceptors are absent in $GABA_{B1}^{-/-}$ mice. Specifically, baclofen no longer depresses excitatory postsynaptic currents or inhibitory postsynaptic currents in CA1 pyramidal neurons through the activation of presynaptic $GABA_B$ receptors *(36)*. $GABA_{B1}^{-/-}$ mice are also devoid of postsynaptic $GABA_B$ receptors in the CA1 pyramidal neurons, as baclofen no longer induces outward K^+-currents mediated by Kir3-type channels *(36,37)*. Electrophysiological $GABA_B$ responses in $GABA_{B1}^{-/-}$ mice are also absent in the dorsal raphe, the arcuate nucleus, and cerebellar granule cells *(37)*. The $GABA_{B1}$-subunit is also critical for $GABA_B$ function in the enteric and peripheral nervous system *(42)*. Taken together, electrophysiological analysis of $GABA_B$ functions in $GABA_{B1}^{-/-}$ mice are consistent with the complete lack of functional $GABA_B$ receptors at both pre- and postsynaptic sites in the nervous system.

The absence of a functional $GABA_B$ receptor in $GABA_{B1}^{-/-}$ mice is corroborated by the lack of baclofen-induced physiological responses. Baclofen is a clinically effective sedative and muscle-relaxant that is well known to induce hypothermia *(43)*. The muscle-relaxant and hypothermic effects of baclofen are completely absent in $GABA_{B1}^{-/-}$ mice *(38,44)*. Analysis of the basal body temperature showed that $GABA_{B1}^{-/-}$ mice are hypothermic by approx 1°C when compared with wild-type mice *(38,44)*. This is likely caused by the loss of $GABA_B$ function in the hypothalamus, where $GABA_B$ receptors are involved in the control of body temperature *(45)*. Consistent with a complete lack of $GABA_B$ activity, injection of baclofen failed to induce the typical δ-wave pattern in the electroencephalogram (EEG) of $GABA_{B1}^{-/-}$ mice *(36)*.

$GABA_{B1}^{-/-}$ mice exhibit overt and characteristic behavioral abnormalities; one such phenotype being the presence of spontaneous epileptic seizures. C57B1/6J $GABA_{B1}^{-/-}$ mice appear normal at birth with spontaneous epileptiform activity observed from postnatal day 12 (P12) onwards *(37)*. The mice develop generalized epilepsies that result in growth retardation and premature death. To gain insight into the cellular mechanisms contributing to the generation and maintenance of this epileptic phenotype, Brown and colleagues *(46)* studied hippocampal slices obtained from $GABA_{B1}^{-/-}$ mice. They conclude that the loss of postsynaptic $GABA_B$ receptor-mediated inhibition enhances *N*-methyl-D-aspartate receptor triggered synaptic processes, which could be the cause of the observed epileptiform activity. Adult Balb/c $GABA_{B1}^{-/-}$ mice exhibit a variety of spontaneous epileptiform activities, including absence type seizures, clonic, and tonic–clonic convulsions *(36)*. Moreover, Balb/c $GABA_{B1}^{-/-}$ mice are very susceptible to audiogenic stimulation, which precipitates tonic–clonic convulsions *(36)*. Absence-type seizures in Balb/c $GABA_{B1}^{-/-}$ mice are "atypical" in that seizures are characterized by rare and long-lasting EEG bursts. Immunohistochemical analysis of seizure-sensitive calcium-binding proteins suggests that the absence of functional $GABA_B$ receptors in Balb/c mice causes epileptiform activity through a mechanism that crucially involves dentate gyrus granule cells *(47)*.

Balb/c $GABA_{B1}^{-/-}$ mice are hyperalgesic and have a reduced latency to respond to the noxious stimulation in acute pain tests. $GABA_{B1}^{-/-}$ mice exhibit pronounced hyperalgesia to noxious heat in the hot-plate and tail-flick tests, as well as a reduced paw withdrawal threshold in response to mechanical pressure *(36)*. $GABA_{B1}^{-/-}$ animals are hyperalgesic in paradigms that involve distinct motor responses and therefore the observed hyperalgesia is unlikely to be related to the contextual hyperlocomotor activity that is also observed with these mice *(48)*. The acute pain data are consistent with a loss of intrinsic $GABA_B$-related inhibitory tone in the nociceptive system of $GABA_{B1}^{-/-}$ mice, which could be exploited for pain management with allosteric modulators *(49,50)*.

Pharmacological experiments indicate a strong $GABA_B$ receptor influence on locomotor activity *(43,51–53)*. In line with this, $GABA_{B1}^{-/-}$ mice exhibit a pronounced hyperlocomotor activity when exposed to novel environments *(36)*. The hyperlocomotor activity is characterized by repeated rapid circling during a 1–2 h time period, followed by hypoactivity *(54)*. A moderate hyperdopaminergic state likely underlies this contextual hyperlocomotor activity of $GABA_{B1}^{-/-}$ mice *(48)*. The extracellular dopamine levels in the striatum of $GABA_{B1}^{-/-}$ mice are increased twofold, whereas tyrosine hydroxylase activity, tissue dopamine content, and dopamine metabolism are not measurably affected. The hyperdopaminergic state of $GABA_{B1}^{-/-}$ mice is therefore not severe enough to inactivate dopamine D2 receptors and disrupt D2-mediated feedback inhibition of tyrosine hydroxylase, as observed in dopamine transporter mice that have fivefold elevated extracellular dopamine levels *(55,56)*. $GABA_B$ mice may therefore serve as an animal model for a moderate hyperdopaminergic state.

Genetic and pharmacological evidence supports that $GABA_B$ receptors modulate anxiety and antidepressant-like behaviors in mice *(57)*. Treatment of wild-type mice with a positive allosteric $GABA_B$ modulator, GS37983, reduces anxiolytic behaviors in the light-box test *(58)*. Consistent with these pharmacological experiments, $GABA_{B1}^{-/-}$ mice display marked increases in anxiety-related behaviors in several paradigms, including the light-dark box, staircase, and elevated zero maze tests *(54)*. Classical benzodiazepine anxiolytics fail to reduce anxiety in $GABA_{B1}^{-/-}$ mice, questioning whether $GABA_{B1}^{-/-}$ mice are a useful model to assess the anxiolytic potential of novel compounds *(54)*. Treatment of wild-type mice with the $GABA_B$-receptor antagonist CGP56433A produces an antidepressant-like phenotype *(54)*. Corroborating with the pharmacological data, $GABA_{B1}^{-/-}$ mice display antidepressant-like behavior in the forced swim test *(54)*, a test that is widely used to assess depression and antidepressant-related phenotypes in genetically altered mice *(59)*. Moreover, $GABA_{B1}$ heterozygote animals showed enhanced prepulse inhibition responses compared with littermate controls, suggesting that $GABA_{B1}$-deficient mice exhibit increased sensorimotor gating *(37)*. These experiments highlight that

$GABA_B$ receptors regulate emotional behaviors and that they represent promising drug targets for the treatment of anxiety and depression *(57)*.

A significant body of literature indicates that $GABA_B$ receptor function influences cognitive performance *(60)*. For example, it was reported that high doses of baclofen induce amnesia *(61)*. Conversely, memory-impairing effects of $GABA_B$ antagonists *(62,63)*, as well as memory-improving effects of baclofen *(62,64)* have also been described. $GABA_{B1}^{-/-}$ mice showed a markedly impaired memory performance in the passive avoidance test *(36)*. The mechanism underlying the memory loss in $GABA_{B1}^{-/-}$ mice is unclear. However, the observed deficiencies in memory functions are in line with the observed memory-impairing effects of $GABA_B$ antagonists.

In summary, the absence of $GABA_{B1}$-subunit expression is sufficient to abolish all typical biochemical and pharmacological $GABA_B$ receptor-mediated responses analyzed. This supports that the $GABA_{B1}$-subunit is obligatory for the formation of functional $GABA_B$ receptors. This implies that $GABA_{B2}$ is not functional by itself or in combination with another protein. In agreement with previous recombinant studies *(40)*, $GABA_{B1}^{-/-}$ mice also confirm that the $GABA_{B1}$-subunit but not the $GABA_{B2}$-subunit, binds GABA and competitive $GABA_B$ ligands.

3. $GABA_{B2}^{-/-}$ Mice

The only published $GABA_{B2}^{-/-}$ mouse was generated on a Balb/c genetic background, whereby exons 8–11 encoding part of the N-terminal extracellular and the first two transmembrane domains were ablated *(65)*. As with Balb/c $GABA_{B1}^{-/-}$ mice, the $GABA_{B2}^{-/-}$ mice are viable and amenable to behavioral testing. The predominantly heterodimeric nature of native $GABA_B$ receptors is emphasized by the substantial down-regulation of $GABA_{B1}$ protein in $GABA_{B2}^{-/-}$ mice. This is analogous to reduced $GABA_{B2}$ protein levels observed in $GABA_{B1}^{-/-}$ mice and indicates that the $GABA_{B1}$- and $GABA_{B2}$-subunits cross-stabilize each other. Mice lacking $GABA_{B2}$ display an unusual subcellular distribution of $GABA_{B1}$ protein in the brain. This is owing to a general redistribution of $GABA_{B1}$ protein from the neuropil to the soma. This redistribution and the failure of $GABA_{B1}$ protein to reach distal neuronal sites in the absence of $GABA_{B2}$ are again consistent with a predominantly heteromeric nature of native $GABA_B$ receptors. However, despite the considerable reduction in $GABA_{B1}$ protein levels in $GABA_{B2}^{-/-}$ mice, $GABA_{B1}$ protein is still detectable in synaptic plasma membranes. Therefore, in contrast to the conventional posit, some $GABA_{B1}$ protein may be able to exit the endoplasmatic reticulum in the absence of the $GABA_{B2}$-subunit in vivo. $GABA_B$ radioligand antagonists, such as [^{125}I]CGP71872 and [^{3}H]CGP62349, still detect GABA binding-sites in

$GABA_{B2}^{-/-}$ mice, consistent with the proposal that GABA binding is exclusively mediated by $GABA_{B1}$-subunits *(40)*. The residual $GABA_{B1}$ protein in $GABA_{B2}^{-/-}$ mice therefore retains significant binding affinity, which is a prerequisite for a putative functional role. [^{35}S]GTPγS binding assays indicate the absence of GABA-induced G protein activation in cortical- or hippocampal membrane preparations of $GABA_{B2}^{-/-}$ mice. This further emphasizes that the predominant native $GABA_B$ receptors form heteromeric assemblies. However, as [^{35}S]GTPγS binding preferentially detects $G_{i/o}$-type G proteins, it is not excluded that the $GABA_{B1}$ protein in $GABA_{B2}^{-/-}$ mice forms functional receptors in association with other G proteins.

Electrophysiological analysis of hippocampal $GABA_{B2}^{-/-}$ neurons surprisingly reveals that $GABA_{B1}$ can be functional in the absence of $GABA_{B2}$. Although $GABA_{B2}^{-/-}$ mice lack functional $GABA_B$ hetero- and autoreceptors, baclofen still induces a G protein-activated inward current in CA1 pyramidal cells. The observed response is atypical, as baclofen normally induces a K^+ outward current. The inward current in $GABA_{B2}^{-/-}$ neurons can be blocked by the $GABA_B$ antagonists CGP55845A and CGP62349, indicating that the response is $GABA_{B1}$-mediated. In agreement with this, the inward current is completely absent in $GABA_{B1}^{-/-}$ mice. It is conceivable that $GABA_{B1}$ receptors couple to G proteins other than $G_{i/o}$ and that inhibition instead of activation of Kir3-type K^+-channels underlies the atypical inward current observed in $GABA_{B2}^{-/-}$ mice. Several studies suggest that various cellular populations in the nervous systems display nonoverlapping expression patterns for $GABA_{B1}$ and $GABA_{B2}$ *(18,66–71)*. The electrophysiological experiments with $GABA_{B2}^{-/-}$ mice thus support that $GABA_{B1}$ could be functional in neurons devoid of $GABA_{B2}$ protein. Interestingly, some scattered hippocampal interneurons in $GABA_{B2}^{-/-}$ mice exhibit a very prominent $GABA_{B1}$ immunostaining, suggesting that these neurons express high levels of $GABA_{B1}$ protein in the absence of $GABA_{B2}$. Unfortunately, it is currently impossible to identify these neurons for electrophysiological recordings. It therefore remains unclear whether the atypical $GABA_{B1}$ responses observed in $GABA_{B2}^{-/-}$ mice are of physiological relevance or represent an artefact of the knockout situation.

Behavioral analysis of $GABA_{B2}^{-/-}$ mice indicates similar phenotypes as in the $GABA_{B1}^{-/-}$ mice *(65)*. $GABA_{B2}^{-/-}$ mice exhibit hyperlocomotor activity, hyperalgesia, impaired passive avoidance learning, and spontaneous epileptiform activity. $GABA_{B2}^{-/-}$ mice also display elevated anxiety-related behavior in the light-dark box paradigm and exhibit an antidepressant-like behavior in force swim test, analogous to $GABA_{B1}^{-/-}$ mice *(72)*. Similarly, prototypic responses to baclofen, including muscle relaxation, hypothermia, and EEG δ-wave induction are missing in the $GABA_{B2}^{-/-}$ mice *(65)*. Taken together, biochemical, pharmacological behavioral, and pharmacological analysis of $GABA_{B2}^{-/-}$ and

$GABA_{B1}^{-/-}$ mice corroborates with current $GABA_B$ receptor models, where inactivation of either subunit prevents receptor function. Small electrophysiological $GABA_B$ receptor responses in $GABA_{B2}^{-/-}$ mice hint at the possibility that $GABA_{B1}$ could be functional in vivo in the absence of $GABA_{B2}$. However, the residual $GABA_B$ responses in $GABA_{B2}^{-/-}$ mice are unable to alter the phenotypes observed in $GABA_{B1}^{-/-}$ mice.

Transgenic mice expressing a $GABA_{B2}$-subunit lacking its C terminal tail (the ΔGB2-Ct mouse) exhibit a complete loss of $GABA_B$ receptor function and produce an epileptic phenotype *(73)*. Presumably, the $GABA_B$ receptor complex is retained intracellularly because the endoplasmatic reticulum retention signal in the $GABA_{B1}$-subunit is no longer masked on dimerization. Loss of $GABA_B$ function in ΔGB2-Ct hippocampal slices promotes depolarizing $GABA_A$ receptor-mediated events, which in turn, leads to the generation of ictal-like events. This may contribute to the epilepsy phenotype observed with these mice.

4. Physiological Effects of GHB at $GABA_B$ Receptors

GHB is a naturally occurring GABA metabolite that has been proposed as a neurotransmitter/neuromodulator that acts through its own receptor *(35)*. The receptors for GHB are postulated to correspond to high-affinity [^{3}H]GHB-binding sites present in the brain, but a physiological significance for endogenous GHB signaling has not been established. Patients suffering from GHB aciduria, a congenital enzyme defect causing GHB accumulation, exhibit a wide variety of symptoms, including psychomotor retardation, delayed or absent speech, hypotonia, ataxia, hyporeflexia, seizures, and EEG abnormalities *(74)*. This shows that elevated levels of GHB influence physiological processes. GHB and its prodrug, γ-butyrolactone (GBL), received public attention as they emerged as popular drugs of abuse. Exogenous administration of GHB elicits central nervous system-dependent effects, such as memory impairment, increase in sleep stages 3 and 4, seizures, and coma. The receptor interactions of exogenously applied GHB are a matter of much debate. GHB clearly activates recombinant $GABA_B$ receptors, albeit at millimolar concentrations *(75)*. In line with this, there is mounting evidence that $GABA_B$ receptors mediate some of the effects of exogenously applied GHB or GBL *(35)*.

$GABA_{B1}^{-/-}$ mice provide the opportunity to study the effects of GHB in the absence of coincident $GABA_B$ responses. Radioligand-binding studies reveal a similar number and a similar spatial distribution of [^{3}H]GHB-binding sites in the brains of $GABA_{B1}^{-/-}$ and wild-type littermates, demonstrating that GHB- are distinct from $GABA_B$-binding sites *(44,76)*. Moreover, analysis of $GABA_{B1}^{-/-}$ mice demonstrates that GHB-induced [^{35}S]GTPγS responses are mediated by

$GABA_B$ receptors and not the high-affinity [^{3}H]GHB-binding sites *(44)*. Following GHB or GBL application, $GABA_{B1}^{-/-}$ mice show neither the hypolocomotion, hypothermia, increase in striatal dopamine synthesis, nor EEG δ-wave induction seen in wild-type mice. $GABA_B^{-/-}$ mice have therefore helped to establish that the vast majority of exogenous GHB actions are mediated by $GABA_B$ receptors. However, the molecular nature and the signaling properties of the specific [^{3}H]GHB-binding sites in the brain remain elusive.

5. Outlook

The work with $GABA_{B1}^{-/-}$ and $GABA_{B2}^{-/-}$ mice support that molecular diversity in the $GABA_B$ system primarily arises from the $GABA_{B1a}$- and $GABA_{B1b}$-subunit isoforms. So far, no unique pharmacological or functional properties could be assigned to $GABA_{B1a}$ or $GABA_{B1b}$. The expression levels for $GABA_{B1a}$ and $GABA_{B1b}$ vary during development and across individual cells, suggestive of a functional specialization. Structurally, the isoforms differ in their amino-terminal ectodomain by a pair of sushi-repeats that is present in $GABA_{B1a}$ but not in $GABA_{B1b}$ *(77)*. Sushi-repeats, also known as complement control protein modules, or short consensus repeats, are found in other GPCRs as well *(78)* and mediate protein interactions in a wide variety of adhesion proteins *(79)*. The presence of sushi-repeats in $GABA_{B1a}$, together with the absence of functional or pharmacological differences in vitro, suggests the existence of auxiliary proteins that modify receptor activity or pharmacology *(11,21)*, precedence for which is found with other GPCRs *(80)*. The lack of selective reagents has not allowed addressing the individual contributions of $GABA_{B1a}$ and $GABA_{B1b}$ to native $GABA_B$ functions. It therefore remains a key question whether $GABA_{B1}$ isoforms exhibit pharmacological and/or functional differences in vivo. A possibly distinct pre- and postsynaptic localization of $GABA_{B1}$ isoforms could help to reconcile conflicting data from recombinant and native work, as pharmacological differences could be the result of differences in the local effector system and/or the receptor reserve. Recently, the authors have taken a genetic approach to dissociate the native functions of $GABA_{B1a}$ and $GABA_{B1b}$ and produced mice that selectively express $GABA_{B1a}$ or $GABA_{B1b}$ under control of their endogenous promoter. Using a combined physiological and morphological approach, currently these mice are being used to address whether $GABA_{B1}$ isoforms localize to distinct subcellular compartments and convey separate functions.

The diverse phenotypes of $GABA_{B1}^{-/-}$ and $GABA_{B2}^{-/-}$ mice indicate that tonic- or phasic GABA activity is present in many diseases relevant to neuronal systems. This could be exploited for therapy with allosteric $GABA_B$ modulators that recently became available *(49,50)*. A spatially and temporally restricted loss of $GABA_B$ function will be necessary to dissociate the circuits underlying the diverse $GABA_{B1}^{-/-}$ phenotypes. To that aim, the authors have generated a conditional *$GABA_{B1}$* allele based on the Cre-loxP system *(81)*. The conditional allele makes it possible to inactivate the *$GABA_{B1}$* gene in defined cellular populations and at defined points in time, making the genetic deletion spatially and temporally restricted. With the genetic tools now available, it will be possible to address the specific functions of $GABA_B$ receptors in different cellular compartments and neuronal systems.

Note Added in Proof

Recently published genetic experiments indicate that $GABA_{B1a}$ predominantly assembles heteroreceptors controlling glutamate release, whereas $GABA_{B1b}$ mostly forms receptors coupled to postsynaptic effector channels *(82–84)*. A selective lack of $GABA_{B1a}$ and $GABA_{B1b}$ in knockout mice significantly impairs synaptic plasticity and behavioral processes, showing that the two $GABA_{B1}$ isoforms cannot fully compensate for each other.

References

1. Hill, D. R. and Bowery, N. G. (1981) 3H-baclofen and 3H-GABA bind to bicuculline-insensitive GABA B sites in rat brain. *Nature* **290,** 149–152.
2. Hill, D. R. (1985) GABAB receptor modulation of adenylate cyclase activity in rat brain slices. *Br. J. Pharmacol.* **84,** 249–257.
3. Bowery, N. G., Bettler, B., Froestl, W., et al. (2002) International Union of Pharmacology. XXXIII. Mammalian gamma-aminobutyric acid(B) receptors: structure and function. *Pharmacol. Rev.* **54,** 247–264.
4. Bettler, B., Kaupmann, K., Mosbacher, J., and Gassmann, M. (2004) Molecular structure and physiological functions of GABA(B) receptors. *Physiol. Rev.* **84,** 835–867.
5. Calver, A. R., Davies, C. H., and Pangalos, M. N. (2002) GABA(B) receptors: from monogamy to promiscuity. *Neurosignals* **11,** 299–314.
6. Thompson, S. M., Capogna, M., and Scanziani, M. (1993) Presynaptic inhibition in the hippocampus. *Trends Neurosci.* **16,** 222–227.
7. Sakaba, T. and Neher, E. (2003) Direct modulation of synaptic vesicle priming by GABA(B) receptor activation at a glutamatergic synapse. *Nature* **424,** 775–778.
8. Lüscher, C., Jan, L. Y., Stoffel, M., Malenka, R. C., and Nicoll, R. A. (1997) G protein-coupled inwardly rectifying K^+ channels (GIRKs) mediate postsynaptic but not presynaptic transmitter actions in hippocampal neurons. *Neuron* **19,** 687–695.
9. Steiger, J. L., Bandyopadhyay, S., Farb, D. H., and Russek, S. J. (2004) cAMP response element-binding protein, activating transcription factor-4, and upstream stimulatory factor differentially control hippocampal GABABR1a and GABABR1b subunit gene expression through alternative promoters. *J. Neurosci.* **24,** 6115–6126.
10. Couve, A., Thomas, P., Calver, A. R., et al. (2002) Cyclic AMP-dependent protein kinase phosphorylation facilitates GABA(B) receptor-effector coupling. *Nat. Neurosci.* **5,** 415–424.

11. Mohler, H. and Fritschy, J. M. (1999) GABAB receptors make it to the top—as dimers. *Trends Pharmacol. Sci.* **20,** 87–89.
12. Bonanno, G. and Raiteri, M. (1993) Multiple GABAB receptors. *Trends Pharmacol. Sci.* **14,** 259–261.
13. Gemignani, A., Paudice, P., Bonanno, G., and Raiteri, M. (1994) Pharmacological discrimination between gamma-aminobutyric acid type B receptors regulating cholecystokinin and somatostatin release from rat neocortex synaptosomes. *Mol. Pharmacol.* **46,** 558–562.
14. Cunningham, M. D. and Enna, S. J. (1996) Evidence for pharmacologically distinct GABAB receptors associated with cAMP production in rat brain. *Brain Res.* **720,** 220–224.
15. Deisz, R. A., Billard, J. M., and Zieglgansberger, W. (1997) Presynaptic and postsynaptic GABAB receptors of neocortical neurons of the rat in vitro: differences in pharmacology and ionic mechanisms. *Synapse* **25,** 62–72.
16. Yamada, K., Yu, B., and Gallagher, J. P. (1999) Different subtypes of GABAB receptors are present at pre- and postsynaptic sites within the rat dorsolateral septal nucleus. *J. Neurophysiol.* **81,** 2875–2883.
17. Kaupmann, K., Huggel, K., Heid, J., et al. (1997) Expression cloning of GABA(B) receptors uncovers similarity to metabotropic glutamate receptors. *Nature* **386,** 239–246.
18. Kaupmann, K., Malitschek, B., Schuler, V., et al. (1998) GABA(B)-receptor subtypes assemble into functional heteromeric complexes. *Nature* **396,** 683–687.
19. Jones, K. A., Borowsky, B., Tamm, J. A., et al. (1998) GABA(B) receptors function as a heteromeric assembly of the subunits GABA(B)R1 and GABA(B)R2. *Nature* **396,** 674–679.
20. White, J. H., Wise, A., Main, M. J., et al. (1998) Heterodimerization is required for the formation of a functional GABA(B) receptor. *Nature* **396,** 679–682.
21. Marshall, F. H., Jones, K. A., Kaupmann, K., and Bettler, B. (1999) GABAB receptors - the first 7TM heterodimers. *Trends Pharmacol. Sci.* **20,** 396–399.
22. Pagano, A., Rovelli, G., Mosbacher, J., et al. (2001) C-terminal interaction is essential for surface trafficking but not for heteromeric assembly of GABA(b) receptors. *J. Neurosci.* **21,** 1189–1202.
23. Margeta-Mitrovic, M., Jan, Y. N., and Jan, L. Y. (2000) A trafficking checkpoint controls GABA(B) receptor heterodimerization. *Neuron* **27,** 97–106.
24. Calver, A. R., Robbins, M. J., Cosio, C., et al. (2001) The C-terminal domains of the GABA(b) receptor subunits mediate intracellular trafficking but are not required for receptor signaling. *J. Neurosci.* **21,** 1203–1210.
25. Galvez, T., Duthey, B., Kniazeff, J., et al. (2001) Allosteric interactions between GB1 and GB2 subunits are required for optimal GABA(B) receptor function. *EMBO J.* **20,** 2152–2159.
26. Robbins, M. J., Calver, A. R., Filippov, A. K., et al. (2001) GABA(B2) is essential for g-protein coupling of the GABA(B) receptor heterodimer. *J. Neurosci.* **21,** 8043–8052.
27. Margeta-Mitrovic, M., Jan, Y. N., and Jan, L. Y. (2001) Function of GB1 and GB2 subunits in G protein coupling of GABA(B) receptors. *Proc. Natl. Acad. Sci. USA* **98,** 14,649–14,654.

28. Duthey, B., Caudron, S., Perroy, J., et al. (2002) A single subunit (GB2) is required for G-protein activation by the heterodimeric GABA(B) receptor. *J. Biol. Chem.* **277,** 3236–3241.
29. Havlickova, M., Prezeau, L., Duthey, B., Bettler, B., Pin, J. P., and Blahos, J. (2002) The intracellular loops of the GB2 subunit are crucial for G-protein coupling of the heteromeric gamma-aminobutyrate B receptor. *Mol. Pharmacol.* **62,** 343–350.
30. Angers, S., Salahpour, A., and Bouvier, M. (2002) Dimerization: an emerging concept for G protein-coupled receptor ontogeny and function. *Annu. Rev. Pharmacol. Toxicol.* **42,** 409–435.
31. Liang, Y., Fotiadis, D., Filipek, S., Saperstein, D. A., Palczewski, K., and Engel, A. (2003) Organization of the G protein-coupled receptors rhodopsin and opsin in native membranes. *J. Biol. Chem.* **278,** 21,655–21,662.
32. Martin, S. C., Russek, S. J., and Farb, D. H. (2001) Human GABA(B)R genomic structure: evidence for splice variants in GABA(B)R1 but not GABA(B)R2. *Gene* **278,** 63–79.
33. Filippov, A. K., Couve, A., Pangalos, M. N., Walsh, F. S., Brown, D. A., and Moss, S. J. (2000) Heteromeric assembly of GABA(B)R1 and GABA(B)R2 receptor subunits inhibits $Ca(2^{+})$ current in sympathetic neurons. *J. Neurosci.* **20,** 2867–2874.
34. Easter, A. and Spruce, A. E. (2002) Recombinant GABA(B) receptors formed from GABA(B1) and GABA(B2) subunits selectively inhibit N-type $Ca(2^{+})$ channels in NG108-15 cells. *Eur. J. Pharmacol.* **440,** 17–25.
35. Crunelli, V., Emri, Z., and Leresche, N. (2006) Unravelling the brain targets of gamma-hydroxybutyric acid. *Curr. Opin. Pharmacol.* **6,** 44–52.
36. Schuler, V., Lüscher, C., Blanchet, C., et al. (2001) Epilepsy, hyperalgesia, impaired memory, and loss of pre- and postsynaptic GABA(B) responses in mice lacking GABA(B(1)). *Neuron* **31,** 47–58.
37. Prosser, H. M., Gill, C. H., Hirst, W. D., et al. (2001) Epileptogenesis and enhanced prepulse inhibition in GABA(B1)-deficient mice. *Mol. Cell. Neurosci.* **17,** 1059–1070.
38. Quéva, C., Bremner-Danielsen, M., Edlund, A., et al. (2003) Effects of GABA agonists on body temperature regulation in $GABA_{B1}^{-/-}$ mice. *Br. J. Pharmacol.* **140,** 315–322.
39. Benke, D., Honer, M., Michel, C., Bettler, B., and Mohler, H. (1999) gamma-aminobutyric acid type B receptor splice variant proteins GBR1a and GBR1b are both associated with GBR2 in situ and display differential regional and subcellular distribution. *J. Biol. Chem.* **274,** 27,323–27,330.
40. Kniazeff, J., Galvez, T., Labesse, G., and Pin, J. P. (2002) No ligand binding in the GB2 subunit of the GABA(B) receptor is required for activation and allosteric interaction between the subunits. *J. Neurosci.* **22,** 7352–7361.
41. Galvez, T., Urwyler, S., Prezeau, L., et al. (2000) $Ca(2^{(})$ requirement for high-affinity gamma-aminobutyric acid (GABA) binding at GABA(B) receptors: involvement of serine 269 of the GABA(B)R1 subunit. *Mol. Pharmacol.* **57,** 419–426.
42. Sanger, G. J., Munonyara, M. L., Dass, N., Prosser, H., Pangalos, M. N., and Parsons, M. E. (2002) GABA(B) receptor function in the ileum and urinary bladder

of wildtype and GABA(B1) subunit null mice. *Auton. Autacoid. Pharmacol.* **22,** 147–154.

43. Badran, S., Schmutz, M., and Olpe, H. R. (1997) Comparative in vivo and in vitro studies with the potent GABAB receptor antagonist, CGP 56999A. *Eur. J. Pharmacol.* **333,** 135–142.
44. Kaupmann, K., Cryan, J. F., Wellendorph, P., et al. (2003) Specific gamma-hydroxybutyrate-binding sites but loss of pharmacological effects of gamma-hydroxybutyrate in GABA(B)(1)-deficient mice. *Eur. J. Neurosci.* **18,** 2722–2730.
45. Pierau, F. K., Yakimova, K. S., Sann, H., and Schmid, H. A. (1997) Specific action of GABAB ligands on the temperature sensitivity of hypothalamic neurons. *Ann. NY Acad. Sci.* **813,** 146–155.
46. Brown, J. T., Gill, C. H., Farmer, C. E., et al. (2003) Mechanisms contributing to the exacerbated epileptiform activity in hippocampal slices of GABAB1 receptor subunit knockout mice. *Epilepsy Res.* **57,** 121–136.
47. Ruttimann, E., Vacher, C. M., Gassmann, M., Kaupmann, K., Van der Putten, H., and Bettler, B. (2004) Altered hippocampal expression of calbindin-D-28k and calretinin in GABA(B(1))-deficient mice. *Biochem. Pharmacol.* **68,** 1613–1620.
48. Vacher, C. M., Gassmann, M., Desrayaud, S., et al. (2006) Hyperdopaminergia and altered locomotor activity in GABAB1-deficient mice. *J. Neurochem.* **97,** 979–991.
49. Urwyler, S., Pozza, M. F., Lingenhoehl, K., et al. (2003) N,N'-Dicyclopentyl-2-methylsulfanyl-5-nitro-pyrimidine-4,6-diamine (GS39783) and structurally related compounds: novel allosteric enhancers of gamma-aminobutyric acidB receptor function. *J. Pharmacol. Exp. Ther.* **307,** 322–330.
50. Urwyler, S., Mosbacher, J., Lingenhoehl, K., et al. (2001) Positive allosteric modulation of native and recombinant gamma-aminobutyric acid(B) receptors by 2,6-Di-tert-butyl-4-(3-hydroxy-2,2-dimethyl-propyl)-phenol (CGP7930) and its aldehyde analog CGP13501. *Mol. Pharmacol.* **60,** 963–971.
51. Colombo, G., Melis, S., Brunetti, G., et al. (2001) GABA(B) receptor inhibition causes locomotor stimulation in mice. *Eur. J. Pharmacol.* **433,** 101–104.
52. Chester, J. A. and Cunningham, C. L. (1999) Baclofen alters ethanol-stimulated activity but not conditioned place preference or taste aversion in mice. *Pharmacol. Biochem. Behav.* **63,** 325–331.
53. Phillis, B. D., Ong, J., White, J. M., and Bonnielle, C. (2001) Modification of d-amphetamine-induced responses by baclofen in rats. *Psychopharmacology* **153,** 277–284.
54. Mombereau, C., Kaupmann, K., Froestl, W., Sansig, G., van der Putten, H., and Cryan, J. F. (2004) Genetic and pharmacological evidence of a role for GABA(B) receptors in the modulation of anxiety- and antidepressant-like behavior. *Neuropsychopharmacology* **29,** 1050–1062.
55. Giros, B., Jaber, M., Jones, S. R., Wightman, R. M., and Caron, M. G. (1996) Hyperlocomotion and indifference to cocaine and amphetamine in mice lacking the dopamine transporter. *Nature* **379,** 606–612.
56. Jones, S. R., Gainetdinov, R. R., Jaber, M., Giros, B., Wightman, R. M., and Caron, M. G. (1998) Profound neuronal plasticity in response to inactivation of the dopamine transporter. *Proc. Natl. Acad. Sci. USA* **95,** 4029–4034.

57. Cryan, J. F. and Kaupmann, K. (2005) Don't worry 'B' happy!: a role for GABA(B) receptors in anxiety and depression. *Trends Pharmacol. Sci.* **26,** 36–43.
58. Cryan, J. F., Kelly, P. H., Chaperon, F., et al. (2004) Behavioral characterization of the novel GABAB receptor-positive modulator GS39783 (N,N'-dicyclopentyl-2-methylsulfanyl-5-nitro-pyrimidine-4,6-diamine): anxiolytic-like activity without side effects associated with baclofen or benzodiazepines. *J. Pharmacol. Exp. Ther.* **310,** 952–963.
59. Cryan, J. F. and Holmes, A. (2005) The ascent of mouse: advances in modelling human depression and anxiety. *Nat. Rev. Drug Discov.* **4,** 775–790.
60. Mondadori, C., Moebius, H. J., and Zingg, M. (1996) CGP 36,742, an orally active GABAB receptor antagonist, facilitates memory in a social recognition test in rats. *Behav. Brain Res.* **77,** 227–229.
61. Sandyk, R. and Gillman, M. A. (1985) Baclofen-induced memory impairment. *Clin. Neuropharmacol.* **8,** 294–295.
62. Castellano, C., Cestari, V., Cabib, S., and Puglisi-Allegra, S. (1993) Strain-dependent effects of post-training GABA receptor agonists and antagonists on memory storage in mice. *Psychopharmacology (Berl.)* **111,** 134–138.
63. Brucato, F. H., Levin, E. D., Mott, D. D., Lewis, D. V., Wilson, W. A., and Swartzwelder, H. S. (1996) Hippocampal long-term potentiation and spatial learning in the rat: effects of GABAB receptor blockade. *Neuroscience* **74,** 331–339.
64. Saha, N., Chugh, Y., Sankaranaryanan, A., and Sharma, P. L. (1993) Effects of post-training administration of (-)-baclofen and chlordiazepoxide on memory retention in ICRC Swiss mice: interactions with GABAA and GABAB receptor antagonists. *Pharmacol. Toxicol.* **72,** 159–162.
65. Gassmann, M., Shaban, H., Vigot, R., et al. (2004) Redistribution of GABAB(1) protein and atypical GABAB responses in GABAB(2)-deficient mice. *J. Neurosci.* **24,** 6086–6097.
66. Kim, M. O., Li, S., Park, M. S., and Hornung, J. P. (2003) Early fetal expression of GABA(B1) and GABA(B2) receptor mRNAs on the development of the rat central nervous system. *Dev. Brain Res.* **143,** 47–55.
67. Li, S. P., Park, M. S., Yoon, H., et al. (2003) Differential distribution of GABA(B1) and GABA(B2) receptor mRNAs in the rat brain. *Mol. Cells* **16,** 40–47.
68. Burman, K. J., Ige, A. O., White, J. H., et al. (2003) GABAB receptor subunits, R1 and R2, in brainstem catecholamine and serotonin neurons. *Brain Res.* **970,** 35–46.
69. Ng, T. K. and Yung, K. K. (2001) Differential expression of GABA(B)R1 and GABA(B)R2 receptor immunoreactivity in neurochemically identified neurons of the rat neostriatum. *J. Comp. Neurol.* **433,** 458–470.
70. Clark, J. A., Mezey, E., Lam, A. S., and Bonner, T. I. (2000) Distribution of the GABA(B) receptor subunit gb2 in rat CNS. *Brain Res.* **860,** 41–52.
71. Kulik, A., Vida, I., Lujan, R., et al. (2003) Subcellular localization of metabotropic GABA(B) receptor subunits GABA(B1a/b) and GABA(B2) in the rat hippocampus. *J. Neurosci.* **23,** 11,026–11,035.
72. Mombereau, C., Kaupmann, K., Gassmann, M., Bettler, B., van der Putten, H., and Cryan, J. F. (2005) Altered anxiety and depression-related behaviour in mice lacking GABAB(2) receptor subunits. *Neuroreport* **16,** 307–310.

73. Thuault, S. J., Brown, J. T., Calver, A. R., Collingridge, G. L., Randall, A., and Davies, C. H. (2005) Mechanisms contributing to the exacerbated epileptiform activity in hippocampal slices expressing a C-terminal truncated GABA(B2) receptor subunit. *Epilepsy Res.* **65,** 41–51.
74. Hogema, B. M., Gupta, M., Senephansiri, H., et al. (2001) Pharmacologic rescue of lethal seizures in mice deficient in succinate semialdehyde dehydrogenase. *Nat. Genet.* **29,** 212–216.
75. Lingenhoehl, K., Brom, R., Heid, J., et al. (1999) Gamma-hydroxybutyrate is a weak agonist at recombinant GABA(B) receptors. *Neuropharmacology* **38,** 1667–1673.
76. Wu, Y., Ali, S., Ahmadian, G., et al. (2004) Gamma-hydroxybutyric acid (GHB) and gamma-aminobutyric acidB receptor (GABABR) binding sites are distinctive from one another: molecular evidence. *Neuropharmacology* **47,** 1146–1156.
77. Blein, S., Ginham, R., Uhrin, D., et al. (2004) Structural analysis of the complement control protein (CCP) modules of GABA(B) receptor 1a: only one of the two CCP modules is compactly folded. *J. Biol. Chem.* **279,** 48,292–48,306.
78. Grace, C. R., Perrin, M. H., DiGruccio, M. R., et al. (2004) NMR structure and peptide hormone binding site of the first extracellular domain of a type B1 G protein-coupled receptor. *Proc. Natl. Acad. Sci. USA* **101,** 12,836–12,841.
79. Lehtinen, M. J., Meri, S., and Jokiranta, T. S. (2004) Interdomain contact regions and angles between adjacent short consensus repeat domains. *J. Mol. Biol.* **344,** 1385–1396.
80. McLatchie, L. M., Fraser, N. J., Main, M. J., et al. (1998) RAMPs regulate the transport and ligand specificity of the calcitonin-receptor-like receptor. *Nature* **393,** 333–339.
81. Haller, C., Casanova, E., Muller, M., et al. (2004) Floxed allele for conditional inactivation of the GABAB(1) gene. *Genesis* **40,** 125–130.
82. Perez-Garci, E., Gassmann, M., Bettler, B., and Larkum, M. E. (2006) The GABA(B1b) Isoform mediates long-lasting inhibition of dendritic Ca(2+) spikes in layer 5 somatosensory pyramidal neurons. *Neuron* **50,** 603–616.
83. Shaban, H., Humeau, Y., Herry, C., et al. (2006) Generalization of amygdala LTP and conditioned fear in the absence of presynaptic inhibition. *Nat. Neurosci.* **9,** 1028–1035.
84. Vigot, R., Barbieri, S., Brauner-Osborne, H., et al. (2006) Differential compartmentalization and distinct functions of GABA(B) receptor variants. *Neuron* **50,** 589–601.

12

$GABA_B$ Receptor as a Potential Therapeutic Target

Norman G. Bowery

Summary

At present, the only therapeutic agent in current use that is known to exert its effects through the GABA-B receptor is baclofen, the β-chlorophenyl analog of GABA. This compound is used extensively as a centrally active muscle relaxant, but it has been shown clinically to exert additional effects, including analgesia, suppression of drug addiction, and cessation of chronic cough. Basic research indicates that there may be many other applications not only for agonists, but also antagonists at the GABA-B receptor, and this chapter seeks to examine the potential significance of these and also of allosteric modulators for the receptor, which have been recognized during the past 5 yr. GABA-B receptors have been implicated in neuronal processing in many brain regions and there is considerable evidence for their pathological involvement in various diseases of the central nervous system, such as absence epilepsy, depression, and even in the etiology of nicotine dependency in humans. Novel chemical entities will surely be the key to exploiting this receptor. At present, the only agonist (baclofen) in the clinic frequently produces unwanted side effects, and this clearly limits its usage. There are no antagonists in clinical therapy, although trials for cognition deficits are underway. Metabotropic receptors, of which the GABA-B receptor is a member, provide most of the targets for our currently used therapeutic agents; thus, there seems to be every reason to believe that the rewards for discovering clinically available GABA-B receptor ligands could be high.

Key Words: $GABA_B$ receptor; G protein; therapeutic target; nociception; cognition; addiction; epilepsy; depression and anxiety.

From: *The Receptors: The GABA Receptors, Third Edition*
Edited by: S. J. Enna and H. Möhler © Humana Press Inc., Totowa, NJ

1. Introduction

In preceding chapters details of the heterodimeric structure of the receptor have been presented and whereas each of the two subunits that make up the receptor might contribute independently to the function of the receptor, the dimer is the form on which the receptor pharmacology is based *(8)*. The two receptor subunits, γ-aminobutyric acid ($GABA_{B1}$) and $GABA_{B2,}$ which provide different functions are mutually dependent on each other. The former contains the GABA-binding domain *(9–12)*, whereas $GABA_{B2}$ provides the G protein coupling mechanism and also incorporates an allosteric modulatory site within its heptahelical structure *(13–15)*. Whereas different functional isoforms of at least $GABA_{B1}$ have been defined *(16–22)* there is, as yet, no unequivocal evidence for distinct $GABA_B$-receptor subtypes *(8)*.

The overall distribution of the receptor in the brain has been delineated using autoradiographical techniques with neurochemical and electrophysiological studies revealing both pre- and postsynaptic locations where they are coupled to Ca^{2+} and K^+ channels, respectively. When receptor activation occurs, not surprisingly, a variety of effects can arise as a consequence of inhibition of transmitter release and/or postsynaptic neuronal hyperpolarization. Having a selective agonist for the receptor together with information derived from "knockout" mice which exhibit hyperalgesia, seizures, hyperlocomotion, impaired learning, loss of responses to baclofen, and lack of $GABA_B$-binding sites throughout the brain *(23–26)*, has provided the basis for much of the speculation about the potential therapeutic benefits of both agonists and antagonists for the receptor. A list of some of the actions of baclofen (in vitro and in vivo) is shown in table 1 and predominant among the in vivo effects are the muscle relaxant, antinociceptive, and antidrug craving effects as well as the reduction in cognitive behavior that have been reported.

2. Mechanisms Associated With the $GABA_B$-Receptor Activation

Both biochemical and ion channel events have been linked to $GABA_B$-receptor activation *(8)*. The common feature of these events is the coupling to G proteins *(27,28)* but the channel events appear not to be dependent on modifications in the generation of cyclic adenosine monophosphate, which occur as a consequence of G protein coupling to adenylyl cyclase *(29–32)*. Instead there would seem to be a direct influence of the G protein on the ion channels. The events that occur following agonist binding to the receptor domain, which is located only in the $GABA_{B1}$-subunit, are mediated by the $GABA_{B2}$-subunit and it is in this subunit within the heptahelical domain where an allosteric receptor modulatory site is also located.

Table 1
Actions Produced by $GABA_B$-Receptor Activation In Vivo and In Vitro and Their Potential Application

Effect	Locus of action	Potential application
Smooth muscle relaxation	Lung, bladder, intestine	Asthma
Smooth muscle contraction	Uterus, oviduct, gall bladder	
Antinociception	Spinal cord, thalamus	Pain
Neuronal hyperpolarization	CNS numerous location	
LTP-modulation	CNS, hippocampus	Cognitive deficits
Generation/exacerbation of absence epilepsy	Thalamus/somatosensory cortex	Absence epilepsy
Enhanced feeding	Higher centers	Food intake modification
Fat intake reduction	Higher centers	Binge eating
Drug addiction suppression	CNS-mesolimbic system	Drug abuse
Vasopressor action	Nucleus tractus solitarius	
Muscle relaxation	Spinal cord	Spasticity
Antitussive action	Cough center in medulla	Coughing
Respiratory depression	Brain stem	
Insulin/glucagon release	Pancreas	Diabetes
Suppression of CRH/MSH release	Pituitary	
Gastrin/gastric acid secretion altered	Vagal center	
Suppression of panic behavior	Dorsal periqueductal Gray	Anxiety/panic disorder
Neutrophil chemotaxis enhanced	Leucocytes	Inflammation
Inhibition of neurotransmitter release	CNS and peripheral nerve terminus	
Inhibition of cognitive function	Higher centers	Cognitive deficits

CRH, corticotropin-releasing hormone; MSH, melanocyte-stimulating hormone.

Two major ionic events can occur following $GABA_B$-receptor activation, an increase in neuronal membrane K^+ conductance or a decrease in Ca^{2+} conductance. More than one type of K^+ channel appears to be linked to $GABA_B$ receptors *(33)* and they are not the same as that opened by 5HT-receptor activation *(34)* even though both receptor systems have been reported to be inactivated by pertussis toxin *(35)*. Modulation of K^+ conductance seems to be primarily, but not exclusively, associated with postsynaptic $GABA_B$ sites, which manifests as a membrane hyperpolarization of long duration. However, presynaptic coupling of $GABA_B$ to a potassium "A" current controlling transmitter outflow has been suggested by Saint et al. *(36)* even though there is overwhelming evidence in favor of a link to Ca^{2+} in presynaptic terminals. Inhibition of terminal Ca^{2+} conductance, first described by Dunlap *(37)* and Desarmenien et al. *(38)*, appears to be the major mechanism by which $GABA_B$ receptors exert their inhibitory control of transmitter outflow. These presynaptic receptors are connected through G proteins to N and P but not L calcium channels. In addition low-threshold T-currents may also be involved indirectly in the response to $GABA_B$-receptor activation *(39,40)* possibly as a consequence of their deinactivation produced by the long-lasting membrane hyperpolarization.

Examination of the list of actions, shown in table 1, not only illustrates the wide distribution of $GABA_B$ receptors, both inside and outside the brain, but also provides the basis for potential therapeutic targeting. Evidence for both pre- and postsynaptic localization of $GABA_B$ sites has been obtained and much data derived from electrophysiological experimentation. Perhaps the most well-documented evidence comes from the generation of the late postsynaptic hyperpolarization in many brain regions *(41–43)* and the involvement of presynaptic $GABA_B$ autoreceptors in paired-pulse inhibition *(44)*. Both of these events are synaptically mediated and appear to be responsible for modulating postsynaptic excitatory processing. Presynaptic $GABA_B$ receptors are not confined to GABAergic nerve terminals as there is good evidence for heteroreceptors on glutamate terminals and these are probably activated by GABA acting in a paracrine-like manner after release from a GABAergic fibre located in close proximity *(45)*. Inhibitory control of transmitter release from a variety of different fibre types has been reported using biochemical techniques, suggesting that synaptically released GABA might well influence numerous transmitter systems. The estimated concentration of GABA within synapses is believed to be in the millimolar range, whereas the affinity of the presynaptic receptor is probably in the nanomolar range. Thus, a millionfold dilution could occur in the extracellular space and there would still be sufficient GABA to activate the receptors on adjacent nerve terminals.

3. Potential Therapeutic Significance of $GABA_B$ Receptor Ligands

The muscle relaxant properties of baclofen that are centrally-mediated, have been well established in the clinic for more than 20 yr. In fact, it is the drug of choice in spasticity associated with cerebral palsy, multiple sclerosis, stiff-man syndrome, and tetanus *(46–70)*. However, the side-effects produced by this drug, which include seizures, nausea, drowsiness, dizziness, hypotension, muscle weakness, hallucinations, and mental confusion, are often poorly tolerated by patients. This is, in part, because of the need for high doses to be administered because of poor brain penetration. This problem has been successfully addressed by the introduction of intrathecal administration using an indwelling pump. There have been numerous clinical reports in which intrathecal application from a pump inserted in the abdomen has proven to be successful over a long period (years) without side-effects or desensitization of the response to the agonist *(46–48,70)*. Presumably, administration in this way, directly to the site of action within the spinal cord, requires only local concentrations that are too low to appear in the systemic circulation, thus avoiding the production of adverse effects. In addition, the reduced extracellular level of the agonist decreases the possibility of receptor desensitization. The response to systemic administration of baclofen is reduced after chronic treatment but this seems not to be the case with intrathecal infusion. The decrease has been attributed to a desensitization of the receptor but the exact mechanism underlying this phenomenon is unclear, but whatever is the cause it may well explain why baclofen has only limited uses, for example, an analgesic.

4. Nociception

The antinociceptive/analgesic action of baclofen administered systemically was first described in laboratory animals more than three decades ago *(71)*. This was subsequently confirmed by different groups. Baclofen has nociceptive activity in acute pain models, such as the tail flick and hot plate tests in rodents at doses below the threshold for muscle relaxation *(72–88)*. Thus, an impairment of locomotor activity can be excluded as a confounding reason for the effect. The antinociceptive action of baclofen in acute pain stems, in part, from a reduction in the release of nociceptive transmitter from primary afferent fibres within the dorsal horn of the spinal cord *(89–94)*. In addition, contribution from an action within higher centers, in particular the thalamus, also occurs *(95,96)*. $GABA_B$-receptor activation in spinal cord slices inhibits the release of substance P, glutamate, and CGRP evoked by electrical stimulation of the dorsal roots *(97–99)*. Each of these substances has been associated with transmission of nociceptive impulses in the spinal cord. In slices from control rats, the application of $GABA_B$-receptor antagonists produces little or no increase in the evoked release of transmitter.

However, in slices obtained from rats with chronic inflammation, produced by complete Freund's adjuvant (monoarthritis), which produces an increase in the concentration of GABA within dorsal horn neurones (up to 25% *[100]*), the evoked release of substance P is dramatically increased by $GABA_B$-receptor antagonism *(101)*. If the antagonist is administered in vivo to monoarthritic rats a significant hyperalgesia occurs. This contrasts markedly with the lack of effect in control rats. These results suggest that an increase in $GABA_B$ innervation to primary afferent terminals occurs during chronic inflammation and this acts as a pathological antinociceptive process to decrease the enhanced sensory input.

Despite the large volume of preclinical data that is now available, the use of baclofen as an analgesic in man has been very limited presumably owing, in part, to rapid tolerance and adverse effects following systemic administration. Possibly the only type of pain for which baclofen has been used systemically is trigeminal neuralgia *(102–105)*. However, when administered intrathecally in the same manner as for the treatment of spasticity, baclofen is able to produce central analgesia such as that as a result of a stroke *(106,107)*. Under these conditions tolerance to the effects of baclofen was not reported.

$GABA_B$ receptors in the spinal cord have also been implicated in the antinoceptive action of spinally administered muscarinic agonists *(108)* and cannabinoid CB1 agonists *(109)*. But as mentioned earlier, the spinal cord is not the only site where a $GABA_B$ agonist exerts its antinociceptive effect. Focal injections into the thalamus can suppress nociceptive processing in chronic inflammation *(96)*. In this study the authors observed that the site at which baclofen acted to reduce the nociceptive response to ankle bending in rats, 14 d after inducing a monoarthritis, was localized within the ventrobasal complex. Injections into adjacent nuclei that unlike the ventrobasal complex are not associated with nociceptive processing, failed to produce an "analgesic" response. Interestingly, it had previously been observed by Ipponi et al. *(110)* that the antinociceptive effect of the GABA uptake inhibitor, tiagabine, in rodents which the authors attributed to the $GABA_B$-receptor activation, was associated with an increase in the extracellular concentration of GABA within the thalamus.

In contrast to models of inflammatory pain the induction of neuropathic pain in rodents does not produce an increase in GABA levels within the spinal cord. However, baclofen produces an antinociceptive effect in models of chronic neuropathy *(81,111–114)*. Moreover, it has been reported that gabapentin, a drug frequently used for the clinical treatment of neuropathic pain, is reputed to be a $GABA_B$-receptor agonist *(115)*. However, this agonist action of GABApentin has not been confirmed by other groups making the potential connection between $GABA_B$-receptor activation and neuropathic therapy rather tenuous. Nevertheless, recent studies by McCarson et al. *(116)* in a rat model of neuropathic pain, suggest that amitriptyline, an antidepressant drug often used clinically in the treatment of neuropathic pain, increases $GABA_B$ receptor function in spinal cord. The implication is that maintenance of $GABA_B$ receptor activity may be crucial in the treatment of neuropathic pain.

Support for a role for $GABA_B$ receptors in pain mechanisms also comes from a series of developmental "knockout" studies performed in mice by Schuler et al.

(24), Prosser et al. *(23)*, and Gassmann et al. *(26)*. In these mice functional $GABA_B$ are not formed, as the mice are deficient in either of the individual $GABA_{B1}$- or $GABA_{B2}$-subunits. In both forms of null mutant mice hyperalgesia was exhibited in acute nociceptive tests, suggesting that functional heteromeric $GABA_B$ receptors are required to maintain pain thresholds.

5. Cognition

Baclofen suppresses cognitive behavior in animals *(117–120)*. In general the cognitive suppression produced by baclofen is similar to that produced by the muscarinic antagonist, scopolamine, but whereas $GABA_B$ mediated cognitive impairment is selectively reversed by $GABA_B$ antagonists but not that produced by scopolamine. The effect of baclofen is, of course, of little consequence to clinical medicine but there is the possibility that $GABA_B$ antagonists might provide a novel opportunity to treat cognitive impairment in man. An enhancement in learning and memory retention in a variety of animal models has been established *(121–128)* and this has led to the first clinical trial of a $GABA_B$ antagonist, SGS742, for mild cognitive impairment *(129)*. The compound is currently in phase II and the indications are that it is progessing satisfactorily. Thus far, SGS742 (600 mg t.i.d for 8 wk) has been administered double-blind to 75 patients and continued assessments of working memory, psychomotor speed, and attention performed. On comparing with a placebo-treated group, significant improvements in all of these parameters were observed with no serious adverse effects emerging.

In order for a competitive antagonist to exert an effect without the addition of the agonist baclofen, the presence of endogenous agonist, presumably, GABA is necessary, i.e., GABAergic "tone" exists. Evidence suggests that the site of action of $GABA_B$ antagonists in relation to cognition may be the hippocampus where an increase in long-term potentiation (LTP) has been implicated *(130,131)*, but the nature of this modification appears to depend on the frequency of stimulation engaged to produce LTP. Presynaptic GABAergic tone within this brain region is reduced leading to enhanced-glutamatergic transmission, which might be responsible for facilitating LTP. If such tone exists elsewhere in the brain then one might expect the production of unwanted effects. In the $GABA_B$ system there would appear to be very little tone under normal conditions. So although one might expect, for example, hyperalgesia to occur as an unwanted effect this does not arise in control animals, as pointed out earlier. Therefore, it would seem unlikely to arise in man.

6. Drug Addiction

Drug addiction is a steadily increasing global problem such that adequate therapeutic treatment of dependence on drugs of abuse is still a major clinical target. Even treatment of dependence on tobacco-related products requires attention. A number of sites have been suggested, including dopamine and glutamate receptors, as potential drug targets to suppress craving and relapse. In addition, $GABA_B$-receptor activation may provide a suitable approach. Baclofen was first shown in 1997, by Roberts and Andrews *(132)*, to reduce the reinforcing effects of cocaine in rats at doses that do not affect locomotion. It soon became clear that other drugs of addiction including nicotine, morphine-related agents, and ethanol were also sensitive to $GABA_B$ agonists, whereas food reinforcement was unaffected *(133–136)*. In fact, centrally or peripherally administered baclofen can even increase food intake in nondeprived rats *(137)*. However, when rats are given access to fatty food they elicit spells of binge eating. Under these conditions baclofen can significantly inhibit the intake of food *(138)*.

The reduction in nicotine self-administration in rats produced by baclofen and other $GABA_B$-receptor agonists, for example, CGP44532, may eventually provide a therapy for smoking cessation *(139)*. However, it has been noted that exposure to cigarette smoke for 10–60 min greatly enhances the expression of $GABA_B$-receptor subunit RNAs in rat prefrontal cortex *(140)*, suggesting that more detailed studies are required to understand the significance of this interaction before the pursuit of therapy. Beuten et al. *(141)* have reported a highly significant association of variants of the human $GABA_{B2}$ gene with nicotine dependence in a cohort of 1276 smoking and nonsmoking individuals, suggesting that the gene plays an important role in the etiology of nicotine addiction.

The finding that baclofen reduces craving for a host of unrelated addictive substances, including heroin, alcohol, and nicotine suggests that there may well be an underlying common mechanism for the $GABA_B$ agonist in all cases. The reward center within the mesolimbic system, possibly the ventral tegmental area, would provide the focus for this action where control of the release/action of dopamine is implicated. Baclofen acts as a mimetic of the natural transmitter GABA, so that if the level of endogenous GABA is raised within this brain region, it would have the same effect. If vigabatrin, an inhibitor of GABA metabolism or the GABA uptake inhibitor, NO-711, is administered centrally in rats they both attenuate heroin and cocaine self-administration and prevent cocaine-induced increases in dopamine in this brain region *(142,143)*. Interestingly, gabapentin, the disputed $GABA_B$ agonist, which might also increase GABA release from stores, has been reported to reduce the craving for cocaine in man *(144)*.

Clinical data indicate that the observations obtained with baclofen on self-administration of cocaine and alcohol in rats, are predictive of man. For example, Ling et al. *(145)* and Haney et al. *(146)* have shown that craving for cocaine can be reduced in man by baclofen and similarly, it can reduce the intake of alcohol, promote abstinence from alcohol, and reduce craving in alcohol-dependent subjects *(147,148)*. Thus, $GABA_B$-receptor activation appears to be an effective and novel approach to the clinical treatment of drug addiction. However, as was alluded to earlier, the administration of baclofen is not without its problems. Apart from the adverse effects produced in some patients when trying to obtain adequate muscle relaxation, the added component of muscle relaxation possibly occurring when treating the drug dependency may detract from any potential benefits. Thus, the systemic use of baclofen might well be limited by the same adverse effects as experienced in the treatment of spasticity. Clearly, focal injection of the agonist, as administered intrathecally in spasticity, is not an option for treating drug addiction, so what might provide an alternate approach? The presence of a positive allosteric modulatory site on the $GABA_{B2}$-subunit may provide the answer.

Allosteric modulation of $GABA_B$ receptors was first described in 2001 by Urwyler and colleagues *(149)*. Subsequent studies then showed that the location of the modulatory site appears to be in the heptahelical domain of the $GABA_{B2}$-subunit *(14)*. Although the $GABA_{B1}$-subunit has the agonist-binding domain it does not have an allosteric site whereas the converse appears to be true for $GABA_{B2}$.Two compounds were originally reported by Urwyler and colleagues as positive modulators, CGP7930 and GS39783 *(149,150)*, and whereas neither of them has any direct agonist activity both accentuate the effects of GABA and baclofen. These compounds have been examined in rat models of addiction and both modulators reduced the self-administration of cocaine and suppressed the acquisition of drinking behavior in alcohol-preferring rats without the need to administer a directly-acting receptor agonist *(151–153)*. These data would suggest that positive allosteric modulation might provide an effective therapy for treating drug dependency, whereas avoiding the possible adverse effects of an agonist, like baclofen.

7. Absence Epilepsy

Absence seizures are primarily associated with juveniles but they frequently disappear during the late teens, although they may be retained in the later years. The seizures do not produce convulsive behavior but have a characteristic electroencephalogram waveform of a 3 Hz spike and wave, which stem from discharges in the thalamic nuclei. It has been believed for many years that the thalamus is the site from which these discharges originate and although an intact thalamocortical network is necessary for generating spike and wave discharges the origin of these discharges appears to lie outside the thalamus. Studies by Meeren et al. *(154)* have shown unequivocally in a genetic rat model of absence epilepsy (WAG/rij) that the site of origin is within the perioral region of the somatosensory cortex. This spreads rapidly across the cortex and initiates a corticothalamic cascade. Injection of a $GABA_B$ agonist into the ventrobasal thalamus or reticular nucleus of a rat, with another genetically based spontaneous form of absence seizures (GAERS), exacerbates the activity *(155)*. By contrast, injection of a $GABA_B$ antagonist into the same regions suppresses the spike and wave discharges *(156)*. Similar result is produced if the $GABA_B$ antagonist is administered systemically. This might indicate that interference with the GABAergic innervation from the reticular nucleus disrupts the thalamocortical loop, which generates the spike and wave activity. However, microinjection of the same selective antagonists into the somatosensory cortex can also suppress the seizure activity indicating the involvement of $GABA_B$ergic mechanisms at this level too *(156)*.

The potential benefits from $GABA_B$ antagonists in the therapy of absence epilepsy seem clear however, their possible introduction into clinical medicine may never occur as absence seizures occur primarily in juveniles. It may depend on previous clinical studies performed in adults for another indication. Preclinical studies with $GABA_B$-receptor agonists suggest that activation of $GABA_B$ receptors in susceptible rodents increases seizure activity. So, in patients with existing absence seizures, $GABA_B$-receptor agonists would be likely to enhance seizure activity and any increase in GABA concentration in the vicinity of the thalamus would be expected to enhance seizure activity. Thus, for example, vigabatrin and tiagabine would be, and are, contraindicated in such individuals.

The mechanism(s) underlying the seizure exacerbation by $GABA_B$ agonists is unclear but an involvement of transient Ca^{2+} T-currents seems possible *(40)*. Synaptic $GABA_B$ mediated late hyperpolarizations might give rise to Ca^{2+} spikes, presumably caused by deinactivation of Ca^{2+} T-currents. Prevention of the hyperpolarization with a $GABA_B$ antagonist would then indirectly prevent the Ca^{2+} spiking and presumably the thalamocortically evoked discharges. No evidence for an increase in the number of $GABA_B$ receptors could be obtained in GAERS *(157)* and WAG/Rj *(158)* rats, although data from another rodent model of absence epilepsy, the Lethargic mouse, indicated that an increase in receptors could contribute to the increase in seizure activity *(159)*. An alternative explanation might be a modest but significant increase in the endogenous extracellular concentration of GABA, which could produce an enhanced hyperpolarization *(160)*. This might then provide an explanation for the action of γ-hydroxybutyric acid (GHB), which produces absence-like seizures in rodents *(161)*. It has been suggested that GHB is a weak $GABA_B$-receptor agonist *(162–167)*. If GHB mimics the effect of the endogenous agonist, GABA, at $GABA_B$ receptors then $GABA_B$-receptor antagonists would be expected to block the spike and wave discharges produced by GHB, which they do *(168,169)*.

8. Depression and Anxiety

A possible link between $GABA_B$ receptors and functional depression was first proposed more than 20 yr ago by Lloyd and colleagues *(170,171)*. These authors demonstrated that an upregulation in $GABA_B$ binding sites occurs in rat frontal cortex after chronic administration of a variety of antidepressant drugs as well as after electroconvulsive therapy. Even though these findings were disputed *(172,173)* there now seems little doubt that $GABA_B$ mechanisms can be associated with depression. Antagonism of $GABA_B$ receptors produces a reversal of depressant-like behavior in recognized animal models *(174)*. For example, after four weeks treatment with the $GABA_B$ antagonist, CGP51176, a clear reversal of depression-like behavior was observed in a mild stress model *(175)*. Similar reversals have been noted in the rodent forced swim test and learned helplessness models of depression. Mombereau et al. *(176)* have observed that mice lacking $GABA_{B1}$- or $GABA_{B2}$-receptor subunits exhibit anti-depressant like behavior but they appear to be more anxious. It has been proposed that $GABA_B$-receptor activation produces anxiolytic activity whereas a loss or blockade of $GABA_B$ receptor function produces antidepressant-like effects.

$GABA_B$-receptor activation in the dorsal periaqueductal gray of rats impairs one-way escape in the elevated T-maze test, which is consistent with an anxiolytic or panicolytic effect *(177)*. Baclofen has also been reported to reduce the incidence of panic attacks in patients following systemic administration *(178)*. However, until clinical studies are performed there is no real way of knowing whether antidepressant activity will emerge from $GABA_B$ receptor blockade. Studies by Heese et al. *(179)* have shown that $GABA_B$-receptor antagonists administered to rats produce a rapid and significant rise in nerve growth factor and brain-derived neurotrophic factor levels, which is also a characteristic of chronic antidepressant drug administration.

9. Other Actions

As indicated in table 1 $GABA_B$-receptor activation produces a variety of effects which do or may have clinical significance. For example, baclofen has antitussive effects that derive, in part, from an action in the brain stem controlling the cough reflex *(180)*. Baclofen is also effective in the treatment of intractable hiccup *(181–186)*. $GABA_B$ agonists also exert an antibronchioconstrictor (antiasthma) action by activating presynaptic receptors on parasympathetic nerve terminals to suppress the release of acetylcholine *(187,188)*. In addition, within the lung, a reduction in nonadrenergic, noncholinergic bronchoconstriction occurs *(189)*. $GABA_B$ agonists also inhibit transient lower oesophageal sphincter relaxations in the dog, which has potential clinical application *(190)*. An interesting effect of $GABA_B$-receptor activation has recently been reported in neutrophils where it appears to act as a chemoattractant receptor and may have a significant role in the inflammatory response *(191)*.

10. Conclusion

The significance of $GABA_B$ ligands as therapeutic agents has yet to be fully realized even though the receptor was discovered nearly three decades ago. However, the potential clinical applications are dependent on the availability of appropriate chemical entities. As the $GABA_B$ receptor system is of predominant, but not exclusive importance within the central nervous system (CNS), it is essential that receptor ligands can gain access to the brain. In addition the lack of demonstrable functional subtyping of the receptor has limited exploitment of any pharmacological specificity. The only receptor agonist in current clinical use is baclofen, which has many limitations, not least of which is its poor ability to penetrate the brain. As a consequence high doses might be needed to achieve a required response. This provoked the idea of local administration as alluled to earlier in the treatment of spasticity. Clearly, this form of application has limited use and so chemical modification would be required to overcome this. As yet little has emerged which is an improvement over baclofen. However, the discovery of allosteric modulators of the $GABA_B$ system may provide the possibility of producing compounds that will readily gain access to the CNS whereas facilitating receptor function. Functional effects have already been demonstrated but further studies are still required.

Perhaps surprisingly, there are no receptor antagonists in therapeutic use and only one antagonist is undergoing clinical trials. CGP 36742 (SGS 742) *(129)* is being tested as a treatment for mild cognitive impairment. This compound has only a low receptor affinity (in the micromolar range) but there are others with subnanomolar affinity, so it will be interesting to follow the progression of SGS742 through the clinical phases as this could open the "flood gates" to the use of more potent antagonists. Clearly, there is still much to do if the full potential of $GABA_B$ receptor ligands as therapeutic agents is to be realized.

References

1. Bein, H. J. (1972) Pharmacological differentiations of muscle relaxants, in *Spasticity: A Topical Survey* (Birkmayer, W. and Huber, H, eds.), Vienna.
2. Keberle, H. and Faigle, J. W. (1972) Synthesis and structure-activity relationship of the gamma-aminobutyric acid derivatives, in *Spasticity: A Topical Survey* (Birkmayer, W. and Huber, H, eds.), Vienna.
3. Bowery, N. G., Hill, D. R., Hudson, A. L., et al. (1980) (-)Baclofen decreases neurotransmitter release in the mammalian CNS by an action at a novel GABA receptor. *Nature* **283,** 92–94.

4. Bowery, N. G., Doble, A., Hill, D. R., et al. (1981) Bicuculline-insensitive GABA receptors on peripheral autonomic nerve terminals. *Eur. J. Pharmacol.* **71,** 53–70.
5. Hill, D. R. and Bowery, N. G. (1981) 3H-baclofen and 3H-GABA bind to bicuculline-insensitive GABAB sites in rat brain. *Nature* **290,** 149–152.
6. Davies, J. and Watkins, J. C. (1974) The action of β-phenyl-GABA derivatives on neurones of the cat cerebral cortex. *Brain Res.* **70,** 501–505.
7. Newberry, N. R. and Nicoll, R. A. (1984) Direct hyperpolarizing action of baclofen on hippocampal pyramidal cells. *Nature* **308,** 450–452.
8. Bowery, N. G., Bettler, B., Froestl, W., et al. (2002) International Union of Pharmacology. XXXIII. Mammalian γ-aminobutyric $acid_B$ receptors: Structure and function. *Pharmacol. Rev.* **54(2),** 247–264.
9. Margeta-Mitrovic, M., Jan, Y. N., and Jan, L. Y. (2000) A trafficking checkpoint controls $GABA_B$ receptor heterodimerization. *Neuron* **27,** 97–106.
10. Calver, A. R., Robbins, M. J., Cosio, C., et al. (2001) The C-terminal domains of the $GABA_B$ receptor subunits mediate intracellular trafficking but are not required for receptor signalling. *Neuroscience* **21,** 1203–1210.
11. Galvez, T., Duthey, B., Kniazeff, J., et al. (2001) Allosteric interactions between GB1 and GB2 subunits are required for optimal $GABA_B$ receptor function. *EMBO J.* **20,** 2152–2159.
12. Pagano, A., Rovelli, G., Mosbacher, J., et al. (2001) C-terminal interaction is essential for surface trafficking but not for heteromeric assembly of $GABA_B$ receptors. *J. Neurosci.* **21**, 1189–1202.
13. Robbins, M. J., Calver, A. R., Fillipov, A. K., Couve, A., Moss, S. J., and Pangalos, M. N. (2001) The $GABA_{B2}$ subunit is essential for G protein coupling of the $GABA_B$ receptor heterodimer. *J. Neurosci.* **21,** 8043–8052.
14. Binet, V., Brajon, C., Le Corre, L., Acher, F., and Pin, J. P. (2004) The heptahelical domain of $GABA_{B2}$ is activated directly by CGP7930, a positive allosteric modulator of the $GABA_B$ receptor. *J. Biol. Chem.* **279,** 29,085–29,091.
15. Uezono, Y., Kanaide, M., Kaibara, M., et al. (2006) Coupling of $GABA_B$ receptor $GABA_{B2}$ subunit to G proteins: evidence from Xenopus oocyte and baby hamster kidney cell expression system. *Am. J. Physiol. Cell. Physiol.* **290,** C200–C207.
16. Kaupmann, K., Huggel, K., Heid, J., et al. (1997) Expression cloning of $GABA_B$ receptors uncovers similarity to metabotropic glutamate receptors. *Nature* **386,** 239–246.
17. Isomoto, S., Kaibara, M., Sakurai-Yamashita, Y., et al. (1998) Cloning and tissue distribution of novel splice variants of the rat $GABA_B$ receptor. *Biochem. Biophys. Res. Commun.* **253,** 10–15.
18. Pfaff, T., Malitschek, B., Kaupmann, K., et al. (1999) Alterantive splicing generates a novel isoform of the rat metabotropic $GABA_BR1$ receptor. *Eur. J. Neurosci.* **11,** 2874–2882.
19. Calver, A. R., Medhurst, A. D., Robbins, M. J., et al. (2000) The expression of $GABA_{B1}$ and $GABA_{B2}$ receptor subunits in the CNS differs from that in peripheral tissues. *Neuroscience* **100,** 155–170.
20. Schwarz, D. A., Barry, G., Eliasof, S. D., Petroski, R. E., Conlon, P. J., and Maki, R. (2000) Characterization of gamma-aminobutyric acid receptor $GABA_{B1(e)}$, a

$GABA_{B1}$ splice variant encoding a truncated receptor. *J. Biol. Chem.* **275,** 32,174–32,181.
21. Wei, K., Eubanks, J. H., Francis, J., Jia, Z., and Snead, O. C. (2001a) Cloning and tissue distribution of a novel isoform of the rat $GABA_BR1$ receptor subunit. *Neuroreport* **12,** 833–837.
22. Wei, K., Jia, Z., Wang, Y. T., Yang, J., Liu, C. C., and Snead, O. C. (2001b) Cloning and characterization of a novel variant of rat $GABA_BR1$ with a truncated C-terminus. *Brain Res. Mol. Brain Res.* **89,** 103–110.
23. Prosser, H. M., Gill, C. H., Hirst, W. D., et al. (2001) Epileptogenesis and enhanced prepulse inhibition in $GABA_{B1}$-deficient mice. *Mol. Cell. Neurosci.* **17,** 1059–1070.
24. Schuler, V., Luscher, C., Blanchet, C., et al. (2001) Epilepsy, hyperalgesia, impaired memory, and loss of pre- and post-synaptic $GABA_B$ responses in mice lacking $GABA_{B1}$. *Neuron* **31,** 47–58.
25. Thuault, S. J., Brown, J. T., Sheardown, S. A., et al. (2004) The $GABA_{B2}$ subunit is critical for the trafficking and function of native $GABA_B$ receptors. *Biochem. Pharmacol.* 2004, **68(8),** 1655–1666.
26. Gassmann, M., Shaban, H., Vigot, R., et al. (2004) Redistribution of $GABA_{B(1)}$ protein and atypical $GABA_B$ responses in $GABA_{B(2)}$–deficient mice. *J. Neurosci.* **24,** 6086–6097.
27. Odagaki, Y., Nishi, N., and Koyama, T. (2000) Functional coupling of $GABA_B$ receptors with G proteins that are sensitive to N-ethylmaleimide treatment, suramin and benzalkonium chloride in rat cerebral cortical membranes. *J. Neural. Transm.* **107,** 1101–1116.
28. Odagaki, Y. and Koyama, T. (2001) Identification of G alpha subtype(s) involved in gamma-aminobutyric $acid_B$ receptor-mediated high-affinity guanosine triphosphatase activity in rat cerebral cortical membranes. *Neurosci. Lett.* **297,** 137–141.
29. Karbon, E. W., Duman, R. S., and Enna, S. J. (1984) $GABA_B$ receptors and norepinephrine-stimulated cAMP production in rat brain cortex. *Brain Res.* **306,** 327–332.
30. Hill, D. R., Bowery, N. G., and Hudson, A. L. (1984) Inhibition of $GABA_B$ receptor binding by guanyl nucleotides. *J. Neurochem.* **42,** 652–657.
31. Xu, J. and Wojcik, W. J. (1986) Gamma aminobutyric acid B receptor-mediated inhibition of adenylate cyclase in cultured cerebellar granule cells: blockade by islet-activating protein. *J. Pharmacol. Exp. Ther.* **239,** 568–573.
32. Enna, S. J. (2000) $GABA_B$ receptor signalling pathways, in *Pharmacology of GABA and Glycine Neurotransmission* (Möhler, H., eds.), Springer-Verlag, Berlin, pp. 329–342.
33. Wagner, P. G. and Dekin, M. S. (1993) $GABA_B$ receptors are coupled to a barium-insensitive outward rectifying potassium conductance in premotor respiratory neurons. *J. Neurophysiol.* **69,** 286–289.
34. Premkumar, L. S. and Gage, P. W. (1994) Potassium channels activated by $GABA_B$ antagonists and serotonin in cultured hippocampal neurons. *J. Neurophysiol.* **71,** 2750–2575.
35. Andrade, R., Malenka, R. C., and Nicoll, R. A. (1986) A G protein couples serotonin and $GABA_B$ receptors to the same channels in hippocampus. *Science (Wash DC)* **234,** 1261–1265.

36. Saint, D. A., Thomas, T., and Gage, P. W. (1990) $GABA_B$ antagonists modulate a transient potassium current incultured mammalian hippocampal neurons. *Neurosci. Lett.* **118,** 9–13.
37. Dunlap, K. (1981) Two types of γ-aminobutyric receptor on embryonic sensory neurons. *Br. J. Pharmacol.* **74,** 579–585.
38. Desarmenien, M., Feltz, P., Occhipinti, G., Santangelo, F., and Schlichter, R. (1984) Coexistence of $GABA_A$ and $GABA_B$ receptors on A and C primary afferents. *Br. J. Pharmacol.* **81,** 327–333.
39. Scott, R. H., Wotton, J. F., and Dolphin, A. C. (1990) Modulation of neuronal T-type calcium channel currents by photoactivation of intracellular guanosine 5′-O(3-thio) triphosphate. *Neuroscience* **38,** 285–294.
40. Crunelli, V. and Leresche, N. (1991) A role for $GABA_B$ receptors in excitation and inhibition of thalamocortical cells. *Trends Neurosci.* **14,** 16–21.
41. Dutar, P. and Nicoll, R. A. (1988) A physiological role for $GABA_B$ receptors in the central nervous system. *Nature* **332,** 156–158.
42. Karlsson, G., Pozza, M., and Olpe, H. R. (1988) Phaclofen: a $GABA_B$ blocker reduces long-duration inhibition in the neocortex. *Eur. J. Pharmacol.* **148,** 485–486.
43. Soltesz, I., Haby, M., Leresche, N., and Crunelli, V. (1988) The $GABA_B$ antagonist phaclofen inhibits the late K^+ dependent IPSP in cat and rat thalamic and hippocampal neurones. *Brain Res.* **448,** 351–354.
44. Mott, D. D. and Lewis, D. L. (1994) The pharmacology and function of $GABA_B$ receptors. *Int. Rev. Neurobiol.* **36,** 97–223.
45. Isaacson, J. S., Solis, J. M., and Nicoll, R. A. (1993) Local and diffuse synaptic actions of GABA in the hippocampus. *Neuron* **10,** 165–175.
46. Ochs, G., Naumann, C., Dimitrijevic, M., and Sindou, M. (1999) Intrathecal baclofen for long-term treatment of spasticity: a multi-centre study. *J. Neurol. Neurosurg. Psychiatry* **52,** 933–939.
47. Ochs, G., Struppler, A., Meyerson, B. A., et al. (1989) Intrathecal baclofen for long-term treatment of spasticity: a multi-centre study. *J. Neurol. Neurosurg. Psychiatry* **52,** 933–939.
48. Penn, R. D., Savoy, S. M., Corcos, D., Latash, M., Gottlieb, G., and Kroin, J. (1989) Intrathecal baclofen for severe spinal spasticity. *N. Eng. J. Med.* **320,** 1517–1521.
49. Penn, R. D. and Mangieri, E. A. (1993) Stiff-man syndrome treated with intrathecal baclofen. *Neurology* **43,** 2412.
50. Becker, W. J., Harris, C. J., Long, M. L., Ablett, D. P., Klein, G. M., and DeForge, D. A. (1995) Long-term intrathecal baclofen therapy in patients with intractable spasticity. *Can. J. Neurol. Sci.* **22,** 208–217.
51. Becker, R., Alberti, O., and Bauer, B. L. (1997) Continuous intrathecal baclofen infusion in severe spasticity after traumatic of hypoxic brain injury. *J. Neurol.* **244,** 160–166.
52. Becker, R., Benes, L. Sure, U., Hellwig, D., and Bertalanffy, H. (2000) Intrathecal baclofen alleviates autonomic dysfunction in severe brain injury. *J. Clin. Neurosci.* **7,** 316–319.
53. Campbell, S. K., Almeida, G. L., Penn, R. D., and Corcos, D. M. (1995) The effects of intrathecally administered baclofen on function in patients with spasticity. *Phys. Ther.* **75,** 352–362.

54. Seitz, R. J., Blank, B., Kiwit, J. C., and Benecke, R. (1995) Stiff-person syndrome with anti-glutamic acid decarboxylase autoantibodies: complete remission of symptoms after intrathecal baclofen administration. *J. Neurol.* **242,** 618–622.
55. Albright, A. L., Barry, M. J., Fasick, P., Barron, W., and Shultz, B. (1996) Continuous intrathecal baclofen infusion for symptomatic generalized dystonia. *Neurosurgery* **38,** 934–939.
56. Azouvi, P., Mane, M., Thiebaut, J. B., Denys, P., Remy-Neris, O., and Bussel, B. (1996) Intrathecal baclofen administration for control of severe spinal spasticity: functional improvement and long-term follow-up. *Arch. Phys. Med. Rehab.* **77,** 35–39.
57. Dressnandt, J. and Conrad, B. (1996) Lasting reduction of severe spasticity after ending chronic treatment with intrathecal baclofen. *J. Neurol. Neurosurg. Psychiatry* **2,** 168–173.
58. Ford, B., Greene, P., Louis, E. D., et al. (1996) Use of intrathecal baclofen in the treatment of patients with dystonia. *Arch. Neurol.* **53,** 1241–1246.
59. Paret, G., Tirosh, R., Ben Zeev, B., Vardi, A., Brandt, N., and Barzilay, Z. (1996) Intrathecal baclofen for severe torsion dystonia in a child. *Acta. Paediatr.* **85,** 635–637.
60. Armstrong, R. W., Steinbok, P., Cochrane, D. D., Kube, S. D., Fife, S. E., and Farrell, K. (1997) Intrathecally administered baclofen for treatment of children with spasticity of cerebral origin. *J. Neurosurg.* **87,** 409–414.
61. Dressler, D., Oeljeschläger, R. O., and Rüther, E. (1997) Severe tardive dystonia: treatment with continuous intrathecal baclofen administration. *Movement Disorders* **12,** 585–587.
62. Dressnandt, J., Konstanzer, A., Weinzierl, F. X., Pfab, R., and Klingelhöfer, J. (1997) Intrathecal baclofen in tetanus: four cases and a review of reported cases. *Intensive Care Med.* **23,** 896–902.
63. François, B., Clavel, M., Desachy, A., Vignon, P., Salle, J. Y., and Gastinne, H. (1997) Continuous intrathecal baclofen in tetanus. A therapeutic alternative. *Presse Med.* **26,** 1045–1047.
64. Francois, B., Vacher, P., Roustan, J., et al. (2001) Intrathecal baclofen after traumatic brain injury: Early treatment using a new technique to prevent spasticity. *J. Trauma-Injury Infect. Crit. Care* **50,** 158–161.
65. Meythaler, J. M., McCary, A., and Hadley, M. N. (1997) Prospective assessment of continuous intrathecal infusion of baclofen for spasticity caused by acquired brain injury: a preliminary report. *J. Neurosurg.* **87,** 415–419.
66. Gerszten, P. C., Albright, A. L., and Johnstone, G. F. (1998) Intrathecal baclofen infusion and subsequent orthopedic surgery in patients with spastic cerebral palsy. *J. Neurosurg.* **88,** 1009–1013.
67. Auer, C., Siebner, H. R., Dressnandt, J., and Conrad, B. (1999) Intrathecal baclofen increases corticospinal output to hand muscles in multiple sclerosis. *Neurology* **52,** 1298–1299.
68. Orsnes, G. B., Sorensen, P. S., Larsen, T. K., and Ravnborg, M. (2000) Effect of baclofen on gait in spastic MS patients. *Acta. Neurol. Scand.* **101,** 244–248.
69. Trampitsch, E., Krumpholz, R., Likar, R., Oher, M., and Gulle, D. (2000) Continuous intrathecal administration of baclofen in severe tetanus. *Anaesthesiol Intensivmed. Notfallmed. Schmerzther.* **35,** 532–533.

70. Krach, L. E. (2001) Pharmacotherapy of spasticity: oral medications and intrathecal baclofen. *J. Child. Neurol.* **16,** 31–36.
71. Cutting, D. A. and Jordan, C. C. (1975) Alternative approaches to analgesia: baclofen as a model compound. *Br. J. Pharmacol.* **54,** 171–179.
72. Levy, R. A. and Proudfit, H. K. (1979) Analgesia produced by microinjection of baclofen and morphine at brain stem sites. *Eur. J. Pharmacol.* 57, 43–55.
73. Wilson, P. R. and Yaksh, T. L. (1978) Baclofen is antinociceptive in the spinal intrathecal space of animals. *Eur. J. Pharmacol.* **51,** 323–330.
74. Liebman, J. M. and Pastor, G. (1980) Antinociceptive effects of baclofen and muscimol upon intraventricular administration. *Eur. J. Pharmacol.* **61,** 225–230.
75. Kendall, D. A., Browner, M., and Enna, S. J. (1982) Comparison of the antinociceptive effect of gamma-aminobutyric acid (GABA) agonists: evidence for a cholinergic involvement. *J. Pharmacol. Exp. Ther.* **220,** 482–487.
76. Sawynok, J. and Dickson, C. (1985) D-Baclofen is an antagonist at baclofen receptors mediating antinociception in the spinal cord. *Pharmacology* **31,** 248–259.
77. Vaught, J. L., Pelley, K., Costa, L. G., Setler, P., and Enna, S. J. (1985) A comparison of the antinociceptive responses to the GABA-receptor agonists THIP and baclofen. *Neuropharmacology* **24,** 211–216.
78. Serrano, I., Ruiz, R. M., Serrano, J. S., and Fernandez, A. (1992) GABAergic and cholinergic mediation in the antinociceptive action of homotaurine. *Gen. Pharmacol.* **23,** 421–426.
79. Aley, K. O. and Kulkarni, S. K. (1991) Baclofen analgesia in mice: a $GABA_B$-mediated response. *Methods Find. Exp. Clin. Pharmacol.* **13,** 681–686.
80. Hammond, D. L. and Washington, J. D. (1993) Antagonism of L-baclofen-induced antinociception by CGP 35348 in the spinal cord of the rat. *Eur. J. Pharmacol.* **234,** 255–262.
81. Smith, G. D., Harrison, S. M., Birch, P. J., Elliott, P. J., Malcangio, M., and Bowery, N. G. (1994) Increased sensitivity to the antinociceptive activity of (±)-baclofen in an animal model of chronic neuropathic, but not chronic inflammatory hyperalgesia. *Neuropharmacology* **33,** 1103–1108.
82. Dirig, D. M. and Yaksh, T. L. (1995) Intrathecal baclofen and muscimol, but not midazolam, are antinociceptive using the rat-formalin model. *J. Pharmacol. Exp. Ther.* **275,** 219–227.
83. Thomas, D. A., McGowan, M. K., and Hammond, D. L. (1995) Microinjection of baclofen in the ventromedial medulla of rats: antinociception at low doses and hyperalgesia at high doses. *J. Pharmacol. Exp. Ther.* **275,** 274–284.
84. Thomas, D. A., Navarrete, I. M., Graham, B. A., McGowan, M. K., and Hammond, D. L. (1996) Antinociception produced by systemic R(+)-baclofen hydrochloride is attenuated by CGP 35348 administered to the spinal cord or ventromedial medulla of rats. *Brain Res.* **718,** 129–137.
85. McCarson, K. E. and Enna, S. J. (1996) Relationship between $GABA_B$ receptor activation and neurokinin receptor expression in spinal cord. *Pharmacol. Rev. Commun.* **8,** 191–194.
86. Wiesenfeld-Hallin, Z., Aldskogius, H., Grant, G., Hao, J. X., Hökfelt, T., and Xu, X. J. (1997) Central inhibitory dysfunctions: mechanisms and clinical implications. *Behav. Brain Sci.* **20,** 420–425.

87. Cui, J. G., Meyerson, B. A., Sollevi, A., and Linderoth, B. (1998) Effect of spinal cord stimulation on tactile hypersensitivity in mononeuropathic rats is potentiated by simultaneous $GABA_B$ and adenosine receptor activation. *Neurosci. Lett.* **247,** 183–186.
88. Przesmycki, K., Dzieciuch, J. A., Czuczwar, S. J., and Kleinrok, Z. (1998) An isobolographic analysis of drug interaction between intrathecal clonidine and baclofen in the formalin test in rats. *Neuropharmacology* **37,** 207–214.
89. Henry, J. L. (1982) Effects of intravenously administered enantiomers of baclofen on functionally identified units in lumbar dorsal horn of the spinal cat. *Neuropharmacology* **21,** 1073–1083.
90. Price, G. W., Kelly, J. S., and Bowery, N. G. (1987) The location of $GABA_B$ receptor binding sites in mammalian spinal cord. *Synapse* **1,** 530–538.
91. Malcangio, M. and Bowery, N. G. (1996a) GABA and its receptors in the spinal cord. *Trends Pharmacol. Sci.* **17,** 457–462.
92. Ataka, T., Kumamoto, E., Shimoji, K., and Yoshimura, M. (2000) Baclofen inhibits more effectively C-afferent than A delta-afferent glutamatergic transmission in substantia gelatinosa neurons of adult rat spinal cord slices. *Pain* **86,** 273–282.
93. Iyadomi, M., Iyadomi, I., Kumamoto, E., Tomokuni, K., and Yoshimura, M. (2000) Presynaptic inhibition by baclofen of miniature EPSCs and IPSCs in substantia gelatinosa neurons of the adult rat spinal dorsal horn. *Pain* **85,** 385–393.
94. Riley, R. C., Trafton, J. A., Chi, S. I., and Basbaum, A. I. (2001) Presynaptic regulation of spinal cord tachykinin signaling via $GABA_B$ but not $GABA_A$ receptor activation. *Neuroscience* **103,** 725–737.
95. Sawynok, J. (1989) GABAergic agents as analgesics, in *GABA: Basic Research and Clinical Applications* (Bowery, N. G. and Nistico, G., eds.), Pythagora, Rome, pp. 383–399.
96. Potes, C. S., Neto, F. L., and Castro-Lopes, J. M. (2006) Administration of baclofen, a $GABA_B$ agonist, in the thalamic ventrobasal complex attenuates allodynia in monoarthritic rats subjected to the ankle-bend test. *J. Neurosci. Res.* **83,** 515–523.
97. Malcangio, M. and Bowery, N. G. (1993) Gamma-aminobutyric $acid_B$ but not gamma-amino butyric and $acid_A$ receptor activation inhibits electrically evoked substance P-like immunoreactivity release from the rat spinal cord in vitro. *J. Pharmacol. Exp. Ther.* **266,** 1490–1496.
98. Malcangio, M. and Bowery, N. G. (1996b) Calcitonin gene-related peptide content, basal outflow and electrically evoked release from monoarthritic rat spinal cord in vitro. *Pain* **66,** 351–358.
99. Teoh, H., Malcangio, M., and Bowery, N. G. (1996) GABA, glutamate and substance P-like immunoreactivity release: effects of novel $GABA_B$ antagonists. *Br. J. Pharmacol.* **118,** 1153–1160.
100. Castro-Lopes, J. M., Tavares, I., Tolle, T. R., Coito, A., and Coimbra, A. (1992) Increase in GABAergic cells and GABA levels in the spinal cord in unilateral inflammation of the hindlimb of the rat. *Eur. J. Neurosci.* **4,** 296–301.
101. Malcangio, M. and Bowery, N. G. (1994) Spinal cord SP release and hyperalgesia in monoarthritic rats: Involvement of the $GABA_B$ receptor system. *Br. J. Pharmacol.* **113,** 1561–1566.

102. Fromm, G. G., Shibuya, T., Nakata, M., and Terrence, C. F. (1990) Effects of d-baclofen and l-baclofen on the trigeminal nucleus. *Neuropharmacology* **29,** 249–254.
103. Fromm, G. H. (1992) Therapeutic applications of baclofen. *Pharmacol. Commun.* **2,** 132–137.
104. Fromm, G. H. and Terence, C. F. (1987) Comparison of L-baclofen and racemic baclofen in trigeminal neuralgia. *Neurology* **37,** 1725–1728.
105. Terence, C. F., Fromm, G. H., and Tenicela, R. (1985) Baclofen as an analgesic in chronic peripheral nerve disease. *Eur. Neurol.* **24,** 380–385.
106. Taira, T., Kawamura, H., Tanikawa, T., Iseki, H., Kawabatake, H., and Takakura, K. (1995) A new approach to control central deafferentation pain: spinal intrathecal baclofen. *Stereotact. Funct. Neurosurg.* **65,** 101–105.
107. Loubser, P. G. and Akman, N. M. (1996) Effects of intrathecal baclofen on chronic spinal cord injury pain. *J. Pain Symptom Manage.* **12,** 241–247.
108. Chen, S. R. and Pan, H. L. (2004) Activation of muscarinic receptors inhibits spinal dorsal horn projection neurons: Role of $GABA_B$ receptors. *Neuroscience* **125(1),** 141–148.
109. Naderi, N., Shafaghi, B., Khodayar, M. J., and Zarindast, M. R. (2005) Interaction between gamma-aminobutyric acid $GABA_B$ and cannabinoid CB1 receptors in spinal pain pathways in rat. *Eur. J. Pharmacol.* **514(2-3),** 159–164.
110. Ipponi, A., Lamberti, C., Medica, A., Bartolini, A., and Malmberg-Aiello, P. (1999) Tiagabine antinociception in rodents depends on $GABA_B$ receptor activation: parallel antinociception testing and medial thalamus GABA microdialysis. *Eur. J. Pharmacol.* **368,** 205–211.
111. Eaton, M. J., Martinez, M. A., and Karmally, S. (1999) A single intrathecal injection of GABA permanently reverses neuropathic pain after nerve injury. *Brain Res.* **835,** 334–339.
112. Patel, S., Naeem, S., Kesingland, A., et al. (2001) The effects of $GABA_B$ agonists and gabapentin on mechanical hyperalgesia in models of neuropathic and inflammatory pain in the rat. *Pain* **90,** 217–226.
113. Sokal, D. M. and Chapman, V. (2003) Inhibitory effects of spinal baclofen on spinal dorsal horn neurons in inflamed and neuropathic rats in vivo. *Brain Res.* **987,** 67–75.
114. Franek, M., Vaculin, S., and Rokyta, R. (2004) $GABA_B$ receptors agonist baclofen has non-specific antinociceptive effect in the model of peripheral neuropathy in the rat. *Physiol. Res.* **53,** 351–355.
115. Ng, G. Y., Bertrand, S., Sullivan, R., et al. (2001) Gamma-aminobutyric acid type B receptors with specific heterodimer composition and postsynaptic actions in hippocampal neurons are targets of anticonvulsant gabapentin action. *Mol. Pharmacol.* **59,** 144–152.
116. McCarson, K. E., Ralya, A., Reisman, S. A., and Enna, S. J. (2005) Amitriptyline prevents thermal hyperalgesia and modifications in rat spinal cord $GABA_B$ receptor expression and function in an animal model of neuropathic pain. *Biochem. Pharmacol.* **71,** 196–202.
117. Nakagawa, Y., Ishibashi, Y., Yoshii, T., and Tagashira, E. (1995) Involvement of cholinergic systems in the deficit of place learning in Morris water maze task induced by baclofen in rats. *Brain. Res.* **683,** 209–214.

118. Tang, A. C. and Hasselmo, M. E. (1996) Effect of long term baclofen treatment on recognition memory and novelty detection. *Behav. Brain Res.* **74,** 145–152.
119. McNamara, R. K. and Skelton, R. W. (1996) Baclofen, a selective $GABA_B$ receptor agonist, dose-dependently impairs spatial learning in rats. *Pharmacol. Biochem. Behav.* **53,** 303–308.
120. Arolfo, M. P., Zanudio, M. A., and Ramirez, O. A. (1998) Baclofen infused in rat hippocampal formation impairs spatial learning. *Hippocampus* **8,** 109–113.
121. Mondadori, C., Jaekel, J., and Preiswerk, G. (1993) CGP 36742: the first orally active $GABA_B$ blocker improves the cognitive performance of mice, rats and rhesus monkeys. *Behav. Neural. Biol.* **60,** 62–68.
122. Brucato, F. H., Levin, E. D., Mott, D. D., Lewis, D. V., Wilson, W. A., and Swartzwelder, H. S. (1996) Hippocampal long-term potentiation and spatial learning in the rat: effects of $GABA_B$ receptor blockade. *Neuroscience* **74,** 331–339.
123. Getova, D., Bowery, N. G., and Spassov, V. (1997) Effects of $GABA_B$ receptor antagonists on learning and memory retention in a rat model of absence epilepsy. *Eur. J. Pharmacol.* **320,** 9–13.
124. Nakagawa, Y. and Takashima, T. (1997) The $GABA_B$ receptor antagonist CGP36742 attenuates the baclofen-and scopolamine-induced deficit in Morris water maze task in rats. *Brain Res.* **766,** 101–106.
125. Getova, D. and Bowery, N. G. (1998) The modulatory effects of high affinity $GABA_B$ receptor antagonists in an active avoidance learning paradigm in rats. *Psychopharmacology* **137,** 369–373.
126. Stäubli, U., Scafidi, J., and Chun, D. (1999) $GABA_B$ receptor antagonism: facilitatory effects on memory parallel those on LTP induced by TBS but not HFS. *J. Neurosci.* **19,** 4609–4615.
127. Farr, S. A., Uezu, K., Creonte, T. A., Flood, J. F., and Morley, J. E. (2000) Modulation of memory processing in the cingulate cortex of mice. *Pharmacol. Biochem. Behav.* **65,** 363–368.
128. Helm, K. A., Haberman, R. P., Dean, S. L., et al. (2005) $GABA_B$ receptor antagonist SGS742 improves spatial memory and reduces protein binding to the cAMP response element (CRE) in the hippocampus. *Neuropharmacology* **48(7),** 956–964.
129. Froestl, W., Gallagher, M., Jenkins, H., et al. (2004) SGS742: The first $GABA_B$ receptor antagonist in clinical trials. *Biochem. Pharmacol.* **68(8),** 1479–1487.
130. Olpe, H. R., Karlsson, G., Pozza, M. F., et al. (1990) CGP 35348: a centrally active blocker of $GABA_B$ receptors. *Eur. J. Pharmacol.* **187,** 27–38.
131. Mott, D. D. and Lewis, D. V. (1991) Facilitation of the induction of long term potentiation by $GABA_B$ receptors. *Science* **252,** 1718–1720.
132. Roberts, D. C. and Andrews, M. M. (1997) Baclofen suppression of cocaine self-administration: demonstration using a discrete trials procedure. *Psychopharmacology* **131,** 271–277.
133. Corrigall, W. A., Coen, K. M., Adamson, K. L., Chow, B. L., and Zhang, J. (2000) Response of nicotine self-administration in the rat to manipulations of mu-opioid and gamma-aminobutyric acid receptors in the ventral tegmental area. *Psychopharmacology* **149,** 107–114.

134. Xi, Z. X. and Stein, E. A. (1999) Baclofen inhibits heroin self-administration behavior and mesolimbic dopamine release. *J. Pharmacol. Exp. Ther.* **290,** 1369–1374.
135. Lobina, C., Pani, M., Reali, R., Addolorato, G., and Gessa, G. L. (2000) Ability of baclofen in reducing alcohol intake and withdrawal severity: I–preclinical evidence. *Alcohol Clin. Exp. Res.* **24,** 58–66.
136. Colombo, G., Agabio, R., Carai, M. A., et al. (2000) Ability of baclofen in reducing alcohol intake and withdrawal severity: I–preclinical evidence. *Alcohol Clin. Exp. Res.* **24,** 58–66.
137. Ebenezer, I. S. and Patel, S. M. (2004) Effects of the $GABA_B$ receptor agonists baclofen and 3-aminopropylphosphinic acid (3-APA) on food intake in rats. *Methods Find. Exp. Clin. Pharmacol.* **26(8),** 627–630.
138. Buda-Levin, A., Wojnicki, F. H., and Corwin, R. L. (2005) Baclofen reduces fat intake under binge-type conditions. *Physiol. Behav.* **86,** 176–184.
139. Paterson, N. E., Froestl, W., and Markou, A. (2005) Repeated administration of the $GABA_B$ receptor agonist CGP44532 decreased nicotine self-administration, and acute administration decreased cue-induced reinstatement of nicotine-seeking in rats. *Neuropsychopharmacology* **30(1),** 119–128.
140. Li, S. P., Park, M. S., Kim, J. H., and Kim, M. O. (2004) Chronic nicotine and smoke treatment modulate dopaminergic activities in ventral tegmental area and nucleus accumbens and the gamma-aminobutyric acid type B receptor expression of the rat prefrontal cortex. *J. Neurosci. Res.* **78(6),** 868–879.
141. Beuten, J., Ma, J. Z., Payne, T. J., et al. (2005) Single- and multilocus allelic variants within the $GABA_B$ receptor subunit 2 ($GABA_{B2}$) gene are significantly associated with nicotine dependence. *Am. J. Hum. Genet.* **76,** 859–864.
142. Ashby, C. R., Rohatgi, R., Ngosuwan, J., et al. (1999) Implication of the $GABA_B$ receptor in gamma vinyl-GABA's inhibition of cocaine-induced increases in nucleus accumbens dopamine. *Synapse* **31,** 151–153.
143. Xi, Z. X. and Stein, E. A. (2000) Increased mesolimbic GABA concentration blocks heroin self-administration in the rat. *J. Pharmacol. Exp. Ther.* **294,** 613–619.
144. Myrick, H., Henderson, S., Brady, K. T., and Malcolm, R. (2001) Gabapentin in the treatment of cocaine dependence: a case series. *J. Clin. Psychiatry* **62,** 19–23.
145. Ling, W., Shoptaw, S., and Majewska, D. (1998) Baclofen as a cocaine anti-craving medication: a preliminary clinical study. *Neuropsychopharmacology* **18,** 403–404.
146. Haney, M., Hart, C. L., and Foltin, R. W. (2006) Effects of baclofen on cocaine self-administration: opioid- and nonopioid-dependent volunteers. *Neuropsychopharmaco*logy **31,** 1814–1821.
147. Addolorato, G., Caputo, F., Capristo, E., et al. (2002) Baclofen efficacy in reducing alcohol craving and intake: a preliminary double blind randomized controlled study, *Alcohol Alcohol.* **37,** 504–508.
148. Flannery, B. A., Garbutt, J. C., Cody, M. W., et al. (2004) Baclofen for alcohol dependence: a preliminary open-label study. *Alcohol. Clin. Exp. Res.* **28(10),** 1517–1523.
149. Urwyler, S., Mosbacher, J., Lingenhoehl, K., et al. (2001) Positive allosteric modulation of native and recombinant γ-Aminobutyric $acid_B$ receptors by

2,6-Di-*tert*-butyl-4-(3-hydroxy-2,2-dimethyl-propyl)-phenol (CGP7930) and its aldehyde analog CGP13501. *Mol. Pharmacol.* **60,** 963–971.

150. Gjoni, T., Desrayaud, S., Imobersteg, S., and Urwyler, S. (2006) The positive allosteric modulator GS39783 enhances $GABA_B$ receptor-mediated inhibition of cyclic AMP formation in rat striatum in vivo. *J. Neurochem.* **94,** 1416–1422.
151. Smith, M. A., Yancey, D. L., Morgan, D., Liu, Y., Froestl, W., and Roberts, D. C. (2004) Effects of positive allosteric modulators of the $GABA_B$ receptor on cocaine self-administration in rats. *Psychopharmacology* **173(1-2),** 105–111.
152. Slattery, D. A., Markou, A., Froestl, W., and Cryan, J. F. (2005) The $GABA_B$ receptor-positive modulator GS39783 and the $GABA_B$ receptor agonist baclofen attenuate the reward-facilitating effects of cocaine: Intracranial self-stimulation studies in the rat. *Neuropsychopharmacology* **30,** 2065–2072.
153. Orru, A., Lai, P., Lobina, C., et al. (2005) Reducing effect of the positive alloseric modulators of the $GABA_B$ receptor, CGP7930 and GS39783, on alcohol intake in alcohol-preferring rats. *Eur. J. Pharmacol.* **525,** 105–111.
154. Meeren, H., van Luijtelaar, G., Lopes da Silva, F., and Coenen, A. (2005) Evolving concepts on the pathophysiology of absence seizures. *Arch. Neurol.* **62,** 371–376.
155. Marescaux, C., Vergnes, M., and Bernasconi, R. (1992) $GABA_B$ receptor antagonists: Potential new anti-absence drugs. *J. Neural. Transm.* **35,** 179–188.
156. Manning, J. P., Richards, D. A., and Bowery, N. G. (2003) Pharmacology of Absence Epilepsy. *Trends Pharmacol. Sci.* **24,** 542–549.
157. Knight, A. R. and Bowery, N. G. (1992) GABA receptors in rats with spontaneous generalized non-convulsive epilepsy. *J. Neural. Transm.* **35(Suppl.)**, 189–196.
158. Parry, K. P., Drinkenburg, W. H. I. M., and Bowery, N. G. (1996) Lack of alteration of $GABA_B$ receptor binding in the absence epileptic WAG/Rij strain of rat. *Br. J. Pharmacol.* **117**, 304.
159. Lin, F. H., Cao, Z., and Hosford, D. A. (1993) Increased number of $GABA_B$ receptors in the lethargic (1h/1h) mouse model of absence epilepsy. *Brain Res.* **608,** 101–106.
160. Richards, D. A., Lemos, T., Whitton, P. S., and Bowery, N. G. (1995) Extracellular GABA in the ventrolateral thalamus of rats exhibiting spontaneous absence epilepsy: a microdialysis study. *J. Neurochem.* **65,** 1674–1680.
161. Snead, O. C. (1995) Basic mechanisms of generalized absence epilepsy. *Ann. Neurol.* **37,** 146–157.
162. Bernasconi, R., Lauber, J., Marescaux, C., et al. (1992) Experimental absence seizures: potential role of gamma- hydroxybutyric acid and $GABA_B$ receptors. *J. Neurol. Trans.* **35(Suppl.),** 155–177.
163. Bernasconi, R., Mathivet, P., Bischoff, S., and Marescaux, C. (1999) Gamma-hydroxybutyric acid: an endogenous neuromodulator with abuse potential. *Trends Pharmacol. Sci.* **20,** 135–141.
164. Xie, X. and Smart, T. G. (1992) γ-Hydroxybutyrate hyperpolarizes hippocampal neurones by activating $GABA_B$ receptors. *Eur. J. Pharmacol.* **212,** 291–294.
165. Engberg, G. and Wissbrandt, H. (1993) Gamma-hydroxybutyric acid (GHBA) induces pacemaker activity and inhibition of substantia nigra dopamine neurons by activating $GABA_B$ receptors. *Naunyn Schmiedebergs Arch. Pharmacol.* **348,** 491–497.

166. Lingenhoehl, K., Brom, R., Heid, J., et al. (1999) Gamma-hydroxybutyrate is a weak agonist at recombinant $GABA_B$ receptors. *Neuropharmacology* **38,** 1667–1673.
167. Snead, O. C. (2000) Evidence for G protein-coupled gamma-hydroxybutyric acid receptor. *J. Neurochem.* **75,** 1986–1996.
168. Snead, O. C. (1992) Evidence for $GABA_B$-mediated mechanisms in experimental generalized absence seizures. *Eur. J. Pharmacol.* **213,** 343–349.
169. Williams, S. R., Turner, J. P., and Crunelli, V. (1995) Gamma-hydroxybutyrate promotes oscillatory activity of rat and cat thalamocortical neurons by a tonic $GABA_B$ receptor mediated hyperpolarization. *Neuroscience* **66,** 133–141.
170. Pilc, A. and Lloyd, K. G. (1984) Chronic antidepressants and GABA "B" receptors: a GABA hypothesis of antidepressant drug action. *Life Sci.* **35,** 2149–2154.
171. Lloyd, K. G., Thuret, F., and Pilc, A. (1985) Upregulation of gamma-aminobutyric acid (GABA) B binding sites in rat frontal cortex: a common action of repeated administration of different classes of antidepressants and electroshock. *J. Pharmacol. Exp. Ther.* **235,** 191–199.
172. Cross, J. A. and Horton, R. W. (1987) Are increases in $GABA_B$ receptors consistent findings following chronic antidepressant administration? *Eur. J. Pharmacol.* **141(1),** 159–162.
173. Cross, J. A. and Horton, R. W. (1988) Effects of chronic oral administration of the antidepressants, desmethylimipramine and zimelidine on rat cortical $GABA_B$ binding sites: A comparison with 5-HT2 binding site changes. *Br. J. Pharmacol.* **93(2),** 331–336.
174. Cryan, J. F. and Kaupmann, K. (2005) Don't worry 'B' happy!: A role for $GABA_B$ receptors in anxiety and depression. *Trends Pharmacol. Sci.* **26(1),** 36–43.
175. Bittiger, H., Froestl, W., Gentsch, C., et al. (1996) $GABA_B$ receptor antagonists: potential therapeutic applications, in *GABA: Receptors, Transporters and Metabolism* (Tanaka, C., and Bowery, N. G., eds.), Birkhaeuser Verlag, Basel, pp. 297–305.
176. Mombereau, C., Kaupmann, K., Gassmann, M., Bettler, B., and van der Putte Cryan, J. F. (2005) Altered anxiety and depression-related behavior in mice lacking $GABA_{B2}$ receptor subunits. *NeuroReport* **16(3),** 307–310.
177. Bueno, C. H., Zangrossi, H. Jr., Nogueira, R. L., Soares, V. P., and Viana, M. B. (2005) Panicolytic-like effect induced by the stimulation of $GABA_A$ and $GABA_B$ receptors in the dorsal periaqueductal grey of rats. *Eur. J. Pharmacol.* **516,** 239–246.
178. Breslow, M. F., Fankhauser, M. P., Potter, R. L., Meredith, K. E., Misiaszek, J., and Hope, D. G. Jr. (1989) Role of gama-aminobutyric acid in antipanic drug efficacy. *Am. J. Psychiatry* **146,** 353–356.
179. Heese, K., Otten, U., Mathivet, P., Raiteri, M., Marescaux, C., and Bernasconi, R. (2000) $GABA_B$ receptor antagonists elevate both mRNA and protein levels of the neurotrophins nerve growth factor (NGF) and brain-derived neurotrophic factor (BDNF) but not neurotrophin-3 (NT-3) in brain and spinal cord of rats. *Neuropharmacology* **39,** 449–462.
180. Dicpinigaitis, P. V. and Dobkin, J. B. (1997) Antitussive effect of the GABA-agonist baclofen. *Chest* **111,** 996–999.

181. Fodstad, H. and Nilsson, S. (1993) Intractable singultus: a diagnostic and therapeutic challenge. *Brit. J. Neurosurg.* **7,** 255–260.
182. Krahn, A. and Penner, S. B. (1994) Use of baclofen for intractable hiccups in uremia. *Am. J. Med.* **96,** 391.
183. Guelaud, C., Similowski, T., Bizec, J. L., Cabane, J., Whitelaw, W. A., and Derenne, J. P. (1995) Baclofen therapy for chronic hiccup. *Eur. Respir. J.* **8,** 235–237.
184. Nickerson, R. B., Atchison, J. W., Van Hoose, J. D., and Hayes, D. (1997) Hiccups associated with lateral medullary syndrome. A case report. *Am. J. Phys. Med. Rehabil.* **76,** 144–146.
185. Kumar, A. and Dromerick, A. W. (1998) Intractable hiccups during stroke rehabilitation. *Arch. Phys. Med. Rehab.* **79,** 697–699.
186. Marino, R. A. (1998) Baclofen therapy for intractable hiccups in pancreatic carcinoma. *Am. J. Gastroenterol.* **93,** 2000.
187. Chapman, R. W., Hey, J. A., Rizzo, C. A., and Bolser, D. C. (1993a) $GABA_B$ receptors in the lung. *Trends Pharmacol. Sci.* **14,** 26–29.
188. Chapman, R. W., Danko, G., Del Prato, M., et al. (1993b) Further evidence for prejunctional $GABA_B$ inhibition of cholinergic and peptidergic bronchoconstriction in guinea-pig with agonists and antagonists. *Pharmacology* **46,** 315–323.
189. Belvisi, M. G., Ichinose, M., and Barnes, P. J. (1989) Modulation of non-adrenergic, non-cholinergic neural bronchoconstriction in guinea-pig airways via $GABA_B$ receptors. *Br. J. Pharmacol.* **97,** 1225–1231.
190. Lehmann, A., Aurell Holmberg, A., Bhatt, U., et al. (2005) Effects of (2R)-(3-amino-2-fluoropropyl)sulphinic acid (AFPSiA) on transient lower oesophageal sphincter relaxation in dogs and mechanism of hypothermic effects in mice. *Br. J. Pharmacol.* **146(1),** 89–97.
191. Rane, M. J., Gozal, D., Butt, W., et al. (2005) γ-Amino Butyric Acid Type B Receptors Stimulate Neutrophil Chemotaxis during Ischemia-Reperfusion. *J. Immunol.* **174,** 7242–7249.

Index

G

R

MIX
Papier aus verantwortungsvollen Quellen
Paper from responsible sources
FSC® C105338

If you have any concerns about our products,
you can contact us on
ProductSafety@springernature.com

In case Publisher is established outside the EU,
the EU authorized representative is:
Springer Nature Customer Service Center GmbH
Europaplatz 3, 69115 Heidelberg, Germany

Printed by Libri Plureos GmbH
in Hamburg, Germany